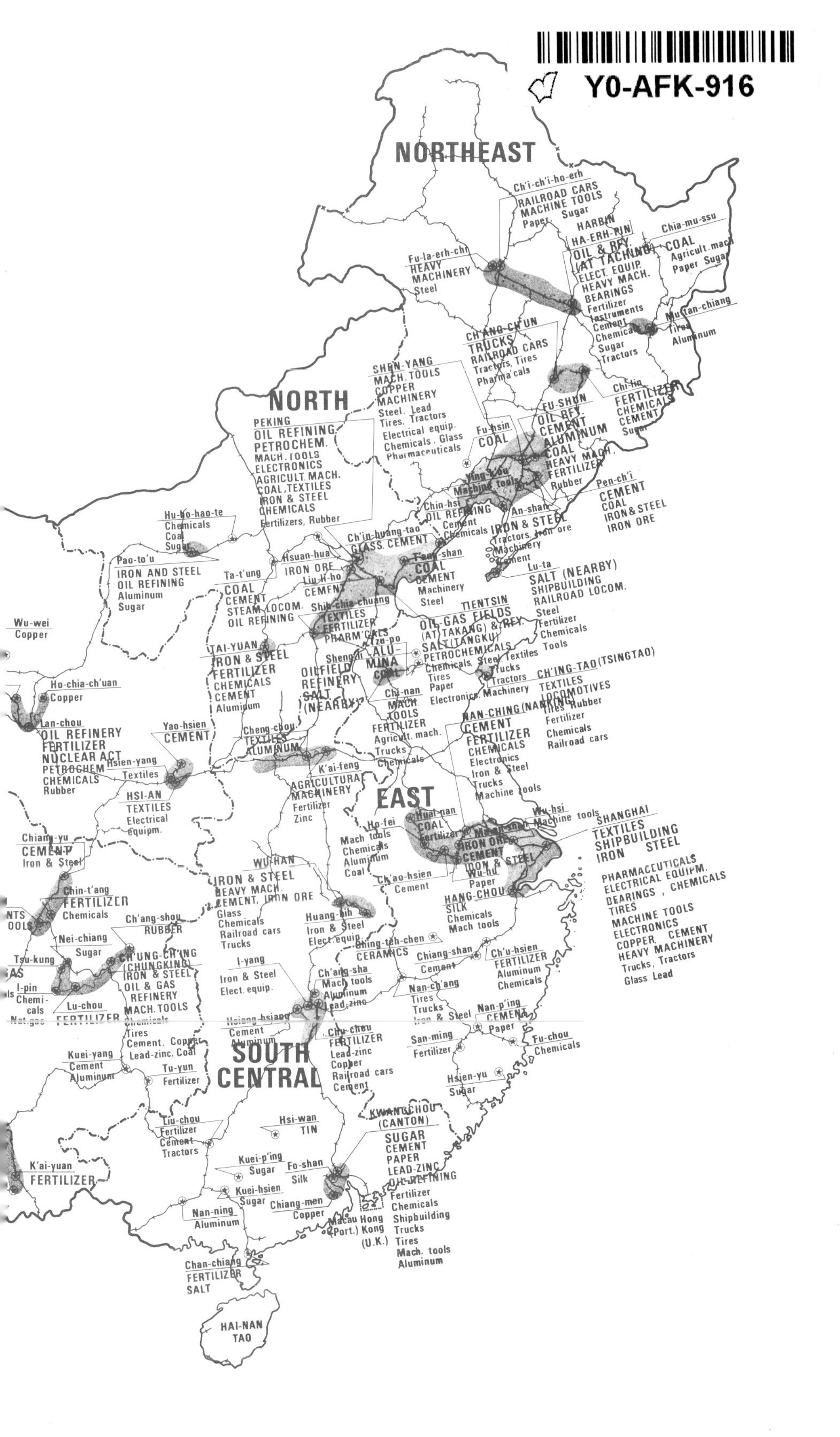
NORTHEAST
Ch'i-ch'i-ho-erh
RAILROAD CARS
MACHINE TOOLS
Paper, Sugar
HARBIN
HA-ERH-PIN
OIL & RFY.
(AT TACHING)
ELECT. EQUIP.
HEAVY MACH.
BEARINGS
Fertilizer
Instruments
Cement
Chemicals
Sugar
Tractors
Chia-mu-ssu
COAL
Agricult. mach.
Paper Sugar
Fu-la-erh-chi
HEAVY
MACHINERY
Steel
Mu-tan-chiang
Tires
Aluminum
CH'ANG-CH'UN
TRUCKS
RAILROAD CARS
Tractors, Tires
Pharma'cals
SHEN-YANG
MACH. TOOLS
COPPER
MACHINERY
Steel, Lead
Tires, Tractors
Electrical equip.
Chemicals, Glass
Pharmaceuticals
Chi-lin
FERTILIZER
CHEMICALS
CEMENT
Sugar
NORTH
PEKING
OIL REFINING
PETROCHEM.
MACH. TOOLS
ELECTRONICS
AGRICULT. MACH.
COAL, TEXTILES
IRON & STEEL
CHEMICALS
Fertilizers, Rubber
FU-SHUN
OIL RFY.
CEMENT
ALUMINUM
COAL
HEAVY MACH.
FERTILIZER
Rubber
Fu-hsin
COAL
Ying-k'ou
Machine tools
Pen-ch'i
CEMENT
COAL
IRON & STEEL
IRON ORE
Chin-hsi
OIL REFINING
Cement
Chemicals
An-shan
IRON & STEEL
Tractors, Iron ore
Machinery
Cement
Hu-ho-hao-te
Chemicals
Coal
Sugar
Ch'in-huang-tao
GLASS, CEMENT
T'ang-shan
COAL
CEMENT
Machinery
Steel
Lu-ta
SALT (NEARBY)
SHIPBUILDING
RAILROAD LOCOM.
Steel
Fertilizer
Chemicals
Tools
Pao-to'u
IRON AND STEEL
OIL REFINING
Aluminum
Sugar
Hsuan-hua
IRON ORE
Ta-t'ung
COAL
CEMENT
STEAM LOCOM.
OIL REFINING
Liu-li-ho
CEMENT
Shih-chia-chuang
TEXTILES
FERTILIZER
PHARM'CALS
TIENTSIN
OIL-GAS FIELDS
(AT TAKANG) & RFY.
SALT (TANGKU)
PETROCHEMICALS
Chemicals, Steel, Textiles
Tires
Paper
Electronics
Trucks
Tractors
Machinery
Wu-wei
Copper
TAI-YUAN
IRON & STEEL
FERTILIZER
CHEMICALS
CEMENT
Aluminum
Tzu-po
ALU-
MINA
COAL
Sheng-li
OILFIELD
REFINERY
SALT
(NEARBY)
CH'ING-TAO (TSINGTAO)
TEXTILES
LOCOMOTIVES
Tires Rubber
Fertilizer
Chemicals
Railroad cars
Ho-chia-ch'uan
Copper
Chi-nan
MACH.
TOOLS
FERTILIZER
Agricult. mach.
Trucks
Chemicals
Lan-chou
OIL REFINERY
FERTILIZER
NUCLEAR ACT.
PETROCHEM
CHEMICALS
Rubber
Yao-hsien
CEMENT
Cheng-chou
TEXTILES
ALUMINUM
NAN-CHING (NANKING)
CEMENT
FERTILIZER
CHEMICALS
Electronics
Iron & Steel
Trucks
Machine tools
Hsien-yang
Textiles
HSI-AN
TEXTILES
Electrical
equipm.
K'ai-feng
AGRICULTURAL
MACHINERY
Fertilizer
Zinc
EAST
Huai-nan
COAL
Fertilizer
Ho-fei
Mach tools
Chemicals
Aluminum
Coal
Ma-an-shan
IRON ORE
CEMENT
IRON & STEEL
Wu-hu
Paper
Wu-hsi
Machine tools
SHANGHAI
TEXTILES
SHIPBUILDING
IRON STEEL
PHARMACEUTICALS
ELECTRICAL EQUIPM.
BEARINGS, CHEMICALS
TIRES
MACHINE TOOLS
ELECTRONICS
COPPER, CEMENT
HEAVY MACHINERY
Trucks, Tractors
Glass Lead
Chiang-yu
CEMENT
Iron & Steel
WU-HAN
IRON & STEEL
HEAVY MACH.
CEMENT, IRON ORE
Glass
Chemicals
Railroad cars
Trucks
Ch'ao-hsien
Cement
HANG-CHOU
SILK
Chemicals
Mach tools
Chin-t'ang
FERTILIZER
Chemicals
Huang-shih
Iron & Steel
Elect. equip.
Ching-teh-chen
CERAMICS
Ch'ang-shou
RUBBER
Nei-chiang
Sugar
Chiang-shan
Cement
Ch'u-hsien
FERTILIZER
Aluminum
Chemicals
Tsu-kung
CH'UNG-CH'ING
(CHUNGKING)
IRON & STEEL
OIL & GAS
REFINERY
MACH. TOOLS
Chemicals
Tires
Cement, Copper
Lead-zinc, Coal
I-yang
Iron & Steel
Elect. equip.
Ch'ang-sha
Mach tools
Aluminum
Lead, zinc
Nan-ch'ang
Tires
Trucks
Iron & Steel
Nan-p'ing
CEMENT
Paper
I-pin
Chemi-
cals
Lu-chou
FERTILIZER
Hsiang-hsiang
Cement
Aluminum
Chu-chou
FERTILIZER
Lead-zinc
Copper
Railroad cars
Cement
San-ming
Fertilizer
Fu-chou
Chemicals
Kuei-yang
Cement
Aluminum
SOUTH
CENTRAL
Tu-yun
Fertilizer
Hsien-yu
Sugar
Liu-chou
Fertilizer
Cement
Tractors
Hsi-wan
TIN
KWANGCHOU
(CANTON)
SUGAR
CEMENT
PAPER
LEAD-ZINC
OIL REFINING
Fertilizer
Chemicals
Shipbuilding
Trucks
Tires
Mach. tools
Aluminum
Kuei-p'ing
Sugar
Fo-shan
Silk
K'ai-yuan
FERTILIZER
Kuei-hsien
Sugar
Chiang-men
Copper
Nan-ning
Aluminum
Macau
(Port.)
Hong
Kong
(U.K.)
Chan-chiang
FERTILIZER
SALT
HAI-NAN
TAO

DOING BUSINESS WITH THE PEOPLE'S REPUBLIC OF CHINA

DOING BUSINESS WITH THE PEOPLE'S REPUBLIC OF CHINA

INDUSTRIES AND MARKETS

BOHDAN O. SZUPROWICZ · MARIA R. SZUPROWICZ
21st Century Research

John Wiley & Sons • New York • Chichester • Brisbane • Toronto

Library of Congress Cataloging in Publication Data:

Szuprowicz, Bohdan O., 1931–
Doing business with the People's Republic of China.

"A Wiley-Interscience publication."
Includes bibliographical references and index.
1. Market surveys—China. 2. China—Industries. 3. China—Commerce. 4. China—Economic condition—1949– I. Szuprowicz, Maria R., joint author. II. Title.
HC427.9.S94 338'.0951 78–2539
ISBN 0–471–03389–8

Printed in the United States of America

10 9 8 7 6 5 4 3 2

To all those who assured us
that a book like this could not be written

"The current international situation is excellent, there is great disorder under heaven. . . ." (from a 1976 New Year's Day editorial in a Peking newspaper)

Preface

Who sells what to China, for how much, and why? This book addresses itself to that question. The objective is to determine China's specific commodity import dependence during the 1970s and to project possible future trends within the framework of its "grand plan" ambitions to become a global superpower by the end of this century.

The book is designed primarily for the international business executive who wants to identify Chinese import markets, determine their size and potential relative to markets in other countries, assess the competition, and make a decision as to whether the opportunities should be pursued.

China's foreign trade reflects imperfections and shortcomings of a large centrally planned economy with autarkic ambitions. Imports are undertaken reluctantly, almost "as a last resort," and the normal profit motivation does not necessarily apply. As a result the direction and the size of China's foreign trade are of political significance in the conduct of its foreign affairs.

Therefore, this book should also be of considerable interest to diplomats, political analysts, international economists, investment bankers, government planners, market researchers, journalists, and editors. It provides a data base of information on the size of Chinese industries and their import dependence and also identifies the market shares of major foreign supplier countries from the capitalist, communist, and the developing world. To the best of our knowledge this type of information has never before been assembled in a single book covering the whole economy of China.

Individual chapters of this book deal with specific economic sectors and import markets in China such as energy, mining and metallurgy, electric power, chemicals, transportation, agriculture, machine building, and communications. Each chapter discusses in detail one or more industries that are a part of that economic sector. For example, the chapter on transportation (Chapter 13) discusses the use, manufacture, and import of motor vehicles, aircraft, ships, and railway rolling stock. Chapter 9 (on mining and metallurgy) deals with production and imports of iron and steel, copper, aluminum, nickel, chromium, cobalt, and other metals and minerals of significance in Chinese foreign trade.

In developing the information about each particular industry we have generally tried to keep to a standard presentation format. Chinese domestic production levels are first determined or estimated and put in global perspective by comparisons with production levels in leading world economies of the United States, the Soviet Union, Japan, and western Europe. In addition, whenever data are available, Chinese performance is compared with other large and populous and developing economies such as those of India, Brazil, Nigeria, Indonesia, and Mexico, as well as with China's neighbors, including North and South Korea, Vietnam, Hong Kong, and Mongolia. Comparisons with yet other countries are also made whenever such countries are involved in production of strategic commodities traded by China, such as chromium, nickel, natural rubber, or petroleum.

Industrial production is then compared with imports of products or equipment in the same commodity category to determine China's import-dependence levels. In addition, individual import market shares of all known supplier countries are determined for each major product category. These market shares are developed for the cumulative period 1970–1975 whenever possible to put in proper perspective any possible distortion created by sudden supplementary Chinese purchases of products or equipment on a short-term basis. China's imports of grains from the United States during 1972–1974 would be a case in point. As further aid to determining the direction and magnitude of trade in specific products, major individual sales in each industry are identified by type of product, manufacturer, and contract value whenever such information is available.

The coverage of individual industries is uneven, partially by design, but sometimes purely from necessity. In formulating our research plans we have been guided by the relative size of Chinese imports in various commodity groups in the first instance. In other words, the first question we tried to answer was on what China was spending most of its foreign currency, rather than with whom it was doing the most trade, or which Chinese industry was the largest employer. Thus we concentrated our efforts primarily on investigation of those industries that showed the largest consistent import dependence or future potential. To make the book a more complete reference text of the Chinese economy, outlines of national infrastructures and other industries have been added, but understandably in less detail.

Another factor contributing to uneven coverage of the various industries is the fact that China does not publish any of the information we were seeking. In some cases we were unable to discover enough data of sufficient accuracy to discuss an industry in as much detail as would be justified by its high import-dependence level. This is particularly true for production and trade in strategic metals and in chemicals.

Once we have established production and import levels as well as market shares of suppliers in a particular industry we also try to determine the stability and future prospects for such trade. Domestic import-substitution capabilities as well as Chinese response to commodity shortages are studied. Trade embargoes, development of alternative sources of supply of strategic materials, and foreign-aid programs of possible future significance are discussed. In this regard the Sino-Albanian connection as an indispensable source of chromium, the Sri Lanka (Ceylonese) rice-for-rubber barter, the Sino-Cuban sugar trade, and the eastern African TAZARA railroad leading to new sources of copper, cobalt, and chromium are examples of Chinese foreign policies clearly subservient to the present and future requirements of the Chinese economy.

Individual industry performance and China's import dependence are studied within the overall framework of China's "grand plan" policy to achieve superpower status by the end of this century. This is done by comparison of China's gross national product (GNP) rates of growth with projected growth of China's major competitors to this title, namely, the United States, the Soviet Union, and Japan. China's potential to become a superpower is also examined in relation to its available energy and raw-materials resources and its vulnerability resulting from its strategic-materials deficiencies.

International relations and Chinese foreign-aid programs are examined in light of China's "grand plan" ambitions and its presumed desire to maintain sufficient stability for the next 25 years to achieve its development objectives. Special attention is given to individual Chinese foreign-aid programs for developing countries that are identified as present or potential suppliers of strategic and other imports to China.

China's foreign trade is less than 5 percent of its GNP in volume and is relatively small in comparison with other developed countries. It is comparable to that of Finland or Hungary today, but its future potential will be considerably larger if China continues its rapid industrialization. However, if China never allows its foreign-trade volume to exceed 5 percent of its GNP, even if it becomes a global superpower by the year 2000, its foreign trade will be no larger than that of Belgium today.

Growth of China's imports is directly linked to its ability to export primarily food, textiles, and petroleum, all of which are commodities in oversupply in world trade. This is also true for oil because OPEC countries possess more than enough surplus production capacity to counteract effectively and rapidly any significant attempt by China to dump its oil at too "friendly" a price. Faced with a limited potential for earning hard currency and the increasing demands of a rapidly growing industry, China may have to reassess its future foreign-trade policies in the light of current economic realities.

For this reason we examine each major Chinese import market to determine its dependence on Soviet or eastern European versus Japanese and Western suppliers. Chinese imports from COMECON countries consist almost exclusively of industrial machinery and transportation equipment. Despite large purchases in the West, the Soviet Union remains China's largest supplier of aircraft and aviation equipment; Poland is leading in coal-mining machinery; Romania and East Germany are major suppliers of locomotives; Czechoslovakia, Romania, and the Soviet Union are among the five leading suppliers of trucks; and Yugoslavia and East Germany are among the top five suppliers of ships to China.

Although China's trade with COMECON represents only 18 percent of the total in absolute volume, it has now grown to a level comparable to its previous peak in 1959. If hard-currency markets for exports do not develop sufficiently, China may well increase further its trade with COMECON in general and with eastern Europe in particular. Predicted Soviet oil shortages in the 1980s and increasing demand for oil throughout COMECON while high OPEC prices continue make China an attractive and most welcome trade partner.

The biggest attraction to both sides in increasing trade is its barter character. Because communist countries conduct trade by bilateral agreements, the problem of hard-currency requirements does not exist. Equipment of COMECON countries may be difficult to sell in Western markets, but it acquires certain irreplaceable qualities when it can be obtained in return for oil, foods, and textiles instead of hard currency.

The Soviet Union is the largest natural market for Chinese export, and the two economies are more complementary today than in previous years. China's dependence on Soviet aid was eliminated long ago. China's oil could go a long way to let the Soviets keep their eastern European clients while they supplement their domestic energy needs with a crash nuclear power program. China's tin could save the Soviet Union $50 million in hard currency every year. In return, the Soviet Union has all the nickel, chromium, cobalt, platinum, and diamonds that China will ever need for its industrial development. Moreover, the Soviet Union is probably the only considerable remaining market capable of absorbing significant Chinese food and textile imports. This may be of importance in the future if most of China's oil is needed for domestic development and free-world export markets for Chinese products remain sluggish.

Our discussions of Chinese industries, foreign aid, imports, and future opportunities focus on economic potentials of China and its trading partners. These we realize are not the only imperatives influencing China's foreign trade, but we believe they are growing in importance as the realities of large-scale industrialization and worldwide interdependence are setting in. Japan's capacity to import Chinese products is not unlimited because it is a much smaller country and its survival depends on exports. Although Japan could increase imports from China, there will come a time when it will be only able to do so at the expense of its older and more predictable trade partners.

We have attempted to present basic economic estimates about China and point out opportunities and problems that it will have to face in the future in its drive toward superpower status. Whether China will follow any particular course of action in its foreign trade as a result of such developments is impossible to predict, nor have we made any attempt to do so.

We have also tried to refrain from quoting Chinese sources, which is a practice rather common particularly among scholarly sinologists. As a result we have been forced to use throughout the book a multitude of qualifiers such as "about," "reportedly," and "believed to be." This does not make us very happy, but at least it prevents us from becoming unwittingly a mouthpiece for Chinese propaganda. We also feel it serves a useful purpose in warning the reader and user of the information about its unconfirmed quality. Our attitude to statistical data about China is very cautious, and we wish to communicate this to our readers at all times. Practically all data are estimates developed by various industry specialists and scholars who, though much better qualified than ourselves to do so, mostly lack the benefits of official Chinese confirmation or assistance. We have made it a point to use statistical data from U.S. government sources whenever possible because we believe theirs to be the largest and most serious effort to monitor China's economy. Any erroneous conclusions that we may have drawn from the use of such data are of course attributable entirely to us.

We must also extend our appreciation to persons and organizations who perhaps even unwittingly were helpful to us in making the writing of this book possible. Among those is the China Export Commodities Fair Authority in Canton who extended to us an invitation to visit China in November 1974. Solomon Manber, who shared the rigors of our China passage and assisted with his incisive evaluation of Chinese equipment and technology, deserves our special thanks. Two unique sinologists and China trade experts whom we always found to be unlimited sources of new information and welcome criticism deserve a special mention. One is Peter Marshall, M.B.E., director of Trade Promotion of the Sino-British Trade Council in London, who probably unwittingly at the time introduced to us the challenge of the Chinese market. The other is Nicholas Ludlow, research director of the National Council for U.S.–China Trade in Washington, whose personal enthusiasm for the China trade was always contagious. We are also thankful to our friend Shohei Kurita, research director of Japan Electronic Computer Company, Ltd. of Tokyo, who was born in China and often gave us the benefit of his advice about high technology markets in Asia and China. Our unnamed helpers include the staffs at the Library of Congress, the Hammarskjöld Library of the United Nations, the New York Central Library, the Graduate Business School Library of New York University, and the United Engineering Libraries in New York, as well as the Teaneck Library in New Jersey. Last, we must thank a large number of officials of foreign-trade organizations and chambers of commerce, diplomats, businessmen, scientists, engineers, booksellers, editors, and sundry "experts" in the United States as well as in Australia, Canada, China, France, Germany, Hong Kong, India, Japan, Macao, Poland, Singapore, Taiwan, the Soviet Union, the United Kingdom, and Zanzibar who bothered to talk to us and whose comments and remarks may have contributed to our understanding of this fascinating subject.

BOHDAN O. SZUPROWICZ
MARIA R. SZUPROWICZ

North Bergen, New Jersey
January 1978

Contents

DOING BUSINESS WITH THE PEOPLE'S REPUBLIC OF CHINA

1

Introduction

As the final paragraphs of this book were being written in late 1977, China finally admitted to the world at large that its population is now in the order of 900 million rather than the 800 million suggested in previous years. This new figure was given to visiting British journalists from the *London Economist* by the deputy prime minister Chi Teng-kuei. It is considered a statistical revelation and is most welcome by economists, businessmen, and scholars throughout the world, not only as a "solid" new statistic from China, but as a sign of possible future normalization of statistical and economic accounting in the most populous country of the world.

Despite questionable rejoicing that another 100 million human beings have suddenly been discovered, many questions remain unanswered. We still do not know exactly when the Chinese realized that their actual population exceeded by at least 12 percent their estimated population. Whether it is now or a year ago does make a difference comparable in size to perhaps all of Canada or Yugoslavia. But if past practice is indicative, we will now see a lot of reassessment work on China using the 900 million population as of 1977 as a base. Not the least of this effort will have to be undertaken by the United Nations, whose estimates for China's population so far favored the lower figures for many years and whose methodology of estimation remained a closely guarded secret to this day.

This is only one example illustrating the frustrations and uncertainties that anyone engaged in research on Chinese economy faces every day. The problem of availability and accuracy of data for several industrial sectors over a period of years is simply magnified so many times.

DEMAND FOR HARD DATA ABOUT CHINA

There are almost no hard data about China anymore. There are, however, numerous estimates painstakingly developed by leading sinologists throughout the world over the last 20 years that provide only approximate magnitudes representing China's economy. Since 1959, when a small statistical handbook entitled *Ten Great Years* had been published in Peking, an almost total statistical blackout was imposed by the government of the most populous nation on earth. Throughout the 1960s economic researchers on China became almost desperate, and after a decade many felt they had reached a point of no return.

It was not until 1970 when Premier Chou En-lai gave a few overall figures to the late Edgar Snow that a small trickle of statistical data resumed, but it was considerably less than before the blackout began. Most data released in recent years are simply percentage increases in production over some year during the 1960s for which no base data were published anyway. Absolute production figures sometimes appear in relation to provinces or municipalities but are more common in data for cities, counties, communes, production teams, and factories.

This scarcity of data coupled with sudden purchases of very large quantities of goods or huge turnkey plants by China in a particular country sends large numbers of businessmen, market researchers, and scholars rushing to discover, confirm, and project the size of Chinese markets, competitive market shares, and possible future sales. Repeatedly they find only very aggregate estimates and varying opinions on validity of sets of alternative figures.

The authors preceeded the writing of this book with a multiclient research project designed to answer some of the questions about production levels, import dependence, market shares of suppliers, and competitive environment in China. Before undertaking that study a questionnaire survey was conducted among 3000 international executives in manufacturing and trading organizations operating in twenty different countries in the world. The largest percentage of those contacted represented companies that were known to have already traded with China, and the remaining were manufacturing products similar to those already imported by China in previous years.

Almost 700 questionnaires were returned from 560 different organizations representing twenty major industries. By far the largest percentage of those responding indicated that lack of data about markets in China was the biggest single obstacle to further expansion of their company interest and trade activities with China. Using the responses of the survey as a guide, the authors conducted research with the objective of filling in that information gap. Additional interviews with executives in the United States, Japan, Hong Kong, France, West Germany, and the United Kingdom, and even those executives attending the Canton Trade Fair in China confirmed a unanimous desire for more meaningful information about China's industries and markets.

SOURCES OF INFORMATION

Original Chinese sources of information about China are primarily domestic radio broadcasts, news releases, and Chinese publications. All other sources are secondary and consist of analyses, opinions, commentaries, estimates, reports, and books published in Hong Kong, Japan, Taiwan, Singapore, England, France, West Germany, and the United States. These are based directly or indirectly on the original sources and to some degree on reports of Chinese refugees and visitors to China. There is also a much smaller body of literature about the economy of China originating from the Soviet Union and eastern European countries. All trade statistics derive from foreign trade statistical yearbooks of individual countries trading with China because no such data are made available by the People's Republic of China (P.R.C.).

Most current information comes from Chinese provincial and international broadcasts, which also includes news releases of the New China News Agency. Translated texts of these broadcasts are available from the BBC Monitoring Service in England, the U.S. Consulate General in Hong Kong, and the Foreign Broadcast Information Service (FBIS) in the United States. Only a small percentage of information in these broadcasts pertains to Chinese industries and trade. A summary of all economic information contained in the broadcasts is published weekly by the BBC Monitoring Service.

Many Chinese newspapers and magazines are translated daily by the U.S. Consulate General in Hong Kong, and these translations are available on a subscription basis. In addition, selected Chinese publications and periodicals are translated by the Joint Publications Research Service (JPRS) in the United States. Both are available from the National Technical Information Service (NTIS) of the U.S. Department of Commerce. Analysis of these publications yields a considerable amount of information about Chinese industries, products, technologies, and the overall economy. There are also Chinese publications in English and other languages published in Peking and Hong Kong that are available on a subscription basis directly there or from agents in other countries.

The most regular sources of original reports by visitors to China are the monthly *China Trade Report* of the Far Eastern Economic Review and daily newspapers in Hong Kong, as well as the internal publications of specialized organizations involved in promotion of trade or scientific exchanges with China. These include *The China Business Review* of the National Council for U.S.–China Trade in Washington, the *Sino-British Trade Newsletter* of the Sino-British Trade Council in London, and the *Republique Populaire de Chine Nouvelles Commercialles* of the Centre Français de Commerce Exterieure in Paris. A *China Exchange Newsletter* published by the National Academy of Sciences in Washington keeps track of scholarly exchanges with the People's Republic of China. All other materials on Chinese economy and trade are based for the most part on data contained in those sources, with some additional statistics from various agencies of the United Nations.

Probably the largest numbers of reports and studies about China are published by the various departments of the U.S. Government, and most of those are available through the Government Printing Office (GPO) in Washington. These publications include comprehensive compendia of papers on Chinese Economy prepared by the Library of Congress for the Joint Economic Committee of the U.S. Congress, a series entitled *Research Aids* concerning Chinese industries published by the Central Intelligence Agency, the periodical *Current Scene* published by the U.S. Information Agency in Hong Kong, several market-assessment reports published by the Bureau of East–West Trade of the Department of Commerce, annual surveys of minerals and metals in China published by the Bureau of Mines of the Department of the Interior, and an annual review of agriculture in China published by the U.S. Department of Agriculture. There are also irregular reports of hearings before the Committees of the Judiciary, on Foreign Relations, on Internal Security, and others that may deal with various aspects of the Chinese economy.

There are several academic establishments engaged in research on contemporary China, and most American sinologists contribute to the various publications of the U.S. Government. Some of their findings are also published in the *China Quarterly* in London and the *Asian Survey* in Berkeley, California. *Contemporary China,* a bimonthly of Columbia University in New York, presents foreign policy aspects, and major policy discussions of China often appear in *Foreign Policy, Foreign Affairs,* and *International Security* quarterlies. There are also several commercial publications often analyzing China's economy such as *Business Asia* in Hong Kong, *Asia Research Bulletin* in Singapore, and *China Trade and Economic Newsletter* in London.

Import and export statistics on Chinese trade are originally derived from statistical reports of countries trading with China because China does not publish its own statistics nor provide such information. These are basically foreign-trade statistical yearbooks for the individual countries. Some export commodity categories are reported in various U.N. publications, usually converted into U.S. dollars. Another source that also publishes basic U.N. trade statistics is the *OECD Commodity Trade Statistics Series B,* but it primarily covers trade with NATO countries.

Interviews with executives, diplomats, or journalists who have had first-hand experience in China generally yield only fragmentary data and personal opinions that often differ widely. This is so because very few persons are able to put their personal observations in proper perspective relative to particular Chinese industry or discipline in the total absence of official statistics. On the other hand, such inputs collected over a long period of time are of definite value and in many cases contribute to the various secondary sources mentioned previously.

METHODOLOGY OF RESEARCH AND ANALYSIS

When undertaking research of the economy of China the authors had the benefit of their previous experience with similar studies of centrally planned economies, industries, and markets of the Soviet Union, Poland, Czechoslovakia, and Hungary in the course of their professional market-research activities. As a result of this work, inadequate statistics and unavailability of data were not novelties, and a method of data collection and classification was developed that proved of value in preparation of meaningful market analyses and reports. Because China is a centrally planned economy originally patterned after the Soviet model, this research and data-collection methodology proved useful in creating a data base of information that in time became comprehensive enough to permit the writing of this book.

The administrative structure of the Chinese economy was identified as an initial outline for research. Individual files were set up for each specific ministry or special agency of the State Council that runs the Chinese government and administers all economic activity in China. This was followed by systematic analysis of radio broadcasts, translations of periodicals, scholarly reports and papers, industry reports, specialized newsletters, newspapers, magazines, and books dealing with China. This activity yielded a large number of fragmentary items about Chinese manufacturing plants, mines, oilfields, ships, construction projects, trade agreements, and other data of economic significance.

All such items are immediately classified by presumed industry affiliation and collected methodically over a period of years before a substantial amount of data collects to describe any one industry. In time, sufficient data accumulates to provide a relatively comprehensive outline of major activity and plants in an industry within each sector of the economy. In some cases it was necessary to analyze back issues of particular publications for the last 10 to 15 years before a satisfactory amount of useful data was found.

It is believed that analysis of the Chinese economy ministry by ministry provides the sharpest outline of the status quo because it reflects an overall picture known only to China's top central planners. It also is a most useful base for evaluation of industrial performance and speculation on future possibilities. As executive organs to the State Council, ministries also outline priorities and development policies for their industries, and their spokesmen often report on planned expansion and plan achievements comparable with the existing data base for each ministry. Even if most such statements are percentage increases once a single base-year production is determined, it is relatively easy to develop a time series for the industry as a whole.

Production estimates for many Chinese industries are being developed in this way in the form of several time series by various sinologists and industry experts who supply their findings to several U.S. Government agencies. From time to time these estimates are published in Joint Economic Committee reports prepared for the U.S. Congress and in various market assessment reports of the U.S. Department of Commerce. These estimates are updated periodically in the *Current Scene* monthly of the U.S. Information Agency and various industry research aids published by the Central Intelligence Agency. Because these estimates form the largest and best researched body of data about the Chinese economy, we have used them whenever they were available.

For some products or industries, and sometimes for specific countries or time periods, figures were not available in any U.S. government publications. In such cases we have resorted to the *Statistical Yearbook of the United Nations* for additional information. If neither of those two sources provided sufficient information, other sources would be consulted until an estimate or other data were found from which an estimate could be derived. In most cases sources used in preparation of tables and figures in the book are identified immediately following the data.

The method of estimating possible future production or imports depended on data available and is usually explained in text. In most cases these are straightforward extrapolations using clearly stated growth factors and reasons for choosing them. Readers with additional insights or different opinions will be able to recompute our projections using their own growth factors whenever they wish with the greatest of ease. None of our projections or conclusions have been presented to or commented on by any representatives of the People's Republic of China.

Import-dependence ratios are developed by comparing the most recent production or consumption estimates in a particular sector with a concurrent annual or average import level. Cumulative import figures are first obtained by identifying all countries exporting a particular commodity to China. This data normally originates from foreign-trade yearbooks of individual countries and certain cumulative reports from the United Nations, OECD, U.S. Department of Commerce, and other agencies. Particular commodity exports to China are extracted from individual country reports and grouped together to form supplier country market shares for 1965–1975 or 1970–1975 periods whenever available. Either average or latest year totals for each commodity are used for comparison with estimated value of domestic production to obtain import-dependence ratios.

RELIABILITY OF DATA

All data presented for China are estimates and must be treated as such. In the absence of complete official statistics several sources may develop varying estimates, and the user then faces a dilemma because more often than not he has no knowledge of the supporting assumptions. One of the best examples is again the discrepancy in China's population estimates. The United Nations estimated 838 million for 1975, whereas U.S. government agencies used an estimate of 934 million for the same year. Other independent sinologists applying different fertility rates and growth factors came up with figures different from either of the above. The Soviet Union and eastern Europe sometimes come up with still different estimates of their own. As was previously mentioned, the Chinese themselves "admitted" that the population is about 900 million. Give or take 100 million the choice is still open.

The problems with any data pertaining to production in plants or to agricultural output is probably less severe than that with China's population. Factory output, acreage yields, and volumes of goods traded are easily measurable quantities. However, whether such data are released by the authorities is another matter. The problem here lies more with data scarcity and irregularity of appearance.

China economy researchers also run the danger of incorrect interpretation of statistics because data released are often qualified by such adjectives as "nearly," "less than," or "more than," and base years for indexes are sometimes not specified. Because we have mostly used well-researched U.S. government estimates we feel this problem has been minimized as far as this text is concerned.

The amount and quality of data available also vary with time, and future data release is unpredictable. However, sinologists in general believe that Chinese statistics are internally consistent, and there is no evidence to suggest that central authorities deliberately practice falsification. Rather, they believe that a vast country such as China could ill afford to keep two sets of national statistics because it could not use false figures for its own planning purposes. China is believed to have adopted a policy of selective publication of data that favorably shows the performance of China's economy. Recent admissions of setbacks by top figures in several ministries, however, augur well for the release in the future of more comprehensive statistics. With patience and ingenuity, meaningful estimates are generally possible on the basis of available statistics published by China and various countries which engage in trade with China.

Because it was necessary to use estimates from a variety of sources in writing this book it is inevitable that errors of definition and interpretation have been introduced. These originate primarily in different methods of valuation of exports or production by various countries as well as from use or omission of unknown adjustment factors in reporting commodity values in different years. While such errors are very difficult to minimize in the absence of adequate Chinese statistics, we believe their magnitude to be tolerable as long as readers regard the information developed in this book as representing approximate relative magnitudes rather than precise accounting figures.

Particularly troublesome is the problem of compiling import market shares, and we believe these may contain the largest percentage of errors. This is so because such reports at specific commodity levels can only be developed from trade statistics that originate from several sources and are not necessarily always compatible. Individual countries report their export statistics usually in their own currencies and often designated by commodity codes that differ from the Standard International Trade Classification (SITC) codes used by the United Nations and the United States. The immediate problem in such cases is to identify to which SITC code categories these commodities belong and convert local export values into U.S. dollars. Because currency conversion rates also vary with time, this in itself is another source of possible discrepancies. Most statistics reported in this manner are also in current dollars, and additional errors can be introduced when cumulating such data over longer periods of time. Differing rounding rules used by different reporting agencies are another source of errors. It should also be kept in mind that unless otherwise stated all import data are derived from official statistical sources and represent shipments of goods in specific years rather than total sales of particular commodities or products which are often reported by the press. Such reports often differ widely from annual statistics because individual sales often are delivered over a period of several years with one or more years of lead time delays, particularly in the case of equipment such as aircraft or ships which are constructed during a long period of time. In order to provide additional insights into Chinese markets the book provides lists of individual sales reported within various industries as an additional report of market shares, contract values, and suppliers which may not be easy to correlate with actual import trade statistics.

In order to keep these problems to a minimum the authors tried to use a single source of production or trade statistics whenever possible, but this was not always practical. The two primary sources used were the United States government publications and United Nations statistics, but in many instances it was necessary to combine data from both sources and yet other documents in order to develop a comprehensive picture of a market or an industry. Such composite tables or figures are probably generally representative of relative positions of countries involved, but it is impossible to define the precision with which they represent individual elements.

Because this book describes the status of China's economy and its foreign trade in the mid-1970s, whenever possible values are expressed in 1975 dollars. On the other hand, it was not always possible to ascertain whether data from all the sources were also available in 1975 currencies. To the extent that some data were not, this is another source of errors equal in magnitude to the different inflation factors involved. In the case of projections of possible future growth in specific industries or markets, extrapolations were made from 1975 data as a base using reported or composite figures for that year.

Clearly these problems are not unique to this particular research. The authors found several instances of reassessment of figures year after year in various sources as additional and more detailed information became available. Nevertheless, even in this imperfect state the book presents some estimates for every sector of the Chinese economy and provides extensive references selected primarily on the basis of their relevance to specific industrial sectors. With those caveats in mind, the reader should be able to use the information to develop a good overall insight into the relative size of Chinese industries and their market potential.

2

Institutional Framework of the People's Republic of China

Any discussion of the economy, the industries, and foreign trade of the P.R.C. sooner or later involves a discussion of the State Planning Commission, the State Council, Ministries, Production Bureaus, and Special Agencies. Hence it is important to develop an overall understanding of the institutional framework of China and the hierarchical relationships between the government, the Chinese Communist Party, and the military as well as scientific establishments. The user of this book is strongly urged to read this chapter before proceeding to more detailed reviews of specific economic sectors.

The People's Republic of China is a communist state in which the real authority lies with the communist party political bureau, better known as the "Politburo." The most recent Constitution, adopted on January 17, 1975, defines the National People's Congress of the People's Republic of China as the "highest organ of state power under the leadership of the Communist Party of China." This formally establishes the Chinese Communist Party as the supreme ruling body in China.

The government of China is accomplished by ministries, special agencies, and revolutionary committees at various administrative levels that parallel the Party committees down to the county, commune, and production brigade levels in rural China. The military establishment is organized by military districts that generally follow provincial administrative divisions but are grouped into eleven distinct Military Regions centered around major municipalities.

Basically the country is divided into six economic regions which contain twenty-nine first-order administrative units all under direct central control: twenty-one provinces, five autonomous regions, and three province-level municipalities. These in turn are subdivided into 289 intermediate level units: prefectures, autonomous prefectures, municipalities, leagues and one administrative district. These in turn are also subdivided into 2233 third order units consisting of 2010 counties, autonomous counties, smaller municipalities, banners, autonomous banners and one town. (see Table 2.1)

Rural counties are subdivided into communes that were first organized in 1958 and function as the basic units of local government. There are now an estimated 50,000 communes in China, and these are further subdivided into production brigades and production teams that form the middle and lower administrative levels controlling collective work. Individual households are permitted to cultivate private plots and keep a limited number of livestock for personal consumption or sale to the state.

CHINESE COMMUNIST PARTY

The Chinese Communist Party is estimated to number at least 30 million members. It is in fact the largest communist party in the world, equal in number to the combined membership of all others in power. This fact may have been one of the contributing causes to the Sino-Soviet dispute about leadership of the world communist movement. However, although it is the largest communist party in the world, its membership represents only 3.1 percent of its country's population. This may reflect the fact that 75 percent of China's population is still rural and a high proportion remain illiterate. Eastern European countries with higher levels of urbanization and literacy show the highest percentage of party membership relative to their populations, higher even than that of the Soviet Union, where the Communist Party members account for 6.2 percent of the total population and 30 percent of the population is still engaged in agriculture (see Table 2.2).

The highest organ of the Chinese Communist Party is the National Party Congress, and the eleventh such Party Congress was held during August 12–18, 1977. The Party Congress, consisting of about 1300 delegates from throughout China, elects a Central Committee, which now consists of 333 persons, an increase of fourteen members since the previous committee was elected in August 1973. The membership of the Central Committee consists of 201 full committee members and 132 alternate members. More than 40 percent of the current membership are new and

Table 2.1 Provincial populations and administrative divisions of China

Economic Region	*First Order Provincial Administration*	*Population in Millions in mid-1976*	*Number of Second Order Intermediate Administrations*	*Number of Third Order County Level Administrations*
Northeast	Liaoning	44.5	15	54
	Kirin	23.0	7	56
	Heilungkiang	32.0	15	83
North	Hopeh	55.2	10	148
	Shansi	23.0	8	106
	Inner Mongolia (A)	8.5	6	47
	Peking (M)	8.5	none	9
	Tientsin (M)	7.2	none	5
	Shantung	78.5	13	111
East	Kiangsu	61.5	14	68
	Anhwei	45.0	15	72
	Chekiang	36.0	11	64
	Kiangsi	29.4	9	85
	Fukien	24.0	9	63
	Shanghai (M)	12.3	none	10
Central South	Honan	67.5	14	120
	Hupeh	43.8	10	77
	Hunan	49.0	12	95
	Kwangtung	54.1	10	106
	Kwangsi Chuang (A)	31.3	14	80
Southwest	Szechwan	100.1	19	189
	Kweichow	24.0	9	82
	Yunnan	28.0	17	123
	Tibet (A)	1.7	6	71
Northwest	Shensi	26.0	10	95
	Kansu	19.8	11	77
	Tsinghai	3.8	7	37
	Sinkiang Uighur (A)	10.0	13	82
	Ningsia Hui (A)	3.0	5	17
	TOTAL	950.7	289[a]	2233[b]

Source: *CIA People's Republic of China Atlas,* March 1975; U.S. Government National Foreign Assessment Center, *China: Economic Indicators,* October 1977.

[a] Includes 174 prefectures, 78 municipalities, 29 autonomous prefectures, 7 leagues, and 1 administrative district.

[b] Includes 2010 counties, 100 municipalities, 66 autonomous counties, 53 banners, 3 autonomous banners, and 1 town.

A = Autonomous region; M = Municipality with provincial status

Table 2.2 Membership of Communist Parties in communist states

Communist-ruled State	*Membership of the Communist Party*	*Party Membership as Percentage of Population*
China	30,000,000	3.1
Soviet Union	16,000,000	6.2
Romania	2,577,534	11.9
Poland	2,437,000	7.0
North Korea	2,000,000	11.3
East Germany	1,900,000	11.2
Yugoslavia	1,500,000	6.9
Czechoslovakia	1,380,000	9.2
Vietnam	900,000	1.8
Bulgaria	788,211	8.9
Hungary	754,000	7.1
Cuba	200,000	2.1
Albania	87,000	3.4
Mongolia	67,000	4.3
Cambodia	NA	
Laos	NA	

Source: *U.S. Government National Basic Intelligence Factbook, July 1977.*

did not serve on the previous Central Committee. This means that over one-third of the members of the old Central Committee were not reelected.

Provincial party officials and civilians constitute most of the membership of the current Central Committee. Representatives from the provinces account for just under two-thirds of the membership. Less than one-third of all the members have a military background, having spent at least half of their careers in the People's Liberation Army (PLA). Only thirty-eight members are women.

The Central Committee in turn elects a Politburo as well as the party chairman, who at present is Hua Kuo-feng, and other top party leaders. The Politburo consists of about twenty-five members, including twenty-one full members and four alternate ones. The Politburo in turn elects from within its membership a Standing Committee of five ranking members who form the most prestigious body, which possesses the ultimate decisionmaking authority (see Fig. 2.1).

The Politburo operates a Military Commission, which directly controls the General Staff Department of the PLA. It may include all military region commanders and political commissars as well as commanders of the service arms of the PLA. The Military Commission also controls the "military–industrial complex," which includes most of the Ministries of Machine Building Industries.

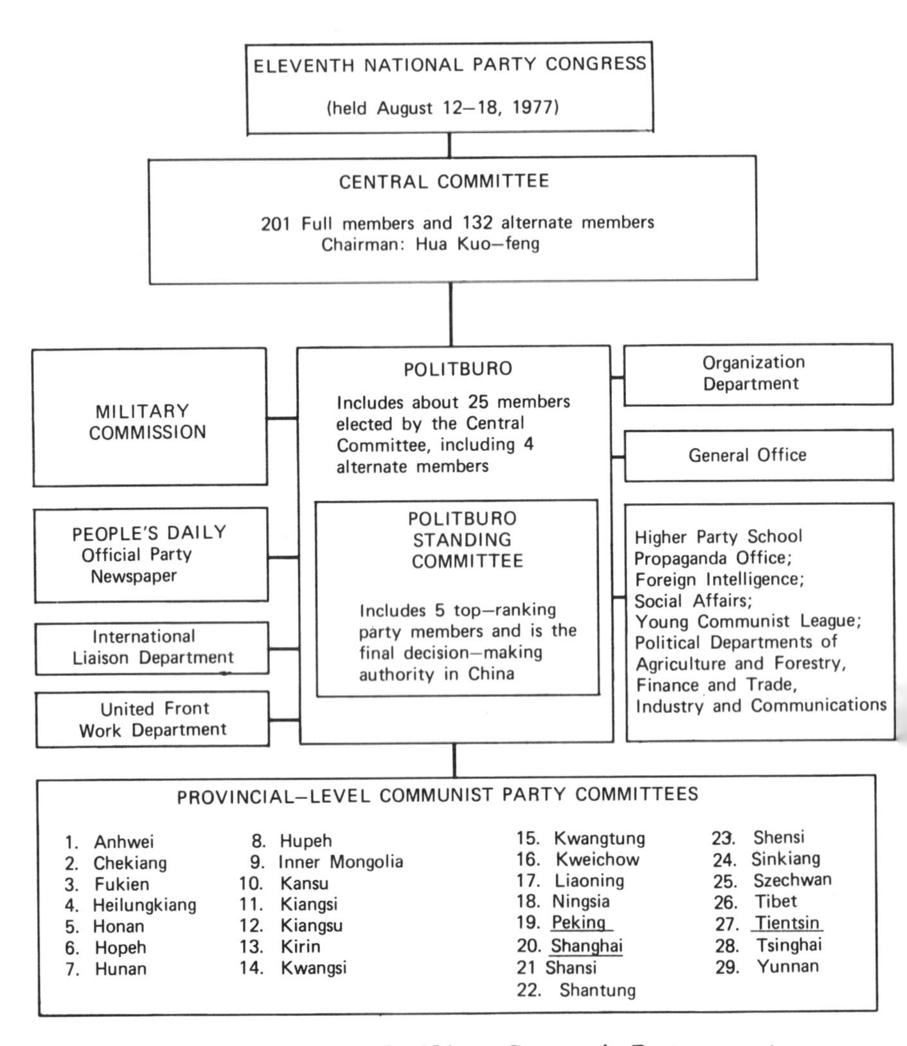

Figure 2.1 Organization of the Chinese Communist Party apparatus.

The Politburo also controls the publication of the *People's Daily*, the official party newspaper with an estimated circulation of 3.4 million, the largest single newspaper circulation in China. The *People's Daily* presents the official Party policy in all domestic and international matters. Several other functional departments support the work of the Politburo.

There are twenty-nine provincial Chinese Communist Party committees, including three municipal organizations in Peking, Shanghai, and Tientsin. Each provincial unit is headed by a first secretary of the provincial communist party, and most of them are also members or alternate members of the Central Committee at the national level. Certain members of the provincial-level party committees of Peking, Shanghai, Kwantung, Shansi, and Sinkiang are also members or alternate members of the Politburo, perhaps underscoring the importance of those areas to the political and economic life of China.

Chinese Communist Party Committees at the commune level are the most important institutions of power in rural China. They interpret policy decisions made in Peking, adapt those policies to meet local conditions, and ensure that policies and production plans are implemented. A branch committee also operates at the brigade level under a party branch secretary.

CENTRAL GOVERNMENT STRUCTURE

The National People's Congress, which is not to be confused with the National Party Congress, is theoretically the highest organ of the government. Its functions include amendments to the constitution, legislation, appointment and removal of the premier and members of the State Council on proposal of the Communist Party Central Committee, approval of the national economic plan, state budget, and final state accounts. In reality, the National People's Congress endorses the programs of the Chinese Communist Party.

There have been four national congresses of this body since its establishment in 1954. The second National People's Congress met in 1959, the third met in 1964, and the first session of the fourth National People's Congress was held on January 13–18, 1975, when a new constitution of the People's Republic of China was adopted.

The Standing (or Executive) Committee of the National People's Congress consists of a chairman, twenty vice-chairmen, and 144 members. Most of the vice-chairmen of the Standing Committee are members or alternate members of the Communist Party Central Committee, and a few are also members of the Politburo.

The State Council, whose members are appointed by the National People's Congress on proposal of the party, exercises leadership over line ministries, commissions, special agencies, and other organs of the state at various levels throughout China. It is responsible for drafting and implementation of the national and economic plans and the state budget and directs all state administrative affairs.

The State Council was formed in 1954 in a reorganization of the Government Administrative Council, which existed since 1949. The council consists of the premier, nine vice-premiers, and twenty-nine ministers. The current State Council is believed to contain twenty-six ministries, three commissions with ministerial status, and at least twenty special agencies and bureaus (see Table 2.3).

Government administration at lower levels is conducted by Revolutionary Committees. These permanent management groups exist in provinces, municipalitics, counties, factories, research institutes, and communes. Revolutionary Committees are not to be confused with local Communist Party Committees, which exist in parallel at the administrative levels of the government and work jointly in governing an area (see Fig. 2.2). In many cases the local party secretary is also the chairman of the Revolutionary Committee, but members of these committees are not all necessarily members of the Communist Party.

COMMISSIONS OF THE STATE COUNCIL

There are three commissions of the State Council that enjoy ministerial status. These are the State Planning Commission, the State Capital Construction Commission, and the Physical Culture and Sports Commission.

The State Planning Commission received ministerial status in 1975 and is headed by minister-in-charge Yu Chiu-li. The commission is engaged in drafting and implementing the national economic plan. It is believed that annual and five-year plans are both developed by a process of data collection from various production units such as factories and communes and that this process is administered at the district, provincial, and national levels.

Once the objectives are set it usually becomes necessary to deal with shortages of equipment and lack of productive capacity. When the specific needs are established, appropriate research institutes may be asked to determine whether the missing products are being manufactured in China and if not, whether they can be manufactured there. Only when the production of necessary elements does not exist or cannot be developed in time will a decision to buy foreign products be made. Thus research institutes exert a certain amount of influence on import decisions and should be kept up to date about the latest available products not manufactured in China.

The State Capital Construction Commission is the top construction authority in China and from 1961 to 1965 formed a part of the State Planning Commission. The activities of the State Capital Construction Commission are described in more detail in Chapter 15. The Physical Culture and Sports Commission was established in 1952 and in 1967 was placed under the control of the General Political Department of the People's Liberation Army. In 1971 it was reaffiliated with the State Council and elevated to ministerial status in 1975.

Table 2.3 Organs of the State Council and their top officials

	Area of Responsibility	Minister/Director
Commissions	State Planning Commission	Yu Chiu-li
	State Capital Construction Commission	Ku Mu
	Physical Culture and Sports Commission	Wang Men
Ministries	Agriculture and Forestry	Sha Feng
	Coal Industry	Hsiao Han
	Commerce	Fan Tzu-yu
	Communications (Land and Water Transport)	Yeh Wei
	Culture	
	Economic Relations with Foreign Countries	Chen Mu-hua
	Education	Liu Hsi-yao
	Finance	Chang Ching-fu
	Foreign Affairs	Huang Hua
	Foreign Trade	Li Chiang
	Light Industry	Chen Chih-kuang
	First Ministry of Machine Building (civilian)	
	Second Ministry of Machine Building (nuclear industry)	
	Third Ministry of Machine Building (armaments)	Li Chi-tai
	Fourth Ministry of Machine Building (electronics)	Wang Cheng
	Fifth Ministry of Machine Building (heavy weapons)	Li Cheng-fang
	Sixth Ministry of Machine Building (shipbuilding)	Pien Chiang
	Seventh Ministry of Machine Building (aerospace)	Wang Yang
	Metallurgical Industry	Tang Ko
	National Defense	Yeh Chien-ying
	Petroleum and Chemical Industries	Kang Shih-en
	Posts and Telecommunications (communications)	Chung Fu-hsiang
	Public Health	
	Public Security	Hua Kuo-feng
	Railways	Tuan Chun-i
	Water Conservancy and Power	Chien Cheng-ying
Special Agencies	Bank of China	Chiao Pei-hsin
	Central Meteorological Bureau	Meng Ping
	China Travel and Tourism Bureau	Yang Kung-su
	Chinese People's Insurance Company	Feng Tien-shun
	Civil Aviation Administration	Liu Tsun-hsin
	Commodity Inspection and Testing Bureau	Chang Ming
	Cultural Relics Administrative Bureau	Wang Yeh-chiu
	Foreign Experts Bureau	Mi Yung
	Foreign Language Publications/Distribution	
	Government Offices Bureau	Kao Fu-yu
	New China News Agency	Chu Mu-chih
	People's Bank of China	Chen Hsi-yu
	Publishing Department	
	Religious Affairs Bureau	
	State Building and Materials Industry Bureau	
	State Geology Bureau	Sun Ta-kung
	State Labor Bureau	Kang Yung-ho
	State Oceanography Bureau	
	State Publications Bureau	
	State Seismological Bureau	Liu Ying-yung
	State Statistics Bureau	
	State Supplies Bureau	Li Kai-hsin
	State Surveying and Cartography Bureau	
	Written Chinese Language Reform Committee	

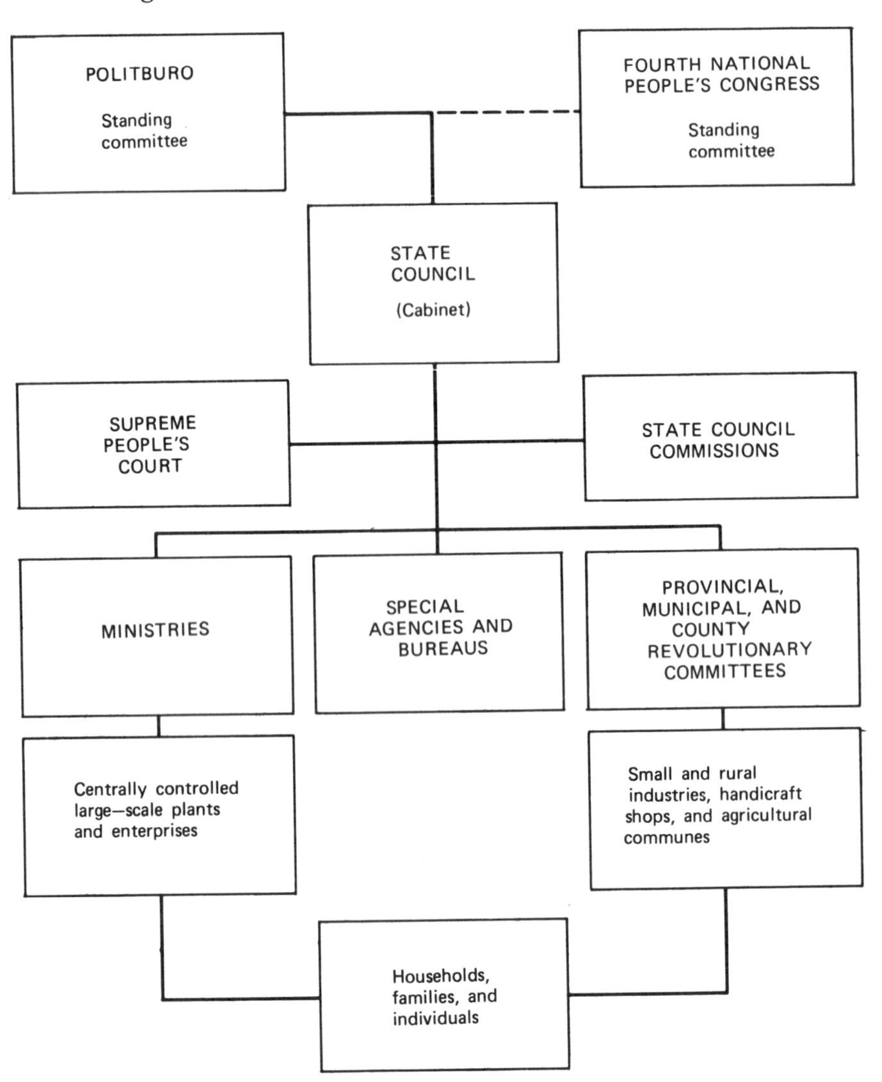

Figure 2.2 Administrative structure of the government.

MINISTRIES OF THE STATE COUNCIL

The State Council, which emerged after the adoption of the new Chinese Constitution in 1975, is a streamlined version of the State Council in existence prior to the Cultural Revolution. This was particularly noticeable in the number of ministries, which was reduced from forty to the present twenty-six.

Industrial ministries are of considerable interest to any executive interested in doing business with China. Most equipment imports are industrial producer goods that are most likely to be used by an enterprise under the supervision of one of the ministries. Although the Ministry of Foreign Trade in China throws up insurmountable barriers between the foreign supplier and the Chinese end user in form of monopolistic Foreign Trade Corporations, it is nevertheless worthwhile to know what industrial potential and domestic product competition do exist in China.

The ministries are large centralized administrations of various branches of all centrally controlled economic activity in China. They supervise the operations of large-scale enterprises throughout China that account for most of the industrial output in all branches of the economy. Chinese industrial ministries can perhaps best be visualized as monopoly conglomerates specializing in production of specific products or provision of particular services on a national scale. In this respect Chinese ministries are quite similar to those controlling the state sectors of the economy in the Soviet Union or other COMECON countries.

The Ministry of Foreign Trade is responsible for the formulation of overall import and export plans that are incorporated into the national Five Year Plans. These are used as a basis for specific plans prepared by the individual Foreign Trade Corporations, which are the only organizations authorized to engage in foreign trade. Activities of this ministry are discussed in detail in Chapter 4.

The Ministry of Foreign Affairs is responsible for the operation of China's diplomatic service and the conduct of Chinese foreign policy and works closely with the Ministry of National Defense. It is probably also responsible for shaping of China's foreign-aid programs to other developing countries, which are believed to be administered by the Ministry of Economic Relations with Foreign Countries. China's foreign policy and its foreign-aid programs are discussed in Chapter 6.

The Ministry of Finance controls the banking, foreign-exchange, and insurance activities through special agencies of the State Council. It is responsible for capital formation and investment allocation. Its control of foreign currency and gold reserves make this ministry an important factor in China's foreign trade. Its activities are covered in Chapter 5.

The Ministry of Commerce came into existence as a result of the division of a former Ministry of Trade in 1952. Since 1958 it has included a department previously known as the Ministry of City Services. It is most probably involved in supervision of consumer goods and supplies and distribution and orderly operation of local food markets.

The Ministry of Communications is responsible for operation, maintenance, and expansion of land and water transportation systems. It administers Chinese ports and its Maritime Administration operates the merchant fleet that engages in foreign cargo shipping and charter services. The Ministry of Railways operates the railroads, which is the most important transportation system in China. Air transport is operated by the Civil Aviation Administration of China, which is a special agency of the State Council. The activities of all those organizations as well as production and imports of motor vehicles, ships, aircraft, and locomotives are discussed in detail in Chapter 13.

The Ministry of Posts and Telecommunications is in charge of China's communications systems. It operates the major part of all telecommunications facilities such as telephones, telex, cable, and satellite communications. Radio and television broadcasting is under the control of a Bureau of Broadcasting Affairs. Operation of China's communication systems as well as production and imports of telecommunications equipment are covered in Chapter 14.

The Ministry of Coal Industry is one of the most important industrial ministries because it is responsible for the production of almost 70 percent of all energy consumed in China. Rapidly expanding production of oil also brings to the front ranks the Ministry of Petroleum and Chemical Industries. Detailed analysis of China's coal, gas, and oil industries as well as production and import of associated equipment are discussed in the comprehensive Chapter 7. The chemical and petrochemical industries and trade are covered in Chapter 10.

The Ministry of Water Conservancy and Power is responsible for the construction of thermal and hydroelectric power plants, transmission systems, and water-conservation projects of regional importance. Chapter 8 discusses in detail the activities of this ministry, as well as the manufacture and import of electric power-generating equipment.

The Ministry of Metallurgical Industry is responsible for the construction and operation of iron ore mines and major steel plants as well as other metal ore mines and refining plants. China has considerable mineral resources of most of the important metals and ores but is deficient in nickel, chromium, and cobalt. Production of metals as well as imports of strategic materials are discussed in detail in Chapters 4, 6, and 9.

The First Ministry of Machine Building is involved in production of equipment for economic development. Its products are generally discussed in chapters describing economic activity in which such products are used. Trucks, locomotives, aircraft, and ships, for example, are all discussed in Chapter 13. However, industrial equipment such as machine tools, engines, instruments, and computers, as well as all military production that is the output of the other six ministries of machine building, are discussed in Chapter 11.

The Ministry of Light Industry controls the supply of equipment and goods for the vast consumer market of China. It supervises textile industry, food processing, and consumer durables such as bicycles, sewing machines, and watches. It is an important ministry not only because of its size, but as an organization producing goods for export. Some industries under its control are discussed in Chapter 16.

The Ministry of Agriculture and Foresty is of paramount importance in China because it is responsible for the productivity in agriculture, which has to provide food for the largest population in the world. Agriculture also contributes a large percentage of China's exports. China is the second largest food-producing country in the world, and constant improvement of agriculture receives top investment priority. Production of major crops, imports and exports of agricultural products, and use of mechanization in agriculture are discussed in detail in Chapter 12.

The Ministry of National Defense is responsible for the activities of the PLA, which is estimated to number 4.3 million persons under arms. In addition, PLA units are heavily involved in economic activity throughout the country and in many areas operate numerous major plants and transportation services. The military–industrial complex of China is responsible for introduction of the most advanced technology into Chinese industry. Most of this activity is discussed in detail in Chapter 11.

The Ministries of Culture, Education, and Public Health provide the social infrastructure to the people of China. These systems are discussed in Chapters 17 and 18. The Ministry of Public Security is responsible for maintaining law and order in the country. It probably controls the armed civilian militia, which is estimated at up to 7 million people organized into divisions and regiments. There are also an estimated 300,000 paramilitary security and border troops stationed near border areas, but it is not clear whether these are under the jurisdiction of this ministry.

SPECIAL AGENCIES OF THE STATE COUNCIL

About twenty Special Agencies of the State Council have been identified. These organizations generally provide a service to various sectors of the economy but do not necessarily operate on a nationwide basis. Some agencies, like the People's Bank of China, Bank of China, and Chinese People's Insurance Company, are under the control of Ministry of Finance, and their activities are discussed in detail in Chapter 5. Similarly, the Commodity Inspection and Testing Bureau is controlled by the Ministry of Foreign Trade.

The Central Meteorological Bureau has operated since 1954, and its function is to coordinate weather research and forecasting activities. The agency collects weather data from regional administrative centers as well as from some 10,000 rural weather posts in communes and work brigades. Electronic computers are used in weather forecasting, and the agency recently purchased two large computers from Hitachi in Japan, valued at $8.6 million to equip its largest weather-forecasting center in Peking.

The State Surveying and Cartography Bureau has existed since 1955, and its known activities include the operation of Institutes of Surveying and Cartography in Peking and Wuhan. The State Oceanography Bureau came into being in 1964 to coordinate oceanographic research in cooperation with shipping and fishing units. The Seismological Bureau is involved in earthquake research and forecasting and involves about 100,000 people throughout China, including some 10,000 professional seismologists.

The Civil Aviation Administration of China exists since 1962 and combines the functions of a national airline, airports operator, and central aeronautical authority. Its activities are discussed in considerable detail in Chapter 13. The China Travel and Tourism Bureau, in existence since 1964, is the agency responsible for all foreign travel in China. In this respect its activities are very similar to those of Intourist of the Soviet Union.

The New China News Agency operates an international network for worldwide information collection and distribution. It often performs functions of semiofficial nature in countries where China has no diplomatic status and is sometimes considered as part of the foreign-affairs establishment of China. Radio and television broadcasting as well as the New China News Agency are discussed in more detail in Chapter 14.

Very little is known about the remaining special agencies of the State Council. Independent existence of these agencies is sometimes short-lived, and they may be incorporated in some other organization that is a large user of the services of a particular bureau. In general, the status of all other Special Agencies of the State Council must be regarded as uncertain.

THE MILITARY ESTABLISHMENT

Extensive involvement of the PLA in economic development of China and its relationship to the political apparatus make it unique among military institutions in the world. Actually the armed forces fulfill a role envisaged by the Chinese Constitution, which states in Article 15 that ". . . the Chinese People's Liberation Army is at all times a fighting force and simultaneously a working force and a production force."

Besides the Chinese Communist Party the PLA is the only other significant power structure in existence throughout China. The ultimate authority over the PLA rests with the Military Commission of the Central Committee of the Communist Party, which includes Chairman Hua Kuo-feng, the Minister of National Defense, and several members of the Politburo (see Fig. 2.3).

There are at least eight service arms of the PLA that parallel several of the ministries of the machine-building industries. For example, the Second Artillery Corps, which is in fact the Chinese strategic missile delivery force, is supported in industry by the Seventh Ministry of Machine Building. A full discussion of the Chinese military–industrial complex is found in Chapter 11.

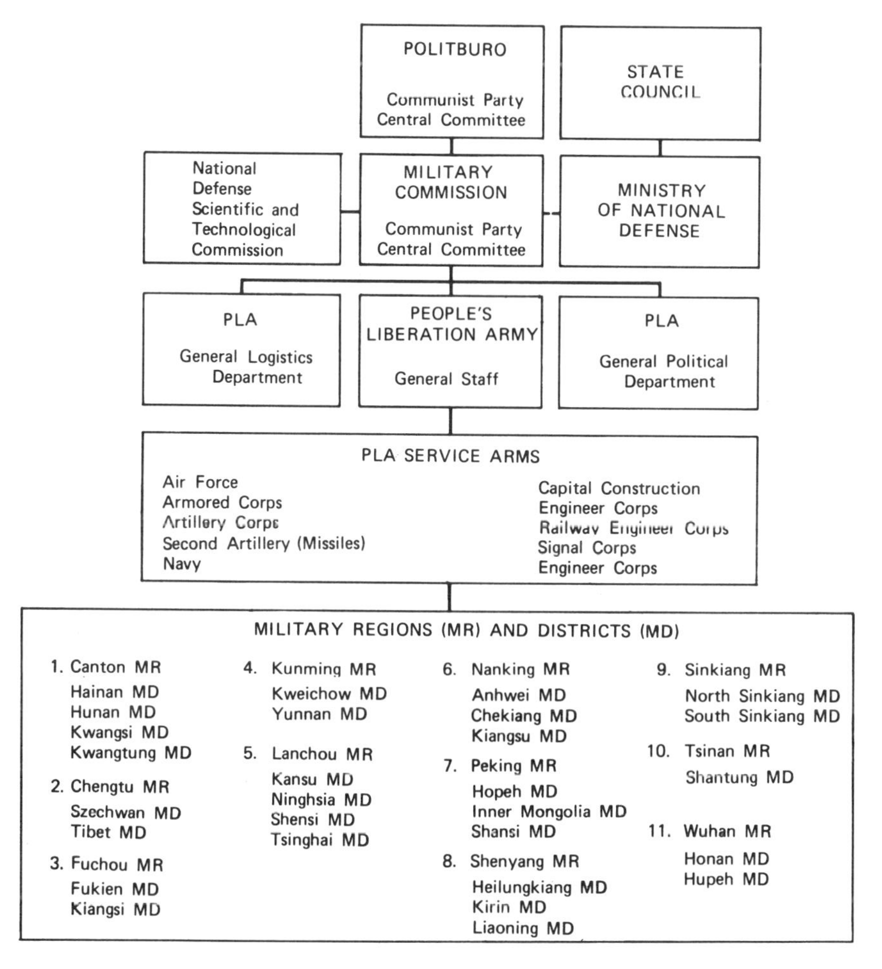

Figure 2.3 Military establishment of the People's Republic of China.

The defense-oriented ministries as a group operate the most advanced technology industries, and at least 50 percent of their total output is believed to be equipment for military armed forces. Such industries as electronics, nuclear power, and aerospace owe their existence in China almost exclusively to the military establishment.

One of the most significant contributions of the PLA is in the development of the Chinese economy. Its railway engineering corps was instrumental in reconstruction of major communications links and building of new lines within both China and East Africa. The PLA was also instrumental in building a large portion of highways and to this date operates transportation systems in remote regions such as Tibet.

The PLA Capital Construction Engineering Corps has massive economic functions, and in some areas of the country PLA units literally run the economy. In the Sinkiang military region the PLA is reported to operate 500 large and small factories and mines. These include iron and steel, power generation, coal and metal mining, chemicals, machine building, textiles, food processing, and agricultural production.

All units of the PLA cultivate their own food and operate their own small plants whose surplus is sold to the state. In addition, the PLA provides assistance to other organizations in construction projects or harvesting, or it is called on to help in cases of emergency. Although the PLA is highly political, it is a very important institution, which permeates the daily economic life of China to a much higher degree than do armed forces of other nations.

REFERENCES

Ashbrook, Arthur G. "China: An Economic Overview, 1975," in *China: A Reassessment of the Economy,* Joint Economic Committee, U.S. Congress, July 1975, p. 20.

Brown, General George S., "*U.S. Military Posture for FY 1978,*" Joint Chiefs of Staff. 20 January 1977, Washington, D.C.

China News Summary. "Fourth National People's Congress," No. 552, January 22, 1975, Hong Kong.

Current Scene. "PRC Provincial Party Committees," April 1976, Hong Kong.

FBIS. EEU-75-20. "Nepszabadsag Views Recent PRC Congress Decisions," Hungarian Radio Broadcast Summary, January 25, 1975.

U.S. Department of State. *World Strength of the Communist Party Organizations, 24th Annual Report,* Bureau of Intelligence and Research, 1972.

U.S. Government. *People's Republic of China Atlas,* November 1971.

U.S. Government. "Directory of Officials of the People's Republic of China," *Reference Aid,* CR77-15208, October 1977.

U.S. Government. "Military Organizations of the People's Republic of China," *Reference Aid,* CR76-12748, June 1976.

U.S. Government. "Politburo of the Chinese Communist Party," *Reference Aid,* CR76-12982, July 1976.

U.S. Government. "The Chinese Communist Party Central Committee," *Research Aid,* CR77-14473, September 1977.

U.S. Government. "Central Government Organizations of the People's Republic of China," *Reference Aid,* CR76-14670, November 1976.

U.S. Government. "China's Economy," *A Research Paper,* ER77-10733, November 1977.

3

Economic Growth and Central Planning

This chapter is intended as an overview of the Chinese economy in the mid 1970s and a discussion of basic trends in various industries. The objective is to develop a comprehensive picture of the existing industrial base and to outline reasonable upper bounds of growth of the Chinese economy in the years ahead.

It is also designed to familiarize the reader with the objectives of the Chinese leadership and their policies developed to attain planned targets. Methodology employed in Chinese economic planning and technological innovation is also outlined in this chapter.

This discussion contains several statements and conclusions that have been developed in other chapters of this book dealing with specific industries in question. The scenarios describing possible Chinese development during the rest of this century are only one plausible set of assumptions and may never materialize. Rates of growth of various countries vary greatly each year, and after a few years trends may set in that could invalidate the assumptions used in this analysis. This chapter is intended more as an exercise for comparison of possible economic magnitudes and should not be regarded as forecasts or projections of actual future growth.

CHINA IN THE MID-1970s

As an international power measured in terms of GNP, China is the sixth largest economy in the world. In 1976 its GNP was estimated at about

$307 billion, about 13 percent smaller than the GNP of France and of the same order as the combined GNP of the six countries of eastern Europe. Besides France, only the economies of the United States, the Soviet Union, Japan, and West Germany are larger than that of China (see Table 3.1).

The per capita GNP, however, is only $318 per year, which ranks China among the less developed countries of the world. Although this GNP level puts China ahead of India and Indonesia, it is still lower than that of Nigeria and well below the per capita GNP of Brazil. However, it is an unusual developing country and although one third of all economic output still comes from agriculture, China has been able to develop several large modern industries. These include significant steel, petroleum, shipbuilding, electronics, and machine-building industries that were able to produce nuclear weapons, missiles, and satellites and provide support for a very large military establishment.

Population

China's population is now estimated at 965 million. The People's Republic of China never published any official population data, but a deputy prime minister admitted in 1977 to a reporter from *The Economist* of London that China's population is 900 million. United Nations estimates, often quoted in the media, put China's population at 835 million in 1975, whereas the U.S. Department of Commerce, Joint Economic Committee, and other government agencies use the higher figure, which was also adopted for use in analysis in this book. Another set of estimates comes from Soviet sources, which put China's population at 820 million in 1972.

China's population is also very young in comparison with those of other large and populous countries. About 57 percent of all Chinese are under 25 years of age. This means that at least half of the population neither experienced nor remember China before the communist takeover in 1949. In the economically active population group aged 15 to 65 years there are an estimated 564 million people, of which at least 420 million, or 44 percent of the total population constitute China's labor force. It is believed that as much as 85 percent of the labor force is engaged in agriculture. About 700 million Chinese are peasants living in the rural areas, but the literacy rate of the whole population is believed to be at least 25 percent. It appears to be comparable to the literacy rate of the population of India, which was reported to be 29 percent in 1971 but is well below that in Indonesia or Brazil, where it is 60 percent and 67 percent, respectively.

Table 3.1 Economic profile of China and selected developed and developing countries (All data for 1976 unless otherwise stated)

Country	*GNP in Billions of $U.S.*	*Population in Millions*	*GNP per capita in $U.S.*	*Electric Power in Billion kW-h*[a]	*1975 Grains Output in Millions of Metric Tons*	*Steel Output in Millions of Metric Tons*	*Foreign Trade in Millions of $U.S.*
China	307	965	318	130	260	23	13.0
United States	1,516	216	7,018	2,000	204	106	211.0
Soviet Union	897	259	4,152	1,109	139	156	75.0
Japan	555	114	4,868	504	17	107	122.0
West Germany	473	61	7,754	353	21	61	190.0
France	353	53	6,660	206	35	23	119.0
Brazil	110	112	982	80	26	10	23.0
India	73	642	113	93	101	7.2	10.3
Indonesia	28	137	204	5.6	27	—	13.5
Nigeria	27	66	409	3.6	—	—	16.3
Vietnam	6.5	51	127	3.0	—	—	1.2

Source: *United Nations Statistical Yearbook, 1976; U.S. Government Handbook of Economic Statistics, 1976; U.S. Government Factbook, 1977.*

[a] kW-h = kilowatt-hours.

The estimated annual growth rate of China's population is now about 1.5 percent, which is more in line with that of Japan than other developing countries, where it is usually well over 2 percent. The actual rate is about 15.3 persons per thousand annually. The Foreign Demographic Analysis Division of the U.S. Department of Commerce projects that this rate will further decrease to about 13–14 per thousand between now and the year 2000. Nevertheless, at the present rate of growth China's population is expected to reach 1007 million by 1980 and be over 1300 million by the year 2000.

Agriculture

China is the largest grains producer in the world, primarily because of its preeminent position as the world's top rice-growing country. It exports rice and imports wheat, and its import dependence on grains is in the order of only 2 percent of total production. China's productivity in agriculture is considerably lower than that of many other countries, and there is still room to increase crops with fertilizer inputs without the need to cultivate additional land.

Grains production on a per capita basis in China is actually considerably higher than in India or Indonesia and almost 20 percent better than in Brazil. But it is still less than half the per capita level in the Soviet Union and almost four times smaller than that in the United States. If China wants only to maintain the present per capita grains production by the year 2000, total production must increase to at least 350 million tons per year, 90 million tons more than in 1975. A doubling of per capita production, whenever it occurs, would probably be reaching the productivity limits of available agricultural land in China.

Energy and Power

China is richly endowed with energy resources in the form of coal, oil, gas, and hydroelectric power potential and is the fourth largest energy-producing country in the world. It ranks third in the production of coal, fifth in the production of natural gas, and already is eleventh in the production of crude oil. Chinese energy resources are comparable in the order of magnitude to those of the United States and the Soviet Union and are considered to be adequate for centuries to come.

Because of its huge population, however, China's per capita energy consumption is only 522 kg (kilograms) of coal equivalent per year.

Whereas this is at least twice as high as in India, Pakistan, or Indonesia, it is well below the energy-consumption level in Brazil. If reasonable energy-production growth rates are maintained and if huge investments are made in this industry, China could achieve production levels of 1000 million tons of coal and 1000 million tons of oil annually by the year 2000. But even at such high production levels Chinese per capita consumption would only then reach the average world consumption level of today.

Electric power production was estimated at 133 billion kW-h in 1976, which makes China the ninth largest electric power-producing country in the world. Power-generation levels are comparable to those existing in the United States about the year 1930 and in the Soviet Union during the early 1950s. Electric power supply remains inadequate even though only 3.5 percent of all the electricity produced is available to the residential and commercial consumers. Special authorization is required for the purchase of a light bulb more powerful than 40 W (watts).

Industry

China manufactures all types of equipment and materials required by a modern economy, but in many industrial sectors the output does not meet the demand. Limited investment resources mandate assignment of priorities that in turn may leave some industries underdeveloped. Recognized or planned deficiencies of production are then met by imports of materials and equipment from abroad, mainly Japan, western Europe and the Soviet Bloc countries.

China appears to have an adequate production of agricultural machinery, internal combustion engines, construction and mining equipment, and simple electronic and telecommunications equipment. Domestic industries are short by at least 10 percent in production of railway vehicles, machine tools, and motor vehicles. Even larger shortages exist in production of steel, ships, electric power-generating equipment, and instruments.

Iron and steel is imported in large quantities, even though China is the sixth largest steel-producing country in the world. For the last several years iron and steel has been the largest single import-product category in Chinese foreign trade. Domestic production of copper and aluminum is inadequate, and large quantities of these metals are also imported. Nickel, chromium, and cobalt are strategic metals because their production in China is negligible and practically all the requirements are met by imports, in many cases from the Third World Countries.

Production of chemicals is also inadequate. During the early 1970s China became the largest importer of chemical fertilizers in the world. It also imports considerable amounts of organic chemicals, synthetic fibers, plastics, and rubber because the domestic production of these materials cannot meet the demand.

To improve its production capacity China has imported a total of at least 70 modern turnkey plants since 1970. Valued at over $2 billion, they include petrochemical, fertilizer, electric power, petroleum exploration and extraction, steel finishing, and other plants whose output is now entering the production stream.

China's textile industry is one of the largest in the world, equal to that of the Soviet Union and India. Consumer durables are limited to such items as bicycles, sewing machines, watches, and radios. Private ownership of automobiles is not allowed.

Foreign Trade

Despite the size of its economy and operation of some of the largest industries in the world, China's foreign trade accounts for only 4.7 percent of its GNP. China ranks as the twenty-seventh in the import market, and its total foreign trade is comparable to that of either South Korea, Hong Kong, Singapore, or Yugoslavia.

About 80 percent of this trade is conducted with Japan, western Europe, Canada, Australia, and the United States. China traditionally has a trade deficit with developed countries that is balanced by surpluses resulting from exports of foods, textiles, and increasingly crude oil and petroleum products to Hong Kong, Singapore, and other developing countries. Trade with communist countries is conducted by bilateral agreements and consists mostly of imports of Soviet and eastern European machinery in return for Chinese raw materials.

Although it is a developing country itself, China also grants foreign aid to other developing countries in Asia, Africa, the Middle East, and Latin America. Chinese foreign aid is in the form of credits repayable in the future by exports of local products. Most countries receiving Chinese foreign aid produce metals and materials that China has to import.

A CHINESE SUPERPOWER BY THE YEAR 2000?

China is now believed to possess most of the natural resources required for development into a future superpower. Most, but not all. In comparison with the existing superpowers, China is self-sufficient in energy, which is the *sine qua non* of superpower status. China's energy situation appears to be superior to that of the United States, although per capita energy consumption is extremely low. It is also superior to that of the Soviet

Union, which is facing the possibility of declining oil production in the future and an obligation to continue oil and gas supplies to eastern and western European clients. China is in fact unaffected by the OPEC oil cartel as far as its domestic energy requirements are concerned. Japan and West Germany, the two next largest economic powers after the United States and the Soviet Union, both must depend on imports of oil. Brazil may have a certain potential in the longer term but so far has not been able to develop energy independence.

Most of the basic raw materials required for the operation of modern industries are also available in China. However, the Soviet Union has a definite advantage because it is practically self-sufficient in all materials, including strategic metals. China lacks significant deposits of nickel, chromium, and cobalt, which are vital to the modern steel industry. In this respect China's deficiency is almost identical to that of the United States. China is also believed to have at present very limited domestic deposits of platinum, cadmium, titanium, diamonds, and gold, all of which are mined in much larger quantities in the Soviet Union.

On the other hand, China has rather limited capital investment resources in comparison with the United States and the Soviet Union. During the 1960s China found it more lucrative to develop petroleum rather than the minerals and metals resources, because oil offered a more readily exportable product that would increase foreign-exchange earnings. As a result, China is considerably underexplored, and some industry observers believe that in the area as large as that of China economic deposits of most strategic materials will sooner or later be found.

The main problem confronting China in its bid for superpower status is agricultural production. This must increase faster than population to improve the present relatively low consumption levels. There is ample evidence that Chinese leaders are trying to improve agricultural productivity, particularly by inputs of chemical fertilizers and mechanization because most of the fertile land in China is already under cultivation. Various other countries, including developing nations such as Egypt or Mexico, have been able to obtain considerably higher yields of rice, wheat, soybeans, or cotton than China has at present. There is every reason to believe that China will improve its agricultural productivity, but even if production of grains is doubled by the year 2000 to at least 500 million tons per year, per capita grain production will still be below that of the Soviet Union today. This may nevertheless eliminate imports of wheat, which appear significant but actually represent only 2 percent of apparent grains consumption. China already claims it is fairly self-sufficient in grains because it exports comparable amounts of rice every year.

Improved productivity in agriculture can also release any additional manpower that may be required by industry in the future. Here China has an advantage over the Soviet Union, where 25 percent of the labor force is still required in the farms and industrial labor shortages are looming ahead. Another advantage of a large reservoir of manpower also lies in its availability to develop labor-intensive modern industries such as electronics, which is also indispensable to superpower status.

Given all those preconditions, whether China becomes a superpower by the year 2000 or not is more a question of definition. If China's GNP increases at an average 6 percent per year from $307 billion in 1976, it will be in the order of $1,250 billion in the year 2000 or about 85 percent of the GNP of the United States today. But its per capita, GNP, given an increase in China's population to about 1300 million will then only reach $954 per year (see Table 3.2). This is comparable to the per capita GNP already achieved by Taiwan or Brazil today.

Table 3.2 Projected growth of Chinese GNP relative to the Soviet Union, Japan, and the United States to the year 2000 computed in billions of 1975 $US

Country	*1976 GNP in Billions of $U.S.*	*Percentage of Assumed Average Growth Rates*	*1985 GNP in Billions of $U.S.*	*2000 GNP in Billions of $U.S.*
China	307	6	518	1,241
United States	1,510	2	1,804	2,428
Soviet Union	897	3	1,170	1,823
Japan[a]	555	6	876	1,822

Source: Developed by the authors.

[a] Japanese growth rate assumed at 5 percent after 1982.

On the other hand, assuming slower GNP growth rates for the Soviet Union (3 percent) and United States (2 percent) over the same time period, their respective GNPs will become $1,823 billion and $2,428 billion. As a result of this scenario, China's GNP would increase from 20 to 50 percent of the American GNP and from 35 to 60 percent of the Soviet GNP while remaining basically in the same ratio to the GNP of Japan. In absolute GNP terms, ignoring per capita comparisons, China appears to have the potential to move from being the sixth largest to fourth or even third largest economy in the world, overtaking West Germany and France in the process. Political stability for the next 20 years and significant development of management and technological capabilities would be required to achieve this goal.

TRENDS IN ECONOMIC GROWTH

Previous sections of this chapter presented a profile of Chinese economy relative to other countries and established an upper boundary of possible economic development until the year 2000. China is now at the halfway point on its road to the superpower status it would like to attain by the end of this century. Hence there is a track record of 27 years since the establishment of the People's Republic of China, which can be examined in the hope of providing some indication of possible future developments.

The average annual GNP growth rate for 1949 to 1975 is now estimated to have been 7 percent, despite short-term GNP declines and stagnation during the Great Leap Forward and the Cultural Revolution periods. This is due to extremely rapid GNP growth during the Rehabilitation and the first Five Year Plan period of 1949–1957, when the GNP increased at an average 11 percent per year. During that period Chinese GNP more than doubled from an estimated $49 billion in 1949 to $115 billion in 1957 (see Fig. 3.1).

The effect of the Great Leap Forward policies was truly disastrous to the economy, and the GNP declined from its 1958 high of $137 billion to a low of $99 billion in 1961. This was a precipitous drop of 27 percent and the situation was worsened by poor weather and bad harvests that forced China to import grains equivalent to 3.5 percent of domestic production. Agricultural production suffered a drop of almost 29 percent and did not recover to 1958 levels until 1964.

Industrial production peaked in 1960 following a ruinous growth tempo of small-scale industrialization throughout the country. Industrial development decreased sharply after nationwide shortages of equipment developed as a result of unplanned and uncoordinated construction programs. Industrial production dropped 42 percent from its peak in 1960 in one year but recovered by 53 percent from its low by 1964. Despite this sharp decline, industrial production still managed a 9 percent average annual growth during 1957–1965. Between 1961 and 1967 industrial production staged a rapid comeback, and its growth rate averaged 16 percent per year.

There were also large declines in construction activity and transportation. Foreign trade was particularly hard hit. China first reduced its imports to improve trade balance by exporting its traditional agricultural products. However, as agricultural production decreased, exports became unavailable and imports were further cut to the bone. In 1961 China had to resort to the sale of silver and gold and obtained short-term commercial credits to finance the vital imports of grains from Australia, Canada, and Argentina.

Despite near starvation, a sharp drop in national morale, and a de-

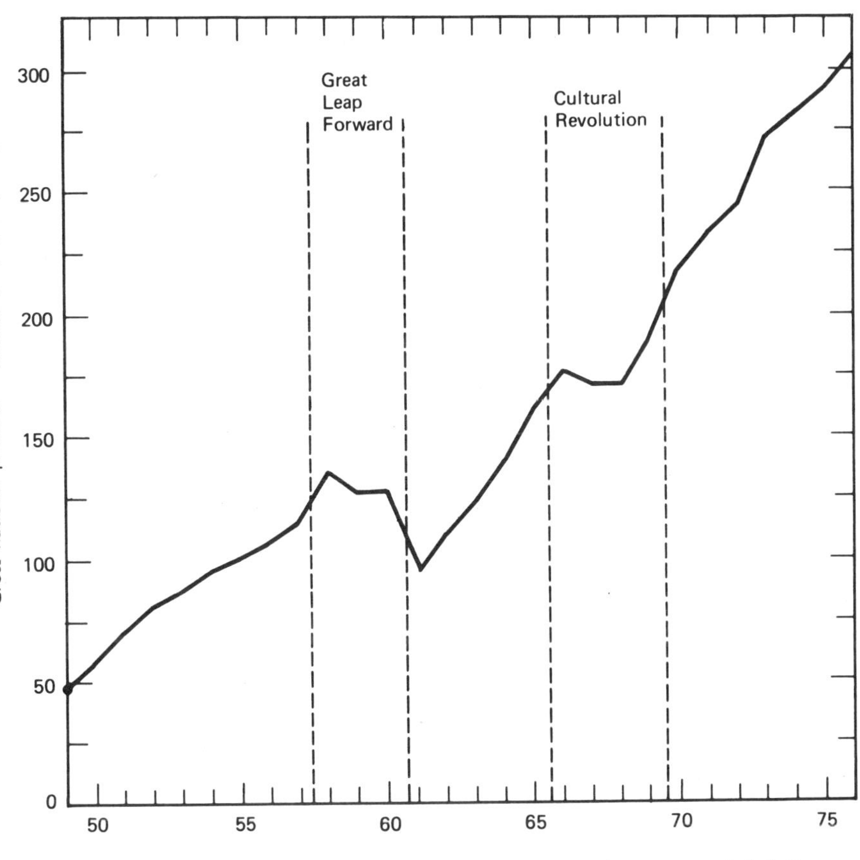

Figure 3.1 Growth of Chinese GNP since 1949. (**Source:** ***Joint Economic Committee of the U.S. Congress, 1975.***)

clining economy, China recovered quickly and successfully from the effects of the Great Leap Forward policies, and by 1964 the GNP reached $142 billion, about $5 billion ahead of its previous high in 1958. What is remarkable is that during the 1957–1965 period of Great Leap Forward implementation and recovery, China's GNP shows an average 4 percent annual growth rate despite three consecutive years of actual decline (scc Table 3.3).

At this point one cannot help but wonder where China would be today were it not for the Great Leap Forward setbacks and the withdrawal of Soviet and eastern European aid. China's GNP was growing at about 7 percent annually in the years preceding the Great Leap Forward and continued to increase at an average growth rate of 6 percent since recovery in 1965. If the GNP of China had continued to grow at an annual average rate of 6 percent uninterrupted since 1958, by 1976 it would have been about $370 billion, or 20 percent higher than it is today. If this growth is projected to the year 2000, China's GNP would then be about 80 percent of the Soviet and Japanese GNP and over 60 percent of the U.S. GNP projected at growth rates discussed in Chapter 2.

Table 3.3 Average annual rates of growth (in percent) of major sectors of the Chinese economy since 1949

Economic Sector	*Rehabilitation (1949–1952)*	*First Five Year Plan (1952–1957)*	*Great Leap Forward and recovery (1957–1965)*	*Cultural Revolution (1965–1970)*	*Fourth Five Year Plan (1970–1975)*	*Projected Fifth Five Year Plan (1976–1980)*
Gross national product	19	7	4	6	6	6
Agriculture	15	4	2	3	2	3
Industry	34	16	9	10	10	8 to 10
Construction	43	20	7	4	9	8 to 10
Transport	35	21	7	7	9	10
Exports	31	12	3	1	27	10[a]
Imports	36	7	3	4	27	5[a]

Source: *U.S. Government Handbook of Economic Indicators for PRC, August 1976; U.S.–China Business Review, July–August 1976;* and authors' estimates.

[a] Initial estimates for 1976 indicate a 16-percent drop in imports and only a 1-percent growth in exports from the previous year. Because China now returned to a positive balance of trade position it is believed that foreign trade will continue to grow during the next few years.

The Cultural Revolution, disruptive as it was and lasting for almost five years, did not produce economic declines as severe as those of the Great Leap Forward. During the revolution China's GNP actually declined by only $5 billion, from $177 billion in 1966 to $172 billion in 1967, or somewhat less than 3 percent. In contrast, it fell by $38 billion between 1958 and 1961, equivalent to at least 30 percent of the average GNP in those years. Thus the Cultural Revolution had a much milder effect on the Chinese economy and despite the political turmoil and various dislocations, the Chinese GNP shows an average 6 percent annual growth rate during 1965–1970. One can venture to say, on the basis of these estimates, that China could probably undergo another political turmoil or two before the year 2000 comparable to that of the Cultural Revolution without seriously impairing its economic growth in the long run. This is particularly true as its economy grows with time and presents an increasingly larger inertia to political disruptions.

Since 1970 China has resumed regular planning, and during the fourth Five Year Plan (1971–1975), GNP growth again averaged 6 percent annually. During that period considerable gains in industrial capacity were made, and industrial production increased by almost 60 percent. China's GNP increased by $80 billion and in 1975 was 36 percent higher than at the beginning of the period.

Initial estimates of China's GNP for 1976 suggest that China once again suffered an economic setback resulting from the combined effects of the Tangshan earthquake and economic disruptions connected with the struggle to neutralize the "gang of four" movement. This time the GNP continued to grow, but at a sluggish 2 to 3 percent rather than the average 6 percent of the preceding years. There was no actual decline in China's GNP as such but only of its rate of growth. However, industrial decline did occur, particularly in the steel industry, whose production dropped by at least 10 percent, primarily as a result of the damage to the largest coal mines and a major steel complex in the wake of the Tangshan earthquake. Increases in oil, agricultural, and equipment production provided an offsetting factor, and the total GNP continued to grow. It is now believed that China has the capacity to increase its industrial development by over 10 percent for a year or two to offset the slower growth in 1976.

Although as the economy becomes larger its rate of growth slows down, China at present appears to have several industries at "takeoff" points, probably entering periods of very rapid growth. Electronics and petrochemicals are two specific examples. Transportation systems also are clearly destined to contribute significantly to future growth. A GNP growth rate above 6 percent would require domestic consumption of all petroleum produced in China and is unlikely to take place if oil exports are required to generate additional hard currency to pay for imports of Western machinery for further industrial expansion.

In this respect China must rely on world markets for its oil and other traditional products, and as its economy grows it may have to do so to a larger degree. These markets are neither stable nor certain, and some OPEC oil-producing countries are operating at only 75-percent capacity, which influences oil prices and may ultimately limit China's oil-export trade. China may find increasing trade with eastern Europe and the Soviet Union more conducive to its economic growth in the future, now that its economy is much larger and less dependent on direct foreign assistance. Complementarity of Soviet and Chinese economies is becoming particularly apparent. China would probably achieve the fastest possible economic growth by dealing with the Soviet Union as its largest natural trading partner. This, of course, is a purely political decision and will probably not come under consideration until a change of present Soviet leadership comes about.

DEVELOPMENT POLICIES AND PRIORITIES

The latest strategy for modernization of China's economy was outlined by its leadership during a series of national conferences dealing with agricultural and industrial development. Agricultural development policies were set out during two National Conferences entitled "Learn from Tachai in Agriculture" held in the autumn of 1975 and December 1976. Overall industrial development policies were outlined during the National Conference on "Learn from Taching in Industry" held from April 20 to May 13, 1977 at the Taching Oilfield in Heilungkiang Province and in Peking. Tachai is a production brigade in Hsiyang County of the Shansi Province with consistent higher-than-average agricultural yields. Taching is the largest oilfield in China. Developed in only 3 years under extremely difficult conditions, it now produces an estimated 50 percent of all crude-oil output in China and runs its own self-supporting establishment.

Those three comprehensive conferences herald a new drive by the leadership to continue national development at a faster pace but along the general policy of "taking agriculture as the foundation and industry as the leading factor." This general policy is paralleled by a strategic policy of "being prepared against war, natural disasters, and doing everything for the people" as well as "digging tunnels deep, storing grain everywhere, and never seeking hegemony."

Besides the major national conferences on agriculture and industry well over twenty conferences were held since November 1976 on specific sectors of the economy that are clearly receiving development priority under the current Five Year Plan (1976–1980) and possibly until 1985. These include petroleum, coal, transportation, iron and steel, national defense, light industry, and science and technology (see Table 3.4).

The long-range objective of China's development policies is now becoming known as the "grand plan," which originated in a 1956 statement by Chairman Mao Tse-tung in which he viewed overtaking of the United States economically in 50 or 60 years as "not only possible, but absolutely necessary and obligatory." This was not necessarily just wishful thinking on the part of Mao Tse-tung, who was at the peak of his achievement at the time and had not yet experienced the failures of the Great Leap Forward experiment. Between 1949 and 1957 China's GNP had grown on the average a remarkable 11 percent per year, and massive technological assistance was pouring in from the Soviet Union and eastern Europe.

Communist sinologists today claim that Chinese leaders at that time were not satisfied with the amount of economic and technical aid received from the Soviet Union and eastern Europe. China apparently demanded increasing levels of assistance proportional to the accelerating growth rate

Table 3.4 Major national conferences outlining development policies and priorities held in China since 1975

Year	*Month*	*Name of Conference*	*Industry*
1975	Fall	First National Tachai Conference	agriculture
		First National Meteorological Conference (Tsingtao)	agriculture
1976	Nov.	National Petroleum Industry Meeting	oil industry
	Dec.	Learn from Taching Conference (Peking)	oil production
		Learn from Taching Conference (Peking)	oil refining
		Second National Tachai Conference (Peking)	agriculture
1977	Jan.	National Coal Industry Conference	coal mining
		Learn from Taching Conference (Peking)	light industry
		Power and Energy Conservation Conference (Canton)	energy
	Feb.	National Defense Conferences (four)	armed forces
		Second National Meteorological Conference (Hangchow)	agriculture
		National Conference on Railway Work	transportation
	Mar.	Forestry and Aquatic Products Conference (Peking)	agriculture
		National Railway Security Conference	transportation
		Metallurgical Enterprises Conference	iron and steel
		Funding of Agriculture Conference (Hsiyang)	agriculture
		Fuel Economy Conference (Tatung)	transportation
	Apr.	Banking and Finance Conference (Taching)	industrial investment
		National Meeting of Academy of Sciences	science/technology
		National Marketing Cooperatives Conference	rural supplies
		National Capital Construction Conference	construction
		National Machine Building Conference	industrial equipment
		National Communications Work Conference	transportation
	May	First National Taching Conference	industry
	Jun.	National Cotton Production Conference	agriculture
		National Farmlands Conference	agriculture
	Jul.	National Foreign Trade Conference	import/export
	Aug.	National Posts and Telecommunications Conference	communications
1978		National Conference on Science (to be held)	science/technology

Source: Current Scene, April/May, Hong Kong.

of China's economy and claimed that it was the responsibility of the Soviet Bloc to provide the largest possible aid to an underdeveloped fraternal socialist state. Because of China's size such action would have quickly made the economies of the Soviet Bloc subordinate to the requirements and demands of China. We may never know what cooperation promises were made by COMECON in the first place, but it appears that when China called their bluff the Sino-Soviet rift became inevitable.

It may well be that the Great Leap Forward programs were introduced in an attempt to force the Soviet Bloc countries to increase their aid to China. Although the damage to the economy has been colossal, it is nevertheless remarkable how quickly China recovered and resumed a relatively rapid growth, particularly in industry. However, any "grand plan" for overtaking the United States that may have been in effect at the time automatically received a setback of about 5 years.

When Premier Chou En-lai discussed China's economic development policy in 1975 he simply indicated that China's goal was to achieve "front rank" status by the end of the century. This is a more modest and plausible objective than overtaking the United States and probably signaled a more pragmatic outlook on economic growth. Chou En-lai also indicated that China was drawing up a 10-year plan for 1975–1985, concurrent with the regular Five Year Plans, and called it "crucial to the attainment of long-term objectives." Chou's statements are in line with possible growth of China's GNP to become the third largest in the world. However, although such an overall objective is within reach, the per capita GNP in China even then will still be comparable to that of some developing countries today.

These policies are being continued by the present leadership with some semblance of a "new leap" attitude toward economic development. Chairman Hua Kuo-feng specifically stated at the First National Taching Conference that it is possible for the Chinese economy to develop faster and better in the next 23 years than it did in the previous 28. This means that Chinese leadership feels confident that China's GNP can increase faster than the already remarkable average growth rate of 6 percent annually since 1949. As it is almost impossible that agriculture can grow faster than 3 percent annually, the overall growth would have to be made up by even faster growth of industry.

Future development is to take place in two phases, with the first crucial stage ending in 1985. Iron and steel, petroleum, coal, electric power, chemicals, and machine building are industries for priority development because their output provides vital inputs to agriculture such as tractors, fertilizers, fuels, and power. Military considerations have not been overlooked, and Vice Chairman and Minister of National Defense Yeh Chien-ying pointed out at the Taching conference that priority development of those industries will also provide modern arms and equipment, transportation, supplies, means of reconnaissance, and advanced command systems vital to a modern national defense establishment.

The "new leap" attitude is particularly apparent in a statement made by Chairman Hua Kuo-feng calling for the construction of ten new oilfields as big as Taching before the end of this century. Since Taching is now producing at least 43 million metric tons of crude oil per year, this means that China hopes or expects to produce at least 500 million tons annually by the end of the century. This is an enormous amount of crude oil, comparable to the total annual production of either Saudi Arabia or the Soviet Union today.

More specifically, the proliferation of national conferences suggests a planned effort to disseminate the expertise of the most successful enterprises in an attempt to raise productivity across the board. One short-term goal is to convert 30 percent of China's industrial units to Taching-like producing entities. The annual target for conversion of large and medium-sized enterprises is reported to be 400 or more units per year. Similarly, in agriculture it was reported that by the end of 1975 about 300 counties advanced their production to a level similar to that of the model Tachai-like Hsinyang county. Recent Tachai conferences called for 400 more counties to achieve such levels, which includes basic mechanization of agriculture. In many cases numbers and names of particular units selected to reach advanced status by 1980 have been specified. Again, one-third, or at least 700, of China's counties are to achieve this status.

The gigantic agroindustrial mobilization is to be administered by provincial, municipal, and autonomous regional party committees. Provincial conferences are to be held annually to report on progress and provide a means of review by central authorities who will assign targets for the next 12 months.

In allowing the provinces to play the leading role, the new leadership is also preparing for another innovation, which is to take place before 1985. China will be divided into six major administrative regions that will continue to develop their economic systems according to varying standards. The reason for this is a plan to further industrialize the interior of China. This will lessen the problem of further overcrowding of the already highly populated existing industrial regions, which would be inevitable under rapid industrial growth policies and strengthen the overall national defense posture.

Finally, consumer-goods production is now receiving more attention than in previous years, and some reports suggest that light industry output increases in 1977 were larger than advances made in heavy industries. This is believed to be part of the "new leap" policy designed to gain the support of the people for the strenuous economic effort that lies ahead. Recent reports also suggest that lowest grade workers will be receiving 15 to 20 percent wage increases as new incentives. This will significantly increase disposable income and consumer spending. Greater sales of high-priced durables such as bicycles, watches, radios, and perhaps even television and mopeds in the future will generate additional capital accumulations (known as "profits" in the capitalist world) which are probably relied on to provide capital for investment in other high-priority sectors of the economy to help in the overall thrust toward the attainment of superpower status before the end of this century.

CHINESE ECONOMIC PLANNING

China's economy is centrally planned by the Planning Commission of the State Council, which works in cooperation with the People's Bank of China and other auxiliary agencies of the State Council. Similar planning groups operate with all Revolutionary Committees, which are the management groups of administrative and economic units in China. These planning groups exist at provincial, county, and municipal levels of the government as well as in industrial enterprises, and rural communes and brigades. The basic model for the central planning system is taken from the Soviet Union, but some modifications have been introduced to reflect differences in economic and political conditions between the two countries.

In China more significance is attached to local industries and their role in providing self-sufficiency to specific geographic and administrative regions. Thus local units appear to carry more importance, which makes for a more decentralized industrial administration.

At the start of the planning cycle the State Planning Commission prepares five-year plans for investment and production, taking into account existing capacity and expected demand. In addition, one-year operating plans are prepared to provide a more manageable budgetary and accounting framework. These plans include all major capital-investment projects and specify output targets for all industrial and agricultural enterprises.

The original plan outlines are then used by large-scale enterprises, small plants, and communes as guidelines for the preparation of detailed plan proposals. These are discussed and revised by the entire staff of a working unit and are then returned to higher administrative levels for approval. If changes are required the plans are returned for further discussion by the originating unit until agreement by both planning levels is reached. This procedure is repeated sometimes two or three times before a final proposal is prepared.

During the next planning phase, officials from province, municipalities, and industrial enterprises hold joint meetings to prepare a final draft of the plan and try to balance proposed production with requirements. At this point, regional and sectoral allocations of investment funds are decided, probably by designating specific production priorities.

The State Council finally submits the resulting plan proposals to the Central Committee of the Communist Party for approval. The approved draft is sent for a formal approval by the Standing Committee of the National People's Congress. Specific targets and tasks are then assigned by provincial and municipal administration to individual plants or communes for implementation and the planning phase is complete. Considerable control over implementation is exercised by the local People's Bank of China branch, which provides investment funds for specific projects.

Although there is general agreement that provinces and economic regions of China enjoy considerable autonomy in determining the size and location of major industrial projects, most overall policy objectives and many specific plan targets are set by Peking. These include decisions on levels of accumulation, regional and industrial investment allocation, production output levels for major product categories, and the retail and accounting prices for raw-material inputs and basic commodities.

Foreign trade is also subject to central planning in China, but it primarily reflects shortages of equipment and capacity to achieve planned targets rather than any particular desire to do business for a profit. When specific needs for new or additional equipment or materials are established, a determination is made as to whether missing inputs are manufactured in China, and if not, whether such production could start in time to meet the demand. Decision to buy foreign products is made after all domestic possibilities have been exhausted. The potential exporter to China who does not know what current "planned deficiencies" are has almost no chance whatsoever of doing business there unless his product is on the current import list or he has been specifically solicited by a Foreign Trade Corporation.

Actual import and export targets are also set by the State Planning Commission, together with the Ministry of Foreign Trade. They are passed to Foreign Trade Bureaus and Foreign Trade Corporations in the provinces for discussion and coordination with local enterprises involved as export manufacturers or end users of imported products. Final decisions are taken after their comments and suggestions have been given consideration.

Western observers and businessmen are not always aware that this process is designed to reduce or even prevent imports if at all possible. In an environment where self-sufficiency is one of the highest virtues, there are various intangible incentives to develop a local solution that will eliminate the need to import. In COMECON countries specific prizes and honors are bestowed on those who succeed in such import-substitution campaigns. This is also one reason why imports of individual advanced product "prototypes" are made. Such imports signify a deficiency but do not necessarily mean that the product is being tested with the objective of quantity purchases. More likely than not, reverse engineering of such equipment will be attempted, and the Chinese openly admit that it is their objective to make "foreign things serve China." Paradoxically, it is part of a planned innovation process, quite characteristic of centrally planned bureaucracies where preoccupation with meeting planned targets often assigns domestic technological innovation efforts a relatively low priority.

TECHNOLOGICAL INNOVATION

Chou En-lai and Teng Hsiao-ping, the original promoters of current Chinese development policies, were fully aware of the need to import advanced equipment and concepts to offset China's relatively brief experience with technological innovation. Recent statements in the Peking press suggests that their policies will be continued to obtain new technologies and achieve scientific breakthroughs. There is also a suggestion that China is willing to consider other useful ideas in politics, economics, literature, and art. Whether this means a possible new attitude toward long-term financing of imports or a willingness to discuss joint natural resources development projects is as yet too early to predict.

Chou En-lai enunciated exact principles for imports of technology: (1) to obtain and use a sample of a new product, (2) criticize its applicability to Chinese conditions, (3) convert and adapt it to perform optimally, and (4) eventually to create an equivalent domestic product eliminating the need for further imports. The Chinese believe this process saves them time and resources, even though they seldom get the advantage of a full technology-transfer licensing procedure. They simply resort to what is commonly known as "prototype purchasing" policies facilitated by the fact that China offers no patent protection on its territory.

The Chinese believe that this procedure speeds up development of their own products. This is probably quite true in the case of various instruments and machines that are basically mechanical in nature and in which the actual construction materials are not vital to the useful service life of a product. The situation is quite different when technological processes are involved that depend on microsecond timing controls, precision instruments, or alloys with very precise characteristics. Reverse engineering is not possible without literally reproducing research and development programs equal in scope to that of the original developers.

This is one reason for purchases of turnkey plants with such processes built in. Builders of such plants usually limit the purchaser of such equipment to a precisely defined market for the output of the plant. There are indications that the Chinese agreed not to export products of several petrochemical plants that they purchased on a turnkey basis during the mid-1970s.

Technology transfer is a difficult process, even when performed with full cooperation of both parties. During the 1960s China obtained licenses for assembly and manufacture of Berliet trucks from France, but for many years it continued to import large quantities of French and later Japanese trucks, which may mean that this particular technology transfer was not yet a total success. Recently China obtained a Rolls Royce license from England to assemble Spey turbofan jet engines. This is a case where domestic research and development simply did not keep pace with aircraft-engine developments. The relative success of this technology transfer will be known only in the future, but China already has a reasonable track record in this area because in the 1950s it obtained licenses for Soviet jet engines and aircraft that it has been building and flying ever since.

The key to the Chinese technological state of the art lies in its imports of machinery and equipment. All Chinese imports fall into two categories from this point of view. Some are imports of raw materials and products simply unavailable or in short supply in China that are imported to meet temporary or permanent deficiencies. Grains, steel, aluminum, copper, ships, trucks, and tank cars probably fall into this category. Other imports, including helicopters, computers, instruments, off-highway trucks, special mining equipment, and other items purchased in relatively small quantities are probably first destined for a program of reverse engineering while performing useful work in the process.

One of the biggest dilemmas facing China at present is the technology for development of its deeper and off-shore oil resources. Rapidly growing industry will continue to require imports of certain raw materials and larger quantities of expensive advanced technological equipment. The Chinese economy will also consume increasingly larger quantities of oil, and experts believe that at current high rates of production China's huge northeastern oilfields in Taching, Shengli, and Takang may be depleted in only 10 years. The solution is to develop offshore oil, but the immensely expensive technology for this is primarily American and western European. If China tries to develop this industry relying on its own technologies, experts believe production will be delayed, with no more than 50 million tons likely to be produced from off-shore areas by 1985. On the other hand, a joint effort with foreign oil companies could produce several times this amount and assure a market for a significant portion of it.

REFERENCES

The China Business Review. "China's Grand Plan," July–August 1977.
China Trade Report. "A Projection of China's Foreign Trade," January 1977, Hong Kong.
Current Scene. "China's Model Production Units," May–June 1975, Hong Kong.
Current Scene. "National Conferences Focus on Economy," April–May 1977, Hong Kong.
The Economist. "China's Great Leap Sideways," November 5, 1977, London.
Hidasi, Gabor. *Ekonomika i Doktryna Maoistowskich Chin,* Panstwowe Wydawnictwo Ekonomiczne, Warsaw, Poland.
Lardy, Nicholas R. "Economic Planning and Income Distribution in China," *Current Scene,* November 1976, Hong Kong.
Lardy, Nicholas R. "Economic Planning in the People's Republic of China: Central–Provincial Fiscal Relations," in *China: A Reassessment of the Economy,* Joint Economic Committee, U.S. Congress, July 1975, p. 94.
Perkins, Dwight H. "Industrial Planning in the PRC," *The China Business Review,* September–October 1977.
Rawski, Thomas G. "Chinese Economic Planning," *Current Scene,* April 1976, Hong Kong.
Shih, Cheng. *A Glance at China's Economy,* Foreign Language Press, Peking, 1974.
U.S.–China Business Review, "A Review of China's Economy and Market Prospects 1971–1980," July/August 1976.
U.S. News and World Report. "Where China is Headed," interview with George Bush, November 14, 1977, Washington, D.C.
Western Investment in Communist Economies. Committee on Foreign Relations, U.S. Senate, August 1974.

4

Import Dependence and Chinese Foreign Trade

Although China is the most populous country in the world and the sixth largest economy in terms of GNP, it ranked only twenty-seventh in the import market in 1975. China's total imports from developed market economies, communist countries, and the Third World amounted to an estimated $7,385 million in that year. This is comparable to the total import markets of such countries as either Finland, Yugoslavia, South Korea, Hong Kong, or India (see Table 4.1).

Chinese imports account for less than 1 percent of total world imports. On a per capita basis, imports in China are only $8 per person and are among the lowest in the world. This is about 25 percent lower than in India and is similar to per capita import levels of Ethiopia or Burma. In contrast, Soviet per capita imports are about $145, and those of Japan are about $520. Hence China's overall imports account for only 2.5 percent of its GNP. On that basis alone China can claim that its economy is more self-sufficient than any other in the world. Even in the Soviet Union, which is as well, if not even better, endowed with all the necessary resources for an autarkic economy, the value of imports already reached 4.3 percent of the GNP. China's import dependence is also well below the world average of 14.7 percent, and only a handful of commodities imported by China can be classified as "strategic goods."

Nevertheless, Chinese imports in certain industries account for a significant percentage of their total output. This is particularly true with regard to the imports of metals and metal ores, which traditionally represent almost 30 percent of all Chinese imports.

Despite extensive metal and mineral resources, China is not com-

Table 4.1 Import dependence of China compared with selected countries in the world during 1975

	GNP in Millions of $U.S.	*Total Imports in Millions of $U.S.*	*Import-dependence Percent*
World	6,130,000	903,200	14.7
China	299,000	7,385	2.46
United States	1,498,000	102,984	6.87
Soviet Union	865,000	36,969	4.27
Japan	484,000	57,881	11.95
West Germany	422,000	74,208	17.58
France	340,000	54,247	15.95
Brazil	97,800	13,658	13.96
India	85,300	6,362	7.46
Finland	28,000	7,607	27.16
Nigeria	25,000	6,041	24.16
South Korea	18,700	7,274	38.89
Taiwan	14,400	5,959	41.38
Hong Kong	7,000	6,767	96.67
Albania	1,200	160	13.00

Source: United Nations Statistical Yearbook, 1976; U.S. Government Handbook of Economic Statistics, 1976.

pletely self-sufficient. This is believed to have resulted partially from a choice of priorities. Additional investment in developing copper and aluminum resources may eventually reduce or even eliminate large imports of those metals that at present are estimated to supply about 50 percent and 25 percent of apparent consumption. Industry observers believe that during the 1960s and early 1970s China's limited mining-development resources were concentrated on expansion of the petroleum industry. Increasing demand for copper and aluminum is being met by imports, especially when world prices appear advantageous. Dependence on imports of copper or aluminum, though relatively high at present, must be regarded as temporary. Some analysts believe that stockpiling, particularly of aluminum, has taken place (see Table 4.2).

Three metals imported by China can be regarded as truly strategic materials: (1) chromium, (2) nickel, and (3) cobalt. No significant production or deposits of those metals have been reported in China, which hence has to import practically 100 percent of its rapidly growing demand for those metals required in production of high-quality stainless steels.

Curiously, Chinese deficiency in chromium, cobalt, and nickel is similar to that experienced by the United States, which also imports about 95 percent of each. China spent about $500 million on imports of these three metals during 1970–1975, of which 85 percent represented purchases of nickel, mostly from Canada. China's dependence on imports of chromium, cobalt, and nickel is probably a significant factor influencing Chinese foreign policy. The Soviet Union is the largest producer of nickel and chromium in the world and the second largest producer of cobalt after Zaire. During the 1950s China imported nickel and chromium from the Soviet Union, but it appears that no such trade is taking place any more. Albania, the third largest chromium producer in the world, has been China's major supplier of chromite ore since the early 1960s. This coincides with the emergence of a common Sino-Albanian political stance directed against the Soviet Union and extensions of Chinese foreign aid to Albania.

China is now also importing chromium from Turkey and Iran and may develop supplies from the Philippines, Madagascar, and even India now that diplomatic relations have been reestablished. In previous years pressures to embargo trade with China throughout Southeast Asia and political inability to trade with Rhodesia and South Africa may have forced China to develop the Albanian connection. If this is so, the price for assuring the supply of chromium from Albania has been high. China extended over $359 million in foreign aid to Albania by 1974 and is believed to have provided large additional credits since then, including assistance for the construction of a 800,000-ton/year ferronickel plant. Considering the size of the Albanian economy, Chinese assistance appears to have been in the order of 2 to 3 percent of the Albanian GNP for more than a decade.

Chinese ventures into eastern Africa, particularly the construction of the TANZAM railway from Tanzania 1100 miles south to Zambia, may also have been an investment to assure China alternate access to chromium as well as copper, cobalt, and platinum. Soviet and Cuban presence in Angola and Mozambique may have partially been prompted by a desire to control any possible future supplies of strategic metals to China from that region. This would be of particular importance to the Soviets if pro-Soviet factions gain strength in Yugoslavia and Sino-Albanian relations become less cordial following the death of President Tito.

Dependence on almost total imports of chromium, nickel, and cobalt is a weakness of the Chinese economy that may become more pronounced as its industry develops and larger inputs of strategic metals are required. Unless China discovers significant domestic deposits of those metals or develops reliable long-term supply sources, it may become increasingly receptive to the consideration of larger and closer trade exchanges with the Soviet Union. As the world's largest producer of nickel and chromium and an exporter of both metals, the Soviet Union may then want to play

Table 4.2 **Value of largest imported commodities and estimated import dependence of various industries expressed as percent of apparent consumption during 1970–1975**

SITC Code	*Import Commodity*	*Total 1970–1975 Imports in Millions of $U.S.*	*Average Annual Imports in Millions of $U.S.*	*Actual 1975 Imports in Millions of $U.S.*	*Imports as Percent of Apparent Consumption (in Percent)*
67	Iron and steel	4,115	685	1,530	15
04	Grains	3,530	588	680	2
56	Fertilizers	1,525	254	405	14
263	Cotton	1,488	248	275	10
682	Copper	956	159	156	50
732	Motor vehicles	954	159	204	15
735	Ships	840	140	377	30
512	Organic chemicals	791	132	163	large
06	Sugar[a]	790	132	185	21
2311/2	Rubber	765	127	150	60
734	Aircraft	546	91	131	
715	Machine tools	456	76	125	10 to 15
684	Aluminum	435	72	261	25
683	Nickel	423	70	92	99
65	Synthetic fabrics	389	65	75	
722/3	Electric power-gen. equip.	351	58	118	up to 30
58	Plastics	296	49	75	large
718.4	Construction/mining equip.	293	49	53	5
266	Synthetic fibers	263	44	94	26
729.5/861	Instruments	235	39	81	30 to 50
221.4	Soybeans	227	38	15	less than 1
64	Paper/paperboard	216	36	64	under 1

731	Railway vehicles	161	27	23	12
711.5	Internal combustion eng.	150	25	27	small
251	Wood pulp	135	22	41	small
712	Agricultural machinery	85	14	41	small
724	Telecommunications equip.	50	8	19	small
283.91	Chromium ores/concentrate	50 (est.)	8	9.5 (1974)	100
	Cobalt	30 (est.)	5	9.6 (1974)	100

Source: Based on data developed in various chapters of this book; originally based on export statistics of countries trading with China.

[a] Derived from volume of sugar traded, extended by average prices in years involved.

its chromium–nickel card in relations with China. If hard-currency markets for Chinese oil do not expand rapidly enough and Soviet oil production starts peaking out in the 1980s, China may also want to barter some of its oil for strategic metals from the Soviet Union. These are some of the reasons that may prompt China to seek closer relations with the Soviet Union if its dealings with the West are not regarded as satisfactory.

China's most visible import dependence is in the iron and steel industry. Imports of iron and steel are the largest single import commodity in China's foreign trade and account for about 15 percent of the apparent consumption of these. Iron and steel also represent 15 percent of the value of all Chinese imports and more than half of all the metal imports. In 1976 imports of steel reached an all-time high of 4.2 million tons, equivalent to about 18 percent of domestic production.

A large percentage of finished steel imports are high-quality steel products containing significant inputs of nickel and chromium. By importing such steel products China is probably able to keep imports of nickel and chromium somewhat lower than they would be if all that steel were domestically produced. However, at this point in time, large imports of steel are necessary anyway because domestic production is inadequate to meet demand. Under the circumstances China might as well import the highest quality steel it can get because this policy lessens its exposure to any threat to its strategic metals supplies.

Lead and zinc, though produced in China, are also imported. Up to 20 percent of lead and about 2 percent of zinc consumption have been supplied by imports in recent years. Domestic deposits are believed adequate and as they are being developed, imports of those metals are expected to decline. Magnesium, titanium, and platinum are also imported, but China's import dependence is difficult to determine as production volumes of titanium and platinum are unknown. Imports of magnesium constitute only 0.3 percent of domestic production (see Chapter 9 for more details).

Another strategic commodity is natural rubber, whose imports represent as much as 60 percent of total rubber consumption. During the 1950s exports of natural rubber from Southeast Asian countries to China were embargoed. Only Ceylon continued to export natural rubber to China in return for rice within a special economic-aid barter agreement. Since then China imported natural rubber from other countries and developed synthetic rubber production. After the new 240,000-ton/year butadiene rubber plant comes on stream, imports of natural rubber may be reduced up to a point.

Grains are the single largest import-product category after iron and steel and account for 13 percent of total Chinese imports value during 1970–1975. However, imports of grains constitute only 2 percent of the total grains production in China and are offset by significant exports of rice. In 1975 China in fact became a net exporter of grains measured in terms of value of total grains traded. Despite considerable publicity, which stems from the fact that China purchases centrally very large quantities at a time, total imports of grains are very small relative to Chinese domestic production.

Imports of sugar, which are also offset by some sugar exports, represent about 21 percent of domestic sugar production. China will probably import and trade in this commodity as long as prices remain very low. There were large imports of soybeans during the early 1970s, but these are irregular and represented only under 1 percent of the estimated production in China. Traditionally, China exports large quantities of soybeans to Japan and has the capacity to increase production. Similarly with cotton, as the second largest cotton-producer in the world, China supplements its output with about 10 percent of imports, depending on demand for Chinese cotton textile exports in the world market. There is also potential to increase cotton production in China. Cotton is the fourth largest import commodity, representing 5.4 percent of all Chinese imports in 1970–1975. Dependence on imports of cotton may lessen as synthetic textile production and exports expand.

The third largest import category is chemical fertilizers, which accounted for up to 15 percent of total fertilizer production. China has been known as the largest fertilizer-importing country in the world, but large new fertilizer plants in China will probably reduce fertilizer imports in the future. Organic chemicals, plastics, synthetic fibers, and fertilizers together represent at least 10 percent of all the imports during 1970–1975.

Of the machinery and transport equipment imported by China, motor vehicles represent the largest import volume. Imports of trucks are estimated to provide about 15 percent of total supply. Equipment markets, which appear to depend on imports for up to 30 percent of total consumption are shipbuilding, electric power-generating equipment, and instruments. China used to import a large percentage of its machine-tool needs, but with expansion of domestic production this is now down to 10 to 15 percent. About 12 percent of China's railway vehicles appear to have been imported in recent years, but probably no more than 5 percent of its construction and mining equipment comes from abroad. Internal combustion engines, agricultural machinery, and telecommunications equipment are also among the largest twenty-five import categories, but these represent a relatively small percentage of domestic output of such products.

TRENDS IN FOREIGN TRADE

China's foreign trade has grown at an average 12 percent per year between 1949 and 1975. During the last 10 years its growth accelerated to an average 15 percent per year, and between 1970 and 1975 the average annual rate of growth was 27 percent. This rapid growth of recent years is not expected to continue, and total 1976 trade actually declined by about 9 percent. Trade is expected to recover somewhat during 1977 but will probably fall short of $14 billion for that year.

Until 1973 China's total foreign-trade volume ranged between 2 percent and 3 percent of the value of its GNP. By 1975 trade volume reached almost 5 percent of the GNP, and China also had a hard-currency trade deficit of $585 million. This was the second largest such trade deficit in China's history, upstaged only by the $1.2 billion deficit of 1974. China may have been prepared to suffer some deficits, expecting to cover them quickly by oil exports that did not increase as rapidly as anticipated. As a result, imports were reduced by about 20 percent, and total trade volume is estimated at only $13.2 billion in 1976. Assuming a possible 6 percent growth in GNP, China's trade volume under those conditions would represent less than 4 percent of its GNP. Some observers believe that China does not want to allow its trade to increase to a level higher than 5 percent of its GNP because it would lose its basis for claiming economic self-sufficiency. Relative magnitudes of 5 percent or less are often regarded as very small or insignificant.

This type of GNP ratio limitation also provides a basis for projecting the possible upper bounds of Chinese trade in the future. Recent average GNP growth rates were in the order of 6 percent per year. This means that China's GNP may reach about $535 billion by 1985. At that level, total trade within the 5 percent GNP boundary could be no more than $26 billion, which is comparable to the total trade of either Switzerland or Spain today.

Some China trade projections made by the Japan External Trade Organization (JETRO) anticipate that China's trade will reach $33 billion by 1985. Their forecast is based on the assumption that Chinese oil and petroleum products exports will increase from $910 million in 1975 to almost $5 billion in 1985. They also believe that trade will increase at growth rates over 10 percent during the early 1980s. If GNP growth continues at an average 6 percent per year, this means that foreign trade would represent more than 6 percent of China's GNP in 1985, a level that China may not want to tolerate.

On the other hand, a more rapid industrial expansion, with the GNP growing at about 7 percent annually or more, imposes a different limitation on China's foreign trade. According to U.S. Government analysts, to maintain an average GNP growth of 7.3 percent China would have to consume domestically most of the oil that it will be producing with practically no surplus for export. This would have the effect of reducing China's trade to about $25 billion in 1985, governed primarily by China's ability to export traditional nonoil commodities. But even if the $33 billion trade level were achieved, it would still be comparable to only the total trade of Sweden today.

The trend for increasing trade with the communist countries must also be taken into account. By 1975 this trade was approaching the levels of its previous high point in 1959, when it amounted to $2.98 billion. It represented only 16 percent of China's total trade in 1975, and remained steady in 1976 while trade with other countries dropped to 12 percent (see Fig. 4.1).

There are important factors that may influence this trade. Recessions and inflation in the West increase the cost of Western products and reduce noncommunist markets for Chinese goods. Unless China can develop a significant hard-currency market for its oil, it may be forced to barter it for eastern European machinery and equipment. This is particularly likely because as Soviet oil production is expected to peak out in the early 1980s, eastern European countries are already looking for alternate sources of oil. Such barter trade should be welcome to both sides since it offers China political advantages and machinery while providing the eastern European countries with oil and foodstuffs without necessitating hard-currency expenditures.

COMMODITY COMPOSITION OF FOREIGN TRADE

Until 1974 Chinese foreign trade was relatively well balanced, with traditional commodity composition patterns prevailing over the last 10 years. In 1974 and 1975 this pattern was disturbed by large trade deficits with the West resulting from massive purchases of turnkey plants in previous years that were being delivered to China. By 1976 foreign imports were reduced and China's trade surplus is now estimated at $945 million of which $745 million is with noncommunist countries. The effect of the massive plant purchases was to increase the share of manufactured goods from 50 percent to over 60 percent of Chinese imports. As a result, the share of foodstuffs and crude materials decreased to 13 percent and 14 percent of China's 1975 imports, respectively. Over the longer range the

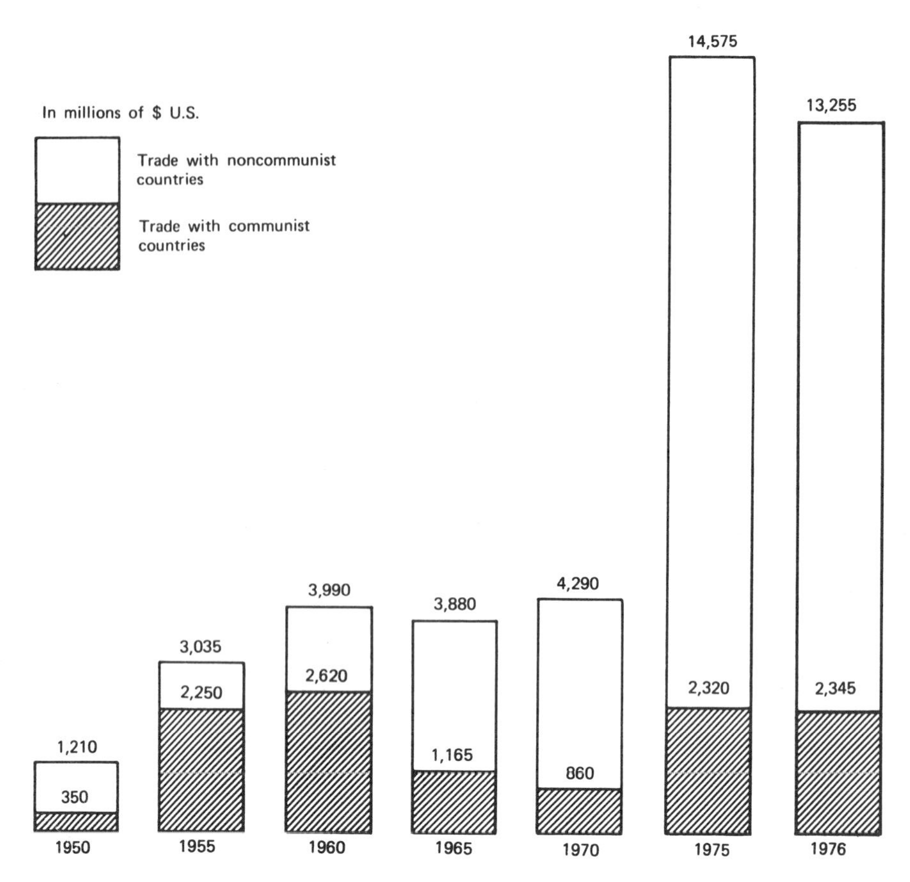

Figure 4.1 Trends in China's foreign trade. **(Source: *Based on data in* U.S. Government People's Republic of China International Trade Handbook, October 1976.)**

pattern will probably continue; however, as agricultural productivity increases in time, imports of manufactured goods may also increase.

Manufactured goods constitute the largest commodity group in Chinese imports as well as exports, but the component products in the export categories are quite different from imported manufactures. About 50 percent of China's manufactured exports consist of textile fabrics, clothing, and footwear. Another 8 percent consists of handicrafts and light manufactures. Minerals, nonferrous metals, iron and steel, nonelectric machinery, and transport equipment account for about 5 percent each. The remaining manufactured goods include paper, leather and skins, metal products, and electric machinery (see Figs. 4.2 and 4.3).

In contrast, imports of manufactured goods are dominated by iron and steel products, which accounted for 37 percent of all manufactures imported in 1976. Nonelectric machinery comprised 28 percent of all manufactured imports, and transport equipment, including trucks, ships, aircraft and railway vehicles made up 12 percent of this import category down from almost 20 percent in 1975. Synthetic textiles, copper, aluminum, and nickel, electric machinery, paper and paperboard, and instruments were some of the other important commodities.

Foodstuffs comprise the second largest export group in China's foreign trade and generally provide China with a substantial trade surplus. In 1975, for example, China exported $2,125 million of foodstuffs, whereas

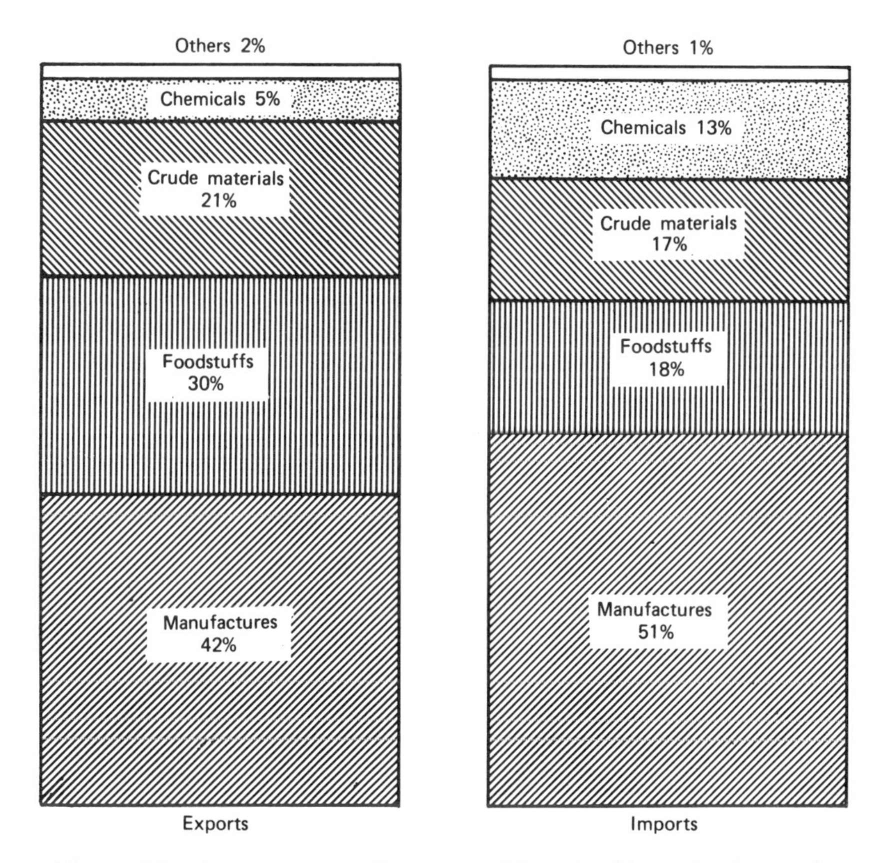

Figure 4.2 Average commodity composition of Chinese foreign trade during 1967–1975.

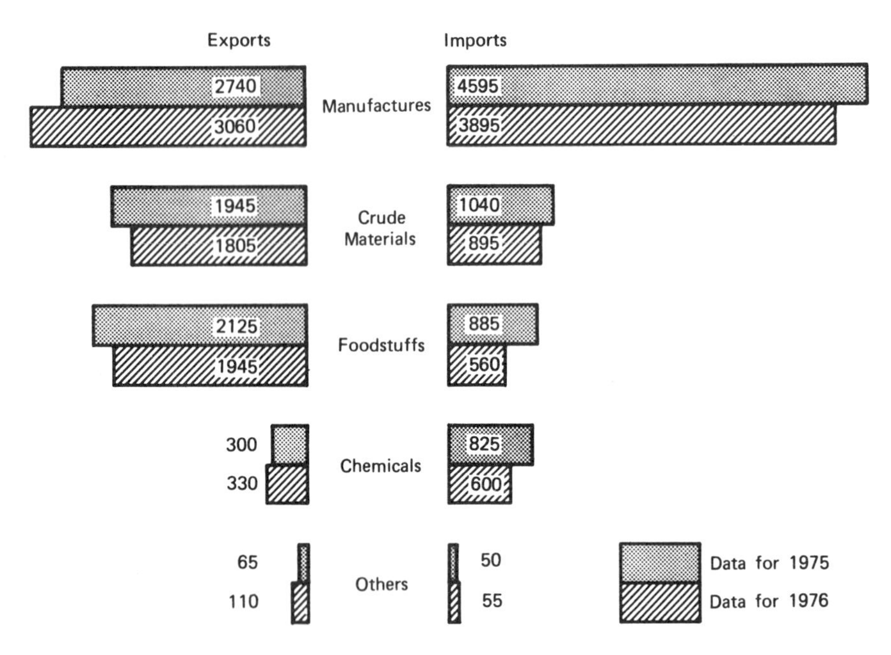

***Figure 4.3** Major commodity groups in foreign trade during 1975 and 1976 expressed in millions of $ U.S.* **(Source: *Based on data in* U.S. Government National Foreign Assessment Center Publication "China: International Trade, 1976–77," November 1977.)**

it imported only $885 million for a net surplus of $1,240 million. Exports in this category consist of rice, meat and fish, fruits and vegetables, and tobacco. Imports consist of wheat, corn, and sugar.

Exports of crude materials also contribute to China's positive trade balance, and if exports of oil and petroleum products increase, this category will provide additional trade surplus. The largest commodity in this category is crude oil, which together with petroleum products amounted to $910 million in 1975, equal to about 43 percent of all exports in this category. Other major crude-materials exported include silk and cotton, oilseeds, crude animal material, coal, and metal ores.

The smallest export category is chemicals, consisting mostly of medicinal products, essential oils, and soap. This category accounts for less than 5 percent of Chinese exports in recent years. In contrast, imports of chemicals are about 13 percent and are dominated by fertilizer products, which account for almost half of all imports in this group. Chemical elements and compounds (predominantly organic chemicals), dyeing materials, plastics, and some pesticides are the remaining imports in this group.

DIRECTION OF FOREIGN TRADE

Since 1969 China's trade with noncommunist market economies averaged over 80 percent of total trade and reached a high of 84 percent in 1975. Developed countries, including Japan, Australia, western Europe, Canada, and the United States account for 51.3 percent of all the trade and remain China's major sources of supply of iron and steel, grains, and machinery. Japan is the leading trade partner, with its 1976 share amounting to 23 percent of all the Chinese trade (see Fig. 4.4).

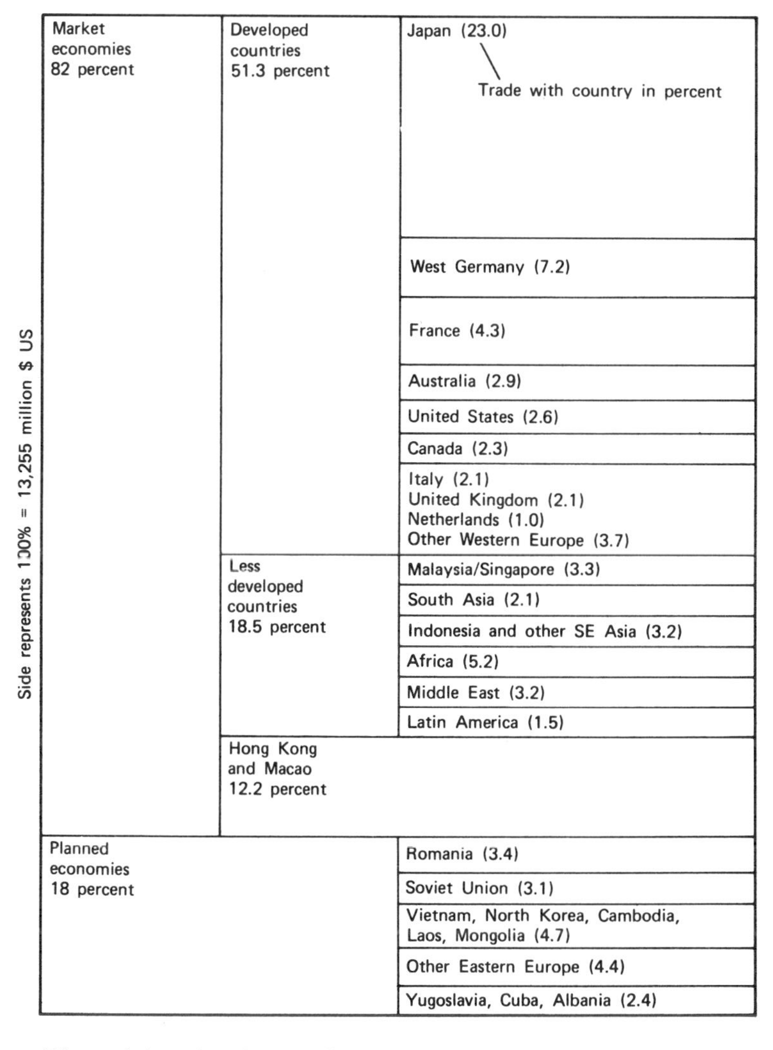

Figure 4.4 Direction of Chinese foreign trade in 1976.

China runs a large trade deficit with developed countries every year. This negative balance was $2,860 million in 1975 and $2,890 million in 1974 and contributed significantly to the overall trade deficits of $215 million in 1975 and $810 million in 1974. Preliminary results for 1976 indicate that China has a positive balance of about $1,115 million in trade with the non-communist countries, although a deficit with developed countries continues at $1,415 million.

Hong Kong, which after Japan is the largest single trading partner, traditionally provides the largest market for Chinese products. Trade with Honk Kong is almost entirely as export and provides China with well over $1 billion in hard-currency earnings every year. China also receives significant earnings from its large exports to Malaysia/Singapore, and Indonesia, which resulted in $262 million and $125 million earnings, respectively, during 1976 (see Table 4.3).

Table 4.3 Trade turnover, exports, imports, and balance of trade of China with its twenty-five largest trade partners during 1976 in millions of $ U.S.

Country	*Trade Turnover*	*Exports*	*Imports*	*Trade Balance*
Total	13,255	7,250	6,005	+ 1,245
Japan	3,052	1,306	1,746	− 440
Hong Kong	1,620	1,590	30	+ 1,560
West Germany	952	236	716	− 480
France	571	169	402	− 233
Romania	453	201	252	− 51
Soviet Union	416	178	238	− 60
Australia	380	102	278	− 176
United States	351	202	149	53
Canada	309	90	219	− 129
Singapore	294	254	40	+ 214
Italy	278	135	143	− 8
United Kingdom	277	136	141	− 5
East Germany	200	96	104	− 8
Cuba	175	75	100	− 25
Albania	150	90	60	+ 30
Malaysia	146	97	49	+ 48
Indonesia	125	125	neg	+ 125
Netherlands	124	78	46	+ 32
Czechoslovakia	120	50	70	− 20
Nigeria	113	108	5	+ 103
Poland	106	40	66	− 26
Egypt	98	39	59	− 20
Iran	95	89	6	+ 83
Belgium/Luxembourg	92	46	46	0
Switzerland	92	32	60	− 28

Source: U.S. Government, National Foreign Assessment Center, *China: International Trade 1976–1977*, November 1977; and authors' estimates.

China has been running trade deficits with Canada and Australia every year for at least the last 15 years. Those countries are primarily suppliers of wheat to China and constitute relatively small markets for Chinese products. Trade with other developed countries such as West Germany, the United Kingdom, and Switzerland more often than not also results in a deficit for China.

Trade is also showing negative balances with Latin American countries such as Brazil, Argentina, Chile, and Peru, which supply China with sugar, grains, and copper. However, trade is considerably more in balance with other less developed countries in the Middle East, Africa, and South Asia.

Trade with communist countries normally results in a positive balance for China of several hundred million dollars every year and, at least on paper, helps to balance total Chinese trade. Much of this trade is conducted by annual agreements between China and its communist trading partners. Commodities to be exchanged and methods of payments are negotiated once a year alternately in Peking or in the capital of the trading country. Reportedly, prices for commodities and goods quoted in Swiss franks are used as a basis for valuation of these exchanges.

Although such virtual barter trade may show surpluses, unless payments agreements between those countries provide for payment in hard currencies, such earnings cannot be used to offset trade deficits with market economies. Most Chinese trade surpluses that arise in dealing with communist countries are with North Korea, Mongolia, Vietnam, Cambodia, and Laos. Except for Mongolia, those countries are also recipients of significant Chinese foreign aid, which further complicates trade accounting.

Trade with communist countries has been steadily declining from 23 percent of total Chinese trade in 1971 to a low of 16 percent in 1975, but it has more than doubled in absolute terms from $1,085 million in 1971 to $2,345 million in 1976. This trade may increase in the future if the rising prices of Western goods and declining markets for Chinese products become intolerable to the Chinese planners. They may look to more barter-type trade with communist countries in particular as a means of obtaining some of the necessary imports of transportation equipment, machine tools, and even electronic equipment in exchange for oil, textiles, or foodstuffs from China.

In summary, whether China is regarded to have a surplus or deficit in its foreign trade is significantly affected by the actual method of account, and three different trade balances can be developed for China every year. The most significant balance is between China and the noncommunist countries in which deficits with developed countries are offset by surpluses in trade with less developed countries. Surpluses in trade with communist countries are only relevant to the extent that they can be applied to offset trade deficits with the noncommunist trading partners (see Table 4.4).

During 1977 China's foreign trade is believed to have recovered from the depressed 1976 levels, but total volume is not expected to reach $14.6 billion, the previous high during 1975. It is also expected to yield another trade surplus in the order of $1 billion despite continuing major repayments of previously imported turnkey plants and grain credits. During the National Foreign Trade Conference in July 1977 China's foreign trade workers were exhorted to do a better job of importing advanced technology and equipment and to improve their exporting capabilities. Plans to

Table 4.4 Recent trends in China's foreign trade in millions of $ U.S.

Trade Area	*1975 Trade*	*1976 Trade*	*Change from Previous Year*	*Percent Change*
Total Trade	14,575	13,225	− 1,320	− 9
Exports	7,180	7,250	+ 70	+ 1
Imports	7,395	6,005	− 1,390	− 19
Trade with Communist Countries	2,390	2,345	− 45	− 2
Exports	1,380	1,240	− 140	− 10
Imports	1,010	1,105	+ 95	+ 9
Soviet Union	279	416	+ 137	+ 49
Exports	150	178	+ 28	+ 19
Imports	129	238	+ 109	+ 84
Eastern Europe	1,010	985	− 25	− 2
Exports	485	435	− 50	− 10
Imports	525	550	+ 25	+ 5
Other Communist Countries	1,095	940	− 155	− 14
Exports	740	625	− 115	− 16
Imports	355	315	− 40	− 11
Trade with non-Communist Countries	12,185	10,915	− 1,270	− 10
Exports	5,800	6,015	+ 215	+ 4
Imports	6,385	4,900	− 1,485	− 23
Developed Countries	8,100	6,805	− 1,295	− 16
Exports	2,620	2,695	+ 75	+ 3
Imports	5,480	4,110	− 1,370	− 25
Less Developed Countries	2,650	2,455	− 195	− 7
Exports	1,780	1,690	− 90	− 5
Imports	870	765	− 105	− 12
Hong Kong and Macao	1,435	1,660	+ 225	+ 16
Exports	1,400	1,630	+ 230	+ 16
Imports	35	30	− 5	− 14

Source: Based on trade statistics in U.S. Government National Foreign Assessment center publication, *China: International Trade 1976–77*, November 1977.

improve packaging of export goods and port-operating efficiencies further indicate longer-range plans to expand foreign trade in the future.

On the other hand, in several developed countries there are mounting pressures to restrict imports of Chinese textiles: Japan is restricting imports of Chinese silk, Australia and Canada have limits on textiles. On the import side China is expected to import about 7 million tons of wheat valued at an estimated $700 million during 1977 out of a total of 11 million tons in contracts already signed. Imports of sugar may reach over 1 million tons. Other agricultural imports such as cotton, soybeans, and edible oils are expected to increase slightly.

Imports of iron and steel, copper, and aluminum are expected to increase, though lower prices will keep total values down. Rubber and fertilizers are likely to increase in price and total imports. Machinery and equipment are expected to drop for the second year to about $1 billion, primarily as a result of completion of deliveries from many turnkey plants purchased in the early 1970s.

Partial trade returns for 1977 (4 to 8 months) suggest that total imports from major noncommunist trading partners and from the Soviet Union are down by about 26 percent, while exports are up by about 13 percent. However, imports from some individual countries are up. Imports from Austria, Norway, Hong Kong, Singapore, New Zealand, and Malaysia rose; imports from Philippines in particular were up 243 percent, reaching $72 million during January–June of 1977. So far, imports from all other major trading partners registered decreases, ranging from 15 percent for Switzerland to 73 percent for Italy. Imports and exports from the Soviet Union for the first half of 1977 were down by 30 percent and 22 percent, respectively. Final overall trade results for 1977 will be affected by changes in trading volume with eastern European countries, Asian communist countries, and the less developed countries for which trading data are not available at the time of writing.

THE FOREIGN-TRADE ESTABLISHMENT

Chinese foreign trade is supervised by the Ministry of Foreign Trade, which is directly responsible to the State Council for the planning and execution of all foreign-trade transactions with China. This ministry came into being in 1952 as a result of division of the former Ministry of Trade. It is headed by Li Chiang, who was identified as minister in October 1973 and became known as China's trade spokesman since then.

Two other ministries are also involved in China's foreign-trade establishment. The Ministry of Finance controls the Bank of China, which is the foreign-exchange bank responsible for the balance of trade. It also supervises the Chinese People's Insurance Company, which in turn controls Chinese insurance companies in Hong Kong. The Ministry of Communications is also involved by virtue of its control of China Ocean Shipping Company (COSCO), which administers Chinese merchant fleet engaged in international commerce and believed to carry over 30 percent of Chinese foreign-trade shipments (see Figure 4.5).

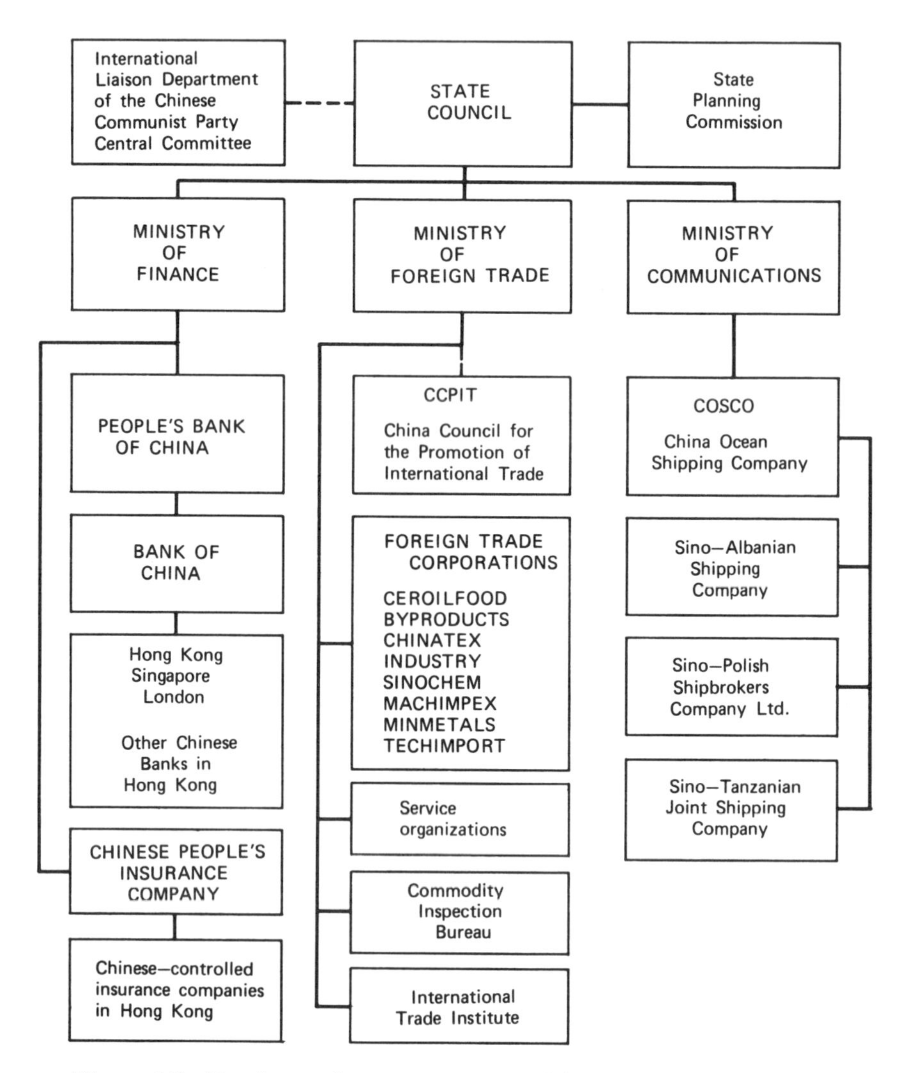

Figure 4.5 Foreign-trade apparatus of People's Republic of China.

The Ministry of Foreign Trade is supervised by a Staff Office of Finance and Trade of the State Council, and its function is to formulate an overall import and export plan that is used as a basis for specific plans prepared by the individual Foreign Trade Corporations and other agencies with foreign-trade authority. The ministry reviews and approves such import and export plans and supervises their execution.

The ministry is organized into four regional bureaus and other departments that supervise customs, commodities inspection, foreign-trade personnel training, market-research institutes, and the trading corporations. It is also involved in the appointment of commercial counselors at Chinese diplomatic posts abroad and is the proper agency for conclusion of trade agreements and exchanges with other countries. An auxillary organization known as China Council for the Promotion of International Trade (CCPIT) performs functions similar to those of an international chamber of commerce.

The four regional bureaus of the Ministry of Foreign Trade supervise trade in specific geographic regions of the world. The First Bureau is responsible for trade with the communist countries, the Second Bureau handles Africa, the Third Bureau supervises trade with western Europe, the Americas, and Oceania (Australia and New Zealand), and the Fourth Bureau is responsible for trade with Asia. These bureaus closely resemble similar geographic departments that exist within the Ministry of Foreign Affairs.

THE FOREIGN-TRADE CORPORATIONS

All trade transactions with China are conducted by eight Foreign Trade Corporations. These can be regarded as huge state trading monopolies acting as agents for manufacturing units and end-user organizations in China. Neither import nor export transactions can be concluded without the participation of the Foreign Trade Corporations.

The Foreign Trade Corporations do not operate according to the "profit motive" but are guided by the Foreign Trade Plan developed by the Ministry of Foreign Trade. This plan is based on the shortfalls and surpluses discovered in the formulation of the national Five Year Plans. Foreign Trade Corporations provide specific planning inputs for imports and exports within their areas of responsibility. Some trade is undertaken to achieve specific foreign policy objectives, but the basic motivation must be regarded as the need to generate sufficient foreign currency to finance imports of strategic and technological goods vital to performance according to planned objectives.

By world standards, Chinese Foreign Trade Corporations (FTCs) are huge trading bureaucracies with an average annual turnover of almost $2 billion each. Judging by the volume of import and export of specific commodities, China National Cereals, Oils, and Foodstuffs Import and Export Corporation (CEROILFOOD) is probably the largest of all Chinese FTCs, with a turnover in the order of $3 billion per year (see Table 4.5).

Table 4.5 Foreign Trade Corporations of People's Republic of China

Name	*Description*
CEROILFOOD	China National Cereals, Oils, and Foodstuffs Import and Export Corporation. Conducts trade in grains, vegetable oils, fruits, feeding stuffs, livestock and meats, dairy products, fish products, vegetable products, sugar, beverages and condiments. It has seventeen branch offices throughout China.
BYPRODUCTS	China National Native Produce and Animal By-products Import and Export Corporation. Principally trades in teas, coffee, cocoa, tobacco, spices, medicinal herbs, timber, and fireworks. Animal by-products include bristles, hides, leather, wool, hair, and feathers. At least thirty-four branches and offices have been identified.
CHINATEX	China National Textiles Import and Export Corporation. Trades in silks, raw cotton and cotton yarn, clothing, woolen goods, man-made fibers, synthetic fabrics, household items, tapestry, and furnishings. Nine branches are known to exist, located in major Chinese cities.
INDUSTRY	China National Light Industrial Products Import and Export Corporation. Principal trade includes general merchandise, paper and paperboards, building materials, electrical appliances, radio and television, consumer hardware, sports goods, toys, pottery and porcelain, pearls, precious stones, ivory and jade, stone and wood carvings, lacquer ware, cloisonné, furniture, and handicrafts. At least nineteen branches are known to exist in China.
SINOCHEM	China National Chemicals Import and Export Corporation. Crude oil and petroleum products are now the principal exports of this FTC. Other products traded include rubber, tires, fertilizers, pesticides, pharmaceuticals, and medicines. It also handles medical instruments, paints, and printing inks. Six major branches of the organization are known to exist in China.
MACHIMPEX	China National Machinery Import and Export Corporation. Specialized trading groups identified within this FTC include: aircraft; instruments and telecommunications; machine tools, electrical, mining, and light industrial equipment; transportation and agricultural equipment. Seven major branches exist in major cities.
MINMETALS	China National Metals and Minerals Import and Export Corporation. Principal trade in various steel products, alloys, iron ore, nickel, chromite, aluminum, copper, rolled materials, sulfur and other minerals as well as steel scrap, welding electrodes, etc.
TECHIMPORT	China National Technical Import Corporation. Specializes in import of complete plants and manufacturing technology.

The Chinese FTCs conduct all negotiations with foreign-trade partners, and contracts are signed in the name of the FTC or one of its branches involved in the transaction. However, when deemed necessary the negotiating terms of the FTC are supplemented by Chinese experts, manufacturers, or end users. As legal entities Chinese FTCs do not possess any assets outside of China and do not maintain any offices abroad, although they do sometimes act through agents in Hong Kong or Macao.

Western businessmen who dealt with Chinese FTCs often observe that little or no salesmanship is involved in dealing with Chinese negotiators. To the buyer of Chinese products FTC officials appear simply as order takers and do not take any initiative to interest a buyer in the products for which he did not ask. In the author's experience the exception may have been Chinese crude oil and petroleum products. Chinese negotiators literally "buttonholed" various visitors at the Canton Trade Fair in the fall of 1974 and tried to interest them in Chinese oil products. The seller to a Chinese FTC may experience extensive questioning about the nature and price of competitive products, but advantages that derive from China's monopoly position are not always utilized.

Each Chinese FTC is basically divided into an import and an export department. Functionally these are subdivided into units that are responsible for specific product groups. In addition, there are departments that specialize in trade with specific geographic areas, presumably when trade with such areas is dominated by a particular product or political objectives. All FTCs maintain headquarters in Peking and regional offices in major Chinese cities, particularly those that are the production centers for the items traded by the FTC.

CHINA COUNCIL FOR THE PROMOTION OF INTERNATIONAL TRADE

Known as CCPIT, this organization is described by the Chinese as "a permanent agency performing duties similar to those of Chambers of International Commerce in other countries." It is funded by the government and directed by persons from the Ministry of Foreign Trade, Ministry of Foreign Affairs, and representatives from the Foreign Trade Organizations.

The organization was created in 1952 and has the authority to engage in agreements and precontract agreements without legal binding force on the parties. It also maintains trade contacts with commercial and paragovernment organizations in countries with which China has no diplomatic relations. Its importance is believed to be decreasing as more countries establish diplomatic relations with China.

The CCPIT has a permanent staff headed by Chairman Wang Yao-ting and at least five Vice-chairmen. The organization is functionally divided into six departments and two commissions. The Legal Affairs Department advises and informs the Foreign Trade Corporations and other trade service organizations about legal developments and requirements in foreign countries. It also analyzes foreign contracts and new legislation that may affect China's foreign trade. Another function of this department is the registration of foreign trademarks.

The Liaison Department develops and maintains contacts with trade associations in other countries. Its major activity is sending Chinese trade delegations to foreign countries to visit plants, research institutes, and manufacturing associations to discover products or technologies of interest to China. There is also a Technical Exchange Department, but its precise role in the organization is not clear.

The Overseas Exhibitions Department is responsible for organizing China's trade exhibitions in foreign countries. This activity may involve participation in an international trade fair or the staging of a solo exhibition of Chinese products and economic achievements in a particular foreign country. There is also a Foreign Exhibitions in China Department, which assists foreign governments, trade associations, or even single companies to stage trade exhibitions and technical symposia in China. The Publicity Department of the CCPIT is probably the closest to being the state advertising agency. It publishes a variety of periodicals that describe China's economic achievements and its foreign-trade activity.

The two commissions of the CCPIT are the Foreign Trade Arbitration Commission and the Maritime Arbitration Commission; both are advised by the Legal Affairs Department. Parties seeking arbitration by any of those commissions are required to select an arbitrator from fifteen to twenty-one members of either commission. Parties may also choose one arbitrator each, and these in turn agree on a third arbitrator to form a panel of three. Most contracts specify Peking as the site of arbitration, and decisions of the panel cannot be appealed.

CHINESE TRADE PROMOTION

The three basic methods employed by the Chinese to "promote" trade with foreign countries are: (1) the Chinese Export Commodities Trade Fair in Canton, (2) Chinese economic and trade exhibitions abroad, and

(3) foreign trade exhibitions held in China. Also, Chinese trade missions abroad combine promotional ventures with the new technology and market research. The precise functions of those events are easier to understand if one keeps in mind the main objective of Chinese foreign trade, which is to sell surplus Chinese production to generate foreign currency for purchases of necessary industrial imports and technological prototypes. Only a negligible amount of consumer products are imported into China.

The semiannual Canton Trade Fair, is primarily an event promoting and displaying Chinese products available for sale. Business observers estimate that 35 to 40 percent of Chinese export business every year is concluded at the fair. During the early 1970s attendance at the fair was estimated to have reached about 25,000 representaives from over 100 different countries. A visit to the fair is only by invitation, which must be obtained from one of the Foreign Trade Corporations. Because the purpose of the fair is to export Chinese products, the most certain way of receiving an invitation is to purchase some Chinese goods in advance and then ask for an invitation to visit the fair. Since 1975 China has also staged several minifairs in other cities. These are staged by individual Foreign Trade Corporations and promote specific products such as carpets, chemicals, arts and crafts, or forest products.

Chinese economic and trade exhibitions in foreign countries are usually miniature reproductions of the Canton Trade Fairs. Their main objective is to sell Chinese products, but emphasis on the type of products varies with location. In developed countries exhibitions stress traditional Chinese goods such as carvings, carpets or embroidery. In African or Latin American countries Chinese exhibitions include agricultural and light industrial products as well as a higher political content. In 1975 China held twelve solo exhibitions; seven of these took place in Africa and Latin America and three in communist countries.

Since 1970 China also participated in over seventy international trade fairs. Of those, thirty-five took place in Third World countries, and seventeen were regular trade fairs in communist countries. Since 1974 Chinese participation declined, particularly in trade fairs staged in western Europe. Countries in whose international fairs China most often participates include Egypt, Algeria, Hungary, Poland, and Bulgaria.

China also organizes national and specialized trade exhibitions, mostly in Peking or Shanghai, where foreign manufacturers can display the products they would like to sell to China. Participation in these fairs is also by invitation only, and usually more firms from a particular country apply than are finally invited to participate. In choosing exhibitors for these events the Chinese also indicate the types of products or technology they would like to see. In addition, they request that exhibitors support their participation with scientific and technical seminars dealing with development, manufacture, and use of products or equipment exhibited.

Exhibiting countries that organize industrial fairs in China found that these are not particularly profitable operations measured against immediate commercial revenues from those events. The cost of organizing such an exhibition from a free-market country is in the order of $1.5 to $2 million and takes up to two years in preparation time. The rewards to individual participants are negligible, more often than not limited to discounted sales of equipment on display.

Some trade observers believe that the primary function of foreign industrial exhibitions in China is education of Chinese manufacturers in the latest and most advanced technologies of interest to China at the time. This is supported by the fact that visitors to those exhibitions are usually selected by invitation and appear to consist mostly of highly qualified specialists from major research centers or manufacturing enterprises.

Foreign industrial exhibitions provide an easy way to perform technology research at practically no cost to China. Display models, which are often sold rather than shipped back to the country of origin, are believed to provide a large number of prototypes for copying and producing in China. Although this method does not always yield the most desirable results, it permits China to perform a considerable amount of technical upgrading of its products.

Analysis of companies invited to exhibit in China and of the products they exhibit does provide one of the best indications of what China's needs are or are likely to be in the near future. An even stronger indication of those needs may be gleaned by following various Chinese trade missions that visit specific research institutes and manufacturing centers outside China. During 1970 to 1975, for example, at least sixty different trade delegations interested in various technological subjects from aviation to telecommunications visited Western countries. It is believed that such missions are performing valid market research in trying to find the best available technology to match their needs. In some cases direct relation between trade-mission visits and later sales is claimed, but these claims are difficult to prove because purchases are made by Foreign Trade Corporations, which seldom identify the end users of purchased goods.

In the final analysis the Chinese so far succeeded in remaining a solid buyer's market. The most promising approach from the Western seller's point of view is a thorough analysis of real present and future Chinese needs and an apparent readiness to assist in solving their anticipated problems in the light of their own planned trade objectives.

REFERENCES

Asia 1977 Yearbook. "China's Economy and Trade," Far Eastern Economic Review, 1977, Hong Kong, p. 159.

Berthoud, Roger. "8% Boost Estimated in China's Foreign Trade," *The Times,* May 3, 1974, London.

Bundesstelle fur Aussenhandelsinformation. *Aussenhandelsgesellschaften und Organizationen in der Volksrepublik China,* April 1975, Cologne, West Germany.

Business Asia. "Future Direction of China's Trade Policies Emerges," June 3, 1977, Business International, Hong Kong.

Business International. "Measuring the Mainland China Market," No. 50, 1970.

Business International. "Selling the Mainland China Market," No. 51, 1971.

Business International. "Snagging an Invitation to Peking; How It's Done," October 5, 1973, New York.

Centre Francais du Commerce Exterieur. *Republique Populaire de Chine Nouvelles Commerciales.* Paris, France, Third Semester, 1977.

Chen, Nai-Ruenn. "China's Foreign Trade 1950–1974," in *China: A Reassessment of the Economy,* Joint Economic Committee, U.S. Congress, July 1975, p. 617.

Chiang, Li. *Nouveau Developpement du Commerce Exterieur de Chine,* CCPIT, Peking, PRC, 1974.

Chinese Exhibition. Brochure published by CCPIT, Peking, 1974.

Clarke, William and Martha Avery. "The Sino-American Commercial Relationship," in *China: A Reassessment of the Economy,* Joint Economic Committee, U.S. Congress, July 1975, p. 500.

Current Scene. "China's Foreign Trade in 1972," October 1973, Hong Kong.

Current Scene. "China's Foreign Trade 1973–1974," December 1974, Hong Kong.

Current Scene. "China's Foreign Trade in 1974," September 1975, Hong Kong.

Current Scene. "China's Foreign Trade in 1975," Part I, September 1976, Hong Kong.

Current Scene. "China's Foreign Trade in 1975," Part II, October 1976, Hong Kong.

Doing Business with the PRC. Summary of Conference in Honolulu, Hawaii Sept. 30 to Oct. 1, 1971. Hawaii International Services Agency, January 1972.

Europa Yearbook 1977, "China–Trade and Industry," 1977, London.

Eckstein, Alexander. "China's Trade Policy and Sino-American Relations," *Foreign Affairs,* 54, 133, October 1975.

First National City Bank. *China Trade Guide,* New York, May 1975.

Hollingworth, Clare. "Chinese Favor Third World and Arabs in Trade Fair Deals," *Daily Telegraph,* May 10, 1974, London.

Industrial Marketing. "What Makes China Buy?," July 8, 1975.

Kraal, Louis. "A High Level Sales Pitch for Shoppers from Peking;" *Fortune,* November 1975.

Kunze, Bernd. "Trade Relations Between the US and China," *Intereconomics,* No. 9, 1971, Hamburg, West Germany.

McDougall, Colina. "A Lull in Capital Goods," *Financial Times,* October 2, 1974.

Ministry of Trade. "*How to Sell to China,*" unpublished notes for British exporters, November 1974, London.

National Council for U.S.–China Trade. *China's Foreign Trade Corporations and Organizations,* Washington, D.C., 1975.

Palladin, A. "Biznes Na Pekinska Modle," *Czerwony Sztandar* (Krasnaya Znamya), November 19, 1974, Vilnius, Lithuanian S.S.R., U.S.S.R.

Price, Robert L. "International Trade of Communist China 1950–1965," in *An Economic Profile of Mainland China,* Joint Economic Committee, U.S. Congress, February 1967.

Smith, Charles. "Trading with China," *Financial Times Survey,* April 29, 1975, London.

Sobin, Julian M. *China as a Trading Partner,* Sobin Chemicals, Boston, Mass.

Sobin, Julian M. "Sino-American Trade Prospects," American Management Association Briefing Conference, *China After Mao,* January 20, 1977, New York.

Stanford Research Institute. *Trade with the People's Republic of China,* Menlo Park, California, 1974.

Ta Kung Pao. "Principle Governing China's Foreign Trade," October 24, 1974, Hong Kong.

Theroux, Eugene A. "Legal and Practical Problems in China Trade," in *China: A Reassessment of the Economy,* Joint Economic Committee, U.S. Congress, July 1975.

U.S.–China Business Review. "Kwangchow Diary—Spring 1975," May–June 1975.

U.S. Department of Commerce. "Trade of the United States with Communist Countries in Eastern Europe and Asia," *Overseas Business Reports* OBR 77–30, June 1977.

U.S. Department of Commerce. *The Chinese Economy and Foreign Trade Perspective 1976,* June 1977.

U.S. Department of State. *Trade with People's Republic of China,* Government Printing Office, Washington, D.C., 1974.

U.S. Government. "Handbook of Economic Statistics 1977," *Research Aid* ER77-10537, September 1977.

U.S. Government. "China: Economic Indicators," *A Reference Aid* ER77-10508, October 1977.

U.S. Government. "China: International Trade, 1976–77," *A Research Paper,* ER77-10674, November 1977.

Usack, A. H. and R. E. Batsavage. "The International Trade of the PRC," in *PRC: An Economic Assessment,* Joint Economic Committee, U.S. Congress, May 1972.

Usine Nouvelle. "Nos Clients de Pekin," June 14, 1973, Paris.

Usine Nouvelle. "Corporations Chinoises d'Importation et d'Exportations," April 11, 1974, Paris.

5

Banking and Finance

Banking and finance in China is a state monopoly operated by the Ministry of Finance. The ministry was established in 1949 and controls the banking, foreign exchange, and insurance activities through special agencies of the State Council. These include the People's Bank of China, the Bank of China, and the Chinese People's Insurance Company. Individual citizens are prohibited to own gold or foreign currencies, but overseas Chinese remittances and investment are encouraged and accepted. The Chinese *renminbi* (RMB; people's money) currency is inconvertible and cannot be brought in or out of China.

The Ministry of Finance executes financial policies of the State Council that are based on the investment requirements of the Five Year Plans drafted and implemented by the State Planning and State Capital Construction Commissions. There is also a direct political influence on China's financial policies through the Finance and Trade Department of the Politburo of the Chinese Communist Party Central Committee (see Figure 5.1).

Through the banking system the ministry plays a very important role in controlling the execution of the national plan and China's foreign trade. The People's Bank of China has extensive powers of supervision over all organizations to which funds are advanced. This prestige enjoyed by Chinese bankers became most evident during the Cultural Revolution when banking and its management were virtually untouched by the disruptions and upheavals of that period.

Besides banking, the ministry also supervises all Chinese insurance activities through the People's Insurance Company of China. Banking and insurance are the two systems responsible for capital formation in China and in a centrally planned economy are understandably under the same control.

The basic currency unit is the *yuan*, and its exchange rate is arbitrarily set at about 1.8 yuan = $1.0 U.S. Until 1974 there was only a single

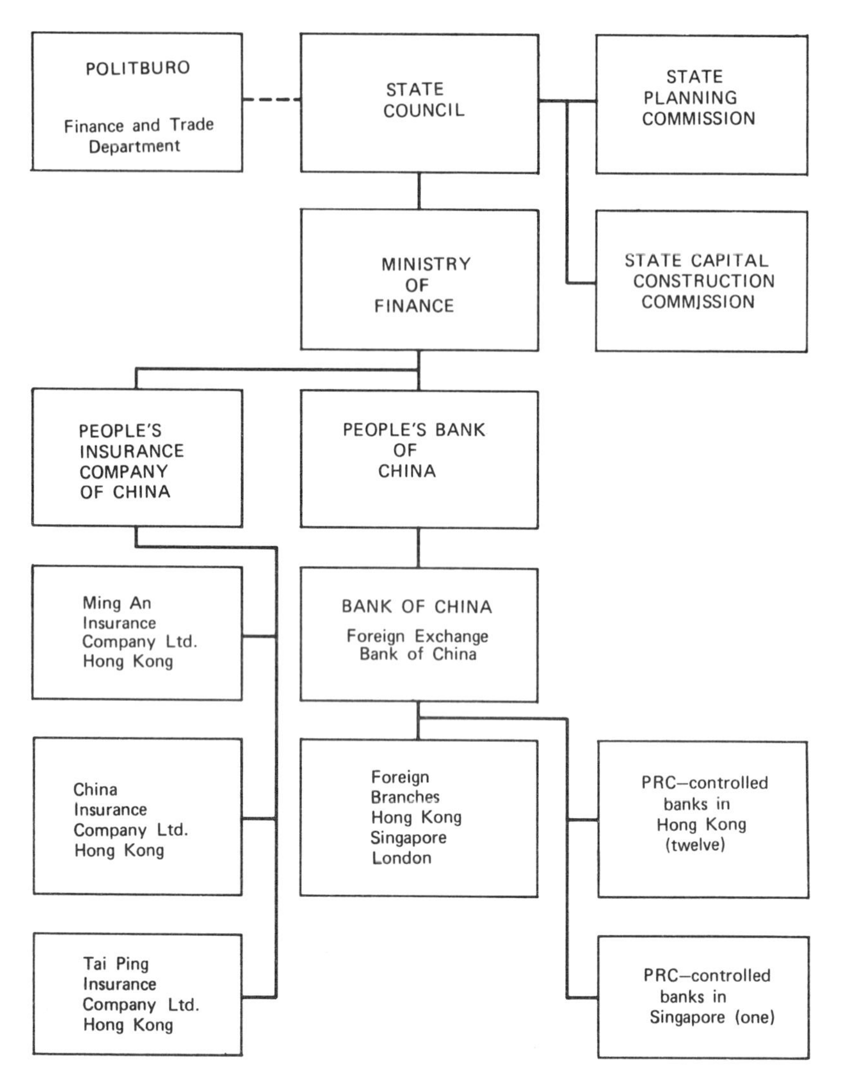

Figure 5.1 Financial establishment of the People's Republic of China.

rate for purchases or sales of the RMB currency, but it is now replaced by announcements of buying and selling rates for RMB with a usual spread between buying and selling rate of about 0.5 percent. The Chinese change the value of the RMB constantly to reflect world monetary conditions.

Western financial specialists have been trying to understand the principle underlying the valuation of the RMB but have been unable to develop a plausible "basket" of foreign currencies, and possibly certain commodities whose prices in the open market would be collectively related to the price of the RMB. This task is even further complicated by the fact that even if the constituent currencies in the "basket" are properly identified, it is impossible to determine the weight assigned to each component or the frequency of changes it undergoes. It is generally agreed that U.S. *dollars*, Japanese *yen*, the British *pound*, Swiss *franc*, German *mark*, Hong Kong *dollar*, French *franc*, Netherlands *gilder*, and Italian *lire* are used in this valuation.

Although at times Chinese foreign currency and gold reserves were very low, China has been able to overcome those situations by drastic cuts in imports while maintaining or even expanding exports. Sales of gold have been necessary to help cover short-term deficits, but availability of surplus oil for export gives China additional flexibility in foreign trade.

Openly, China refuses credits to finance its purchases of grains or plants. On the other hand, some contract terms include deferred payments for up to 80 percent of total cost for 5 years at 6 percent interest after the start-up of plants. Because such financing is standard procedure in this type of contract, China does not consider this as an extension of credit. Technically this is correct because financing is arranged by the seller with banks in his country, but China benefits from such credits as much as if it obtained financing directly from Western banks.

THE PEOPLE'S BANK OF CHINA

Starting out as a Special Agency of the State Council, the People's Bank of China now operates under the control of the Ministry of Finance as China's central bank and the commercial banking system of the country. The bank issues currency, maintains savings for individuals, provides credit to commercial enterprises, controls the money supply, and assists in execution of the national plan. It is believed that the bank supplies over 75 percent of the working capital of the state-owned sector of the economy and oversees all the investment, making sure that unplanned development does not take place. By its direct control of the Bank of China it is also responsible for all international transactions and the balance of foreign trade.

In carrying out the state plan, the guiding principles of the People's Bank of China are to maintain stable prices and balanced accounts throughout the economy. Deficit spending is basically forbidden and, whenever possible, great efforts are made to settle transactions on a noncash basis. Reportedly as much as 85 percent of all settlements are handled that way.

The People's Bank of China was established in 1950 as a state monopoly and absorbed all foreign and domestic private banking systems operating in China. Its immediate task at that time was to arrest a runaway inflation that had plagued China since 1937 and is now considered to have been even more severe than that experienced by Germany between World Wars I and II. Government efforts and programs arrested the inflation by 1950, and in 1955 the People's Bank of China exchanged banknotes in circulation for new currency. For every 10,000 yuans turned in, one new yuan was issued.

Because the People's Bank of China is a monopoly operating within a planned economy it is able to control the performance of individual units by allocation of resources. Whenever possible, settlements are made without any transfer of funds, and the bank in fact continuously acts to reduce the supply of money. The only accumulation of currency results from wages, and the bank is absorbing that in the form of savings. Because it is the ultimate financial authority without competition, the bank is also in a position to make all the collected funds immediately available for capital formation. Unplanned investment and advance payments to ensure supplies of raw-materials in industry are discouraged. The bank appears to have the authority to inspect the financial status of Chinese enterprises and require that surplus capital not required for immediate operating expenses be deposited in the bank.

People's Bank of China is believed to operate at least 15,000 branches throughout China, although some estimates put that number at 39,000. The bank is one of the few end-user organizations in China known to have purchased a HIS 61/62 data-processing system from Honeywell–Bull of France in 1974. Several Western computer firms, including IBM and NCR, have been active in trying to computerize the operations of Chinese banks in Hong Kong. It is believed that success in providing a system for the Bank of China and the twelve Chinese-controlled banks would pave the way toward computerization of the People's Bank of China. The ultimate value of such a system could be worth about $200 million.

THE BANK OF CHINA

The Bank of China is a subsidiary of the People's Bank of China and operates as the foreign exchange bank for the PRC. It has the status of a Special Agency of the State Council within the government framework of China, and its main objective is to maintain a balanced international trade. As such it plays an important if not often decisive role in Chinese foreign trade.

The bank was acquired from an established banking organization in 1949. It is only 66 percent owned by China and still has some of the original shareholders who receive a fixed 5 percent dividend on their investment. Its headquarters are in Peking but it operates at least thirty-two branches in major Chinese cities. In addition it has three foreign branches, one in Hong Kong, one in Singapore, and one in London. The Hong Kong and the Singapore branches play the most important role because those areas are the sources of Chinese foreign-trade surpluses. The Hong Kong branch also supervises twelve other Chinese-controlled banks in Hong Kong. The London branch manages China's hard-currency portfolio and arranges Chinese transactions with European partners (see Table 5.1).

Table 5.1 Chinese-controlled banks in Hong Kong

Name of Bank	*Number of Branches*	*Location of Head Office*
Bank of China	2	Peking
Bank of Communications	13	Peking
China and South Sea Bank	10	Shanghai
China State Bank	8	Shanghai
Chiyu Banking Corporation	5	Hong Kong
Hua Chiao Commercial Bank	6	Hong Kong
Kincheng Banking Corporation	9	Shanghai
Kwangtung Provincial Bank[a]	11	Canton
Nanyang Commercial Bank	12	Hong Kong
National Commercial Bank	6	Shanghai
Po Sang Bank	2	Hong Kong
Sin Hua Trust, Savings & Commercial Bank	15	Shanghai
Yien Yieh Commercial Bank	9	Shanghai

Source: U.S. Government Research Aid No. CR77-13-37, July 1977.
[a] This bank also has a branch in Singapore.

For settling trade contracts with noncommunist countries the Bank of China maintains correspondent relationships with over twenty banks in Japan and various banks in other countries. These correspondent relations are usually arranged by opening an RMB account at the Bank of China in Peking, which in turn opens a hard-currency account with its Western correspondent. The Bank of China exercises very strict control over purchases or sales of Chinese (RMB) currency by its correspondent banks. In general, such transactions are only allowed when documented by specific contracts with Chinese Foreign Trade Corporations.

Trade with communist countries is conducted on a bilateral basis. All transactions are processed through specially set up clearing accounts that are coordinated with annual trade agreements. The agreements specify desired trade levels and outline commodity composition of imports and exports of what in fact amounts to barter trade activity. The clearing accounts are denominated in "trade rubles" or in mutually agreed on Western currency, which in the case of China is often in Swiss francs. The value of bartered commodities is determined by world market prices, and trade is balanced usually by changes in plans for the next trading period.

Since 1970 the Bank of China had to expand its operations to meet the needs of expanding Chinese foreign trade. Its total assets increased from 7 billion RMB in 1971 to 17 billion RMB in 1975, equivalent to almost $9 billion.

The Bank of China is regarded as cautious and conservative by any Western standards. For example, it maintains minimum-risk assets of such proportions that if every private depositor of the Bank of China decided to withdraw his deposit, the bank could satisfy all of them without having to recall any outstanding loans.

In late 1975 the Bank of China in Hong Kong began offering savings accounts as well as fixed deposit accounts denominated in U.S. dollars at competitive interest rates ranging from 3.5 percent for 3 months to 5.75 percent for short-term (12-month) deposits. In addition, Bank of China deposits are not subject to Hong Kong 15 percent interest withholding tax. It was estimated early in 1977 that the Bank of China and all other Chinese-controlled banks in Hong Kong retained the equivalent of about $15 billion in deposit accounts. Western businessmen and tourists may also open RMB accounts in China, but the utility of such accounts is extremely limited. In most respects the Chinese currency remains untraded and inconvertible.

CHINESE INSURANCE ORGANIZATIONS

All insurance activity in China is supervised by the People's Insurance Company of China, which is based in Peking and enjoys the status of a Special Agency of the State Council. Established in 1949, it is a state-owned insurance company and operates directly under the control of the Ministry of Finance. It is headed by General Manager Feng Tien-shun, who is assisted by several deputy general managers. The company underwrites international trade and marine-risk insurance at competitive rates and covers some of its foreign-exchange business by reinsurance in foreign countries.

The terms and conditions of Chinese insurance policies and certificates have been judged similar to those used in international practice, and all valid claims are honored. However, Chinese insurance does not cover strikes, riots, and civil disorders. This makes it impractical to use for insuring foreign trade to countries where longshoremen strikes occur.

The People's Insurance Company of China operates seventeen branch and subbranch offices in China's principal ports and major cities. It also controls three joint state–private insurance companies that maintain foreign branches of their own organizations in Hong Kong, Singapore, and other Asian cities.

The China Insurance Company Ltd. with headquarters in Peking is a joint state–private enterprise that underwrites ocean marine transportation, fire, life, personal accident, workmen's compensation, and motor-vehicle insurance. The company operates branches in Hong Kong, Singapore, Kuala Lumpur, Macao, and Penang. Another Chinese-controlled insurance company is the Tai Ping Insurance Company Ltd., which is also based in Peking. This company appears to offer similar coverage to that available from China Insurance Company Ltd. Its main branches and subbranches are also located in Hong Kong, Singapore, Kuala Lumpur, Penang, and Ipoh. A third company, the Ming An Insurance Company, also operates an office in Hong Kong, but further details about its activities are not available.

In foreign trade, communist countries often insist on purchasing goods on a free-on-board (FOB) basis and selling cost, insurance, and freight (CIF) whenever they can. This is seen as another way to save hard currencies because in some transcontinental shipments insurance and freight charges may add up to 15 percent of the total price of imports.

In 1973 China made a policy decision to carry out insurance of its foreign trade on a "mutually advantageous basis" permitting buyers of Chinese products to arrange their own insurance if they so desire. Some Chinese exports such as oil and coal can be obtained on a FOB basis in Chinese ports, which reflects the shortage of tankers and colliers in China's merchant fleet.

The People's Insurance Company of China has been quite active in the worldwide reinsurance markets. According to statements made by the company China has ten outward reinsurance treaties under which it accepts a certain proportion of the risks of direct insurers in other countries. The Chinese insurance company also maintains relationships with 400 insurance firms in eighty different countries. In 1974 British, French, and Swiss insurers visited China to discuss provision of specialized insurance to cover aviation equipment, marine insurance and third-party passenger insurance. In 1976 an agreement including reinsurance treaty was also signed with the American International Group (AIG) in New York.

CURRENCY AND GOLD RESERVES

Although China started out with minimal foreign-exchange reserves in 1950, it is estimated to have been able to accumulate about $645 million in foreign exchange by 1957. This was made possible partially through rapid expansion of exports, but long-term credits amounting to $1.4 billion granted by the Soviet Union were particularly helpful in accomplishing this task.

It is believed that possibly 50 percent of the original currency and gold reserves was accumulated by nationalization of private holdings and sales of gold produced in China. The remainder was generated from net earnings in foreign trade. These earnings were primarily achieved because of strict and effective controls of foreign trade, which resulted in exports of domestic products and allocation of earnings to the purchase of capital goods at the expense of increased consumption. Resulting currency accumulations may have been even higher if it were not for the fact that China extended as much as $475 million in foreign aid to other communist countries. Domestic collection campaigns are believed to have yielded as much as $250 million of foreign exchange held by private individuals. In addition, remittances from overseas Chinese are estimated to have amounted to $855 million by 1957.

During the Great Leap Forward China's currency and gold reserves declined rapidly. Large volumes of exports to the Soviet Union were destined for repayment of debt and did not contribute to earnings. Disruption of agricultural production led to reduction of exports to other markets where they could earn foreign currency. The only course of action left to China was to reduce imports, particularly from the market economies.

In 1961 acute grain shortage forced China to further reduce its imports of other commodities, increase sales of precious metals, and accept short-term credits to finance grain imported during 1961–1962. Sales of silver, which amounted to $22 million in 1960, increased to over $50 million in 1961 and $47 million in 1962. In 1961 China also exported 7600 troy ounces of gold to the London bullion market. The Soviet Union and eastern European countries provided credits to help China's balance of payments. The Soviet Union extended $320 million to finance most of China's accumulated clearing debt. It also advanced $46 million in credit to finance China's imports of 500,000 tons of Cuban sugar.

By maintaining a high level of exports to the Soviet Union, China was able to accumulate a surplus of $425 million during 1963 and 1964, which was sufficient to cover most of its debt to the Soviet Union. By the end of 1964 $100 million of indebtedness with eastern Europe was also eliminated, and China's foreign currency reserves increased to about $345 million. The worst decline in China's foreign-currency reserves was over.

During 1965 China's gold and convertible currency holdings probably increased by $100 to $150 million. China also purchased $115 million worth of gold for sterling, a move probably prompted by a fear of devaluation of the British pound as well as growing unrest throughout Asia. Other gold purchases included amounts of $35 million worth in 1966, $70 million in 1968, and $20 million in 1969. Throughout the 1960s China's domestic gold production also provided additional reserves. In 1960 China's gold production was estimated at about $35 million current dollars and from then on continued at about $25 million annually until 1970. That year Chinese foreign-currency reserves were estimated at only $160 million, but its gold holdings accumulated to about $560 million. By 1973 some estimates put China's gold reserves at $900 million.

There are also reports that China undertakes a certain amount of gold trade through Macao and that unofficial gold imports from throughout Asia are in the order of $100 to $300 million annually. Most Chinese dealings in gold and foreign currencies, however, are educated guesses because no data have ever been released about such transactions.

The disruption of exports during the Cultural Revolution was another setback for the balance of payments with the noncommunist countries, and in 1968 imports were sharply reduced. The increase of imports by 1970 again created a deficit, but this was quickly offset by expansion of exports in 1971. By 1974 Swiss banking executives reassessed China's foreign currency position and concluded that it was probably much higher than was previously believed. They estimated that it was more likely to be in the order of $2,000 million and that even a $4,000-million level could be regarded as "defensible" because of various other invisible earnings of China as well as a sharp increase in the price of gold. China appears to have made a shrewd and timely investment by buying gold during the 1960s (see Table 5.2).

Remittances of foreign currency by overseas Chinese who number at least 20 million, were estimated at about $100 million annually. These estimates are very imprecise and some are well below this level, but others go as high as $1 billion per year. China encourages such remittances and even provides formal vehicles to invest in China. Two special corporations, the Fukien Overseas Chinese Investment Corporation and Kwangtung Overseas Investment Corporation, reportedly offer 8 percent interest on hard currency investments in China.

Additional sources of foreign exchange are the profits of numerous PRC-owned businesses and investments in Hong Kong. There are also re-

Table 5.2 Estimated reserves, including gold and foreign exchange in selected countries in 1977

Country	*Gold and Currency Reserves in Millions of $U.S.*	*Country*	*Gold and Currency Reserves in Millions of $U.S.*
China[a]	3,000+	Soviet Union[b]	7,200
West Germany	35,100	Brazil	5,873
United States	19,100	Canada	5,100
Japan	17,800	Nigeria	4,663
United Kingdom	17,200	India	4,559
Iran	11,592	South Korea	3,502
Italy	10,500	Mexico	1,501
France	9,900	Taiwan	1,411
		Egypt	405

Source: U.S. Government National Foreign Assessment Center, *Economic Indicators Weekly Review,* October 13, 1977.

[a] Based on estimate in China Trade Report, October 1977. [b] Based on estimate of 1800 tons of gold reserves in the Soviet Union extended by an average price of $4 million per ton.

payments of Chinese loans made to communist and developing countries and other invisible transfers. These include receipts from insurance and freight charges believed to be in the order of $10 million and to a lesser degree tourism, which remains a large, unexploited source due to lack of tourist facilities and inadequate transportation in China at present.

As a result of massive purchases of turnkey plants during 1972 to 1974, China developed hard-currency trade deficit of $1.2 billion by 1974. Many observers now accept that China's total currency and gold holdings were in the region of $3,000 million and are currently down to about $2,000 million.

By 1976 unconfirmed reports began circulating about Chinese gold sales. In November 1976 China reportedly disposed of 15 to 20 tons of gold in Europe, valued at about $73 to 98 million. In December 1976, 81 tons of gold worth $350 million was sold in London. In January 1976, 20 tons of gold was supposedly sold on the Zurich and London exchanges. All those sales could be well over $500 million mark, indicating that China considerably reduced its gold holdings to balance trade. This activity was coupled again with reduction of imports and a return to the trade surplus position in 1976.

Since 1973 China also became an exporter of crude oil, which earned $760 million in 1975 and $665 million in 1976. This added considerable flexibility to China's foreign currency reserves position. Surplus oil production available for export is now in the order of 10 million tons, which is annually equivalent to about $1,000 million. Exports to Japan, Thailand, and the Philippines are already taking place, and in some cases China is offering its oil at a lower "friendship price." Although oil contributes significantly to China's foreign-curency earnings, there is a limit to its utility for this purpose. Many of the OPEC oil-producing countries are operating only at 75 percent capacity, and China must move cautiously in this market recognizing its limitations. It may find bartering its oil for manufactured goods from eastern European countries much more profitable in the long run.

USE OF CREDITS

China not only openly rejects foreign credits in international trade but boasts that it is free as a country of any external and domestic debts as well. This posture is important to the People's Republic of China in projecting the impression of "self-sufficiency," but it is incompatible with China's recent history and even current practice. It is also contrary in principle to the considerable foreign-aid credits that China itself extends to at least fifty less developed countries (LDCs) in Asia, Africa, and Latin America.

Open rejection of foreign credits does not appear to be inconsistent with the purchases of commodities on "commercial credit" of as much as 12 to 24 months. Complete turnkey plants are almost always purchased on deferred, medium-term (up to 5 years) repayment schedules with very explicit interest charges. However, it must be pointed out that the Chinese have seldom, if ever, engaged in various schemes practiced or proposed by the Soviet Union and eastern European countries. The self-reliance policy seems to have kept them out of joint ventures, coproduction, counterpurchase, and product-compensation deals, many of which must depend openly on Western long-term financing.

During the 1950s China received long-term credits from the Soviet Union amounting to over $1.4 billion, and estimates of total loans received from the Soviet Union range up to $5.7 billion. After 1961 China demonstrated its ability to repay its total debt to the Soviet Union in 1965 by maintaining large export surpluses. This procedure was possible primarily because the Soviet Union has an enormous capacity for absorption of foodstuffs and textile goods exported by China. In contrast, China would not be in a position to engage in such trade with Western countries where markets for consumer goods are saturated and protected by various interest groups.

Since 1961 China in fact used short-term Western credits, estimated at $910 million by 1965. Most of these were only short- or medium-term financing arrangements to cover the purchases of Western grains. About $100 million involved medium-term arrangements pertaining to the purchases of at least thirteen industrial plants, but most of those credits have not been used by China. Because such credits are standard business procedure and financing is often arranged by the seller and his bank, China does not consider such transactions to be formal credit extensions, as would be the case when such arrangements are negotiated and directly guaranteed by the governments of the countries involved. Nevertheless, the effect of such financing is identical to that of an equivalent credit granted to China and appears to be quite welcome by the Chinese.

The turnkey plants valued at over $2 million, purchased by China during the 1970s, are the best examples of credit accepted and probably expected by China. Of the sixty plants purchased between 1972 and 1976, at least seventeen involved financing, mostly by Japanese Eximbank and commercial banks. Several plants were purchased for cash on a progress payment basis, but others included specific terms for deferred payments after plant completion and start-up.

For example, the $228-million sale of a hot-rolled steel mill by Nippon Steel consortium in 1974 was arranged on a deferred-payment basis. The contract was denominated and payable in Japanese yen and required a 10-percent cash deposit. Another 20 percent was to be paid on shipments of plant and a further 10 percent, on commencement of trial operations. The remaining 60 percent, which would amount to $137 million, was reportedly to be paid on a deferred basis, presumably after expected plant start-up in 1977.

A typical sale involving financing by Japanese Eximbank was the $72-million 500-megawatt (MW) electric power plant sold by Hitachi Engineering in 1973. Terms called for 10 percent down payment on signing of contract and 15 percent more on shipment of equipment. Another 5 percent was due on completion of the guarantee period (or 28 months after shipment). The balance of 70 percent, equivalent to $50 million was to be repaid in ten installments at 6 percent interest over a 5½-year period.

A fertilizer complex sold by Heurtey Industries of France in 1974 for $118 million required 35 percent down payment, but the remaining 65 percent amounting to $76 million was repayable over 5 years at 6 percent annual interest. Another large French sale of a 2,000,000-ton-capacity petrochemical complex for Shenyang valued at $300 million was made in 1973. Terms of contract included deferred payments over 5 years after delivery of the plant scheduled in 1978. Financing was arranged by Credit

Lyonnais, Banque de l'Union Europeenne, and Banque Fançais pour la Commerce Exterieur. Output of this plant was restricted to internal Chinese consumption only, with no export.

An ethylene plant of 300,000-ton capacity valued at $46 million was financed by Japanese Exim/Commercial Bank in 1972. Terms called for 20 percent down payment FOB, and 80 percent of the cost was deferred over 5 years at 6 percent on completion of plant in 1975. Another ethylene and polyvinyl alcohol plant of 120,000-ton capacity sold by Mitsubishi Petrochemicals in 1973 was valued at $34 million and was also financed by the Japanese Exim/Commercial Bank. Terms called for 20 percent down payment, and 80 percent was to be repaid over 5 years at 6 percent interest per annum.

Japanese Exim/Commercial Bank also financed an ethylene vinyl plant sold by Kuraray Industries in 1973 for $26 million. This contract required 30 percent down payment, but 70 percent was to be repaid in ten semiannual installments in Chinese yuans over a period of 5 years at 6 percent interest, beginning after final shipment had taken place. Previously, Kuraray sold China a vinylon plant in 1963. Terms then required 25 percent down and repayments over 5 years at 6 percent interest with an export restriction on output of the plant imposed on the People's Republic of China for 10 years after start-up.

REFERENCES

Berger, Roland. "Currency and Gold Reserves Shrouded in Mystery," *The Times*, October 2, 1974, London.

Central Intelligence Agency. "Communist China's Balance of Payments, 1950–1965," in *An Economic Profile of Mainland China*, Joint Economic Committee, U.S. Congress, February 1967.

Denny, David L. "International Finance in the People's Republic of China," in *China: A Reassessment of the Economy*, Joint Economic Committee, U.S. Congress, July 1975.

Denny, David L. and Frederic M. Surls. "China's Foreign Financial Liabilities," *The China Business Review*, March–April 1977.

Dziennik Polski. "Chinskie Zloto," March 12, 1977, London.

"First US Insurance Agreements with PRC Established a Good Precedent," *U.S.–China Business Review*, March–April 1976.

Jackson, Howell and Dick Wilson. "The Bank of China in the 1970's," *The China Business Review*, January–February 1977.

Sergeant, Patrick. "Bankers Wait at China's Gates," *Euromoney*, July 1975.

Short, Eric. "China May Insure in London," *Financial Times,* July 16, 1974, London.
Triplett, William. "The Banking Industry," in William W. Whitson, ed., *Doing Business with China,* Praeger, New York, 1974.
U.S. Government, National Foreign Assessment Center. "China: International Trade, 1976–77," *A Research Paper,* ER77-10674, November 1977.
Wilson, Dick. "Another Good Year for the Bank of China," *U.S.–China Business Review,* May–June 1976.

6

International Relations and Chinese Foreign Aid

Since the mid-1960s China has made a remarkable effort to develop diplomatic relations with numerous countries. It has succeeded in emerging from near isolation to a position of a major power playing an increasingly active role in international politics. In 1966 China had diplomatic ties with only forty-six countries. During the Cultural Revolution diplomatic relations with four countries were broken off, and forty-one Chinese ambassadors were recalled to Peking. For a while the most populous nation in the world was represented by only one ambassador, Huang Hua in Cairo. It was not until 1969 that Chinese ambassadors began returning to other diplomatic posts abroad.

By 1971 the People's Republic of China was admitted to the United Nations, and numerous countries began extending diplomatic recognition to the new member. On the Chinese side, Premier Chou En-lai personally oversaw the development of Chinese foreign relations until his illness in 1974. By 1977 China was maintaining diplomatic relations with 115 countries and sent a special ambassador to the European Economic Community (Common Market), a move that the Soviet Union thus far failed to match.

More than half of China's diplomatic ties were established since 1970. The decision to expand international relations was made at the highest levels of the Chinese Communist Party and marked not only the end of the Cultural Revolution but a decisive defeat of any "antiforeign" policies within China.

Big increases in foreign trade and the "ping-pong" diplomacy leading to a dialogue with the United States and admission to the United Nations followed rapidly. What is often overlooked is the fact that the year 1970 also saw China's biggest effort to date in extension of foreign aid. It is only now realized that in 1970 as much as $1,124 million was pledged by China in grants, loans, and credits amounting to 0.5 percent of its GNP at that time. This percentage was higher than either the Soviet Union or the United States granted annually, and even in absolute terms Chinese economic aid extended that year was far more than that given by the Soviets.

These foreign-aid initiatives are all the more remarkable because they followed immediately the disruptions of the Cultural Revolution. Seemingly, China could ill afford such largesse at a time when it was in dire need of capital for investment in its own domestic recovery and development. One cannot escape the impression that China was in a hurry to buy itself into the international trading community. On the other hand, 1970 was also the year when Chinese oil production increased by an almost incredible 8 million tons, moving China into the big league of oil producers.

This emerging energy base of the first order also permitted a new and more adventurous political stance in the international arena. China was well aware of the effect on its own economy of the Soviet aid it received during the 1950s and the problems it experienced after its sudden withdrawal. It realized fully the potential of foreign aid as an economic weapon and lost no time in joining the competition. It also must have realized by then its resource deficiencies and the need to develop several independent sources of certain strategic materials.

When Chou En-lai toured Africa in 1964 he continued to belittle Soviet foreign aid, explaining that it was costly and carried dangerous ties. It appeared as if China were competing with the Soviet Union trying to discourage Soviet foreign-aid clients. Actually, at that time China did not have the resources to do so, and even today it remains the only developing country extending foreign aid to other developing countries.

Since 1970 the extension of Chinese aid and expansion of diplomatic ties with developing countries accounted for a large portion of Chinese diplomatic activity. It is the belief of this author that this activity will continue and even intensify because the paramount objective of Chinese foreign policy must be to secure long-term sources of strategic materials that are unavailable or limited in China. Without assured supplies of chromium, cobalt, nickel, and rubber China cannot continue its industrial expansion, even with abundant supplies of domestic energy. The Soviet Union, which has adequate domestic supplies of most strategic materials, is well aware of China's weakness, and Soviet moves in eastern and southern Africa suggest it does not want to lose that advantage over China. An analysis of major existing or potential exports from countries receiving significant Chinese foreign aid is presented later in this chapter. It confirms that most of these countries possess the vital raw materials that China has to import from abroad.

Whereas foreign policy toward developing countries is based on China's need to secure future supplies of strategic materials, its policy toward developed countries in Europe can be regarded as an attempt to develop a counterforce against any Soviet threat to its own security. The lesson of German–Japanese alliance during the World War II, which forced Stalin to maintain a huge army in the Far East against a possible Japanese attack, has not been lost on the Chinese.

Equally important to China is eastern Europe. It is seldom remembered that in 1956, during the periods of considerable unrest in Poland and Hungary, Chou En-lai made personal appearances in Warsaw and Budapest. There have been reports that Polish diplomats and technicians dealing with the Chinese are sometimes addressed as "old neighbors." There are also accounts of Chinese diplomats at the Warsaw embassy showing their Polish counterparts whom they consider as friendly a wall map of the thirteenth-century world when the Kingdom of Poland and the Empire of Kublai Khan shared a common though very turbulent frontier.

China cannot loosen the Soviet grip on most eastern European countries to any significant degree because those countries are almost completely dependent on the Soviet Union for their petroleum supplies. The greater degree of independence shown by Romania and Albania may have been possible because both these countries have domestic petroleum resources. In addition, Albania's resources include chromium and nickel. When the Soviet shipments of these strategic materials to China were reduced in the early 1960s, Albania became China's major supplier of chromium and nickel and clearly merits China's attention.

In 1968, when the Soviet Union organized the invasion of Czechoslovakia, China condemned the "Brezhnev doctrine" openly and more emphatically than did other governments. China also often points out that there is only one German people though there are two states. This situation is not unlike the question of Taiwan and the "two Chinas" proposition. It is the Chinese belief that German reunification is inevitable.

China considers the emergence of the Common Market in Europe as one of the most important political developments in recent years. Since 1970 it has given much attention to the development of trade and diplomatic relations with all European countries. The key to a strong and powerful western Europe in the eyes of China is, of course, West Germany, which is also the fourth largest economy in the world. China is also well aware that a united Germany would immediately become the third largest economy in the world, displacing Japan from that position and possibly even threatening the Soviet Union in the future.

The Chinese also express their understanding to Europeans for the need to maintain strong economic and military tics with thc Unitcd Statcs. They perecive any reduction in Europe's defense posture as an eventual threat to their own security. The Chinese believe that U.S.S.R. desires to extend its hegemony over western Europe. Neutralization of that area and free access to its vast technology base would enable the Soviets to stage the final showdown with China.

Because much of western Europe and even the United States depend heavily on strategic materials in southern Africa, Soviet and Cuban involvement in Africa has the effect of drawing China and Europe even closer together on that issue. Both parties feel particularly uneasy because the Soviet Union is itself a leading producer of most strategic metals and does not need to import. Its moves, therefore, are perceived as purely political, leading toward strategic materials control. Attempts to counteract will govern much of Chinese foreign policy in the future.

THE FOREIGN-AFFAIRS ESTABLISHMENT

The conduct of Chinese foreign affairs is the responsibility of the Ministry of Foreign Affairs headed by Minister Huang Hua, the onetime Chinese ambassador in Cairo. The ministry operates the 115 diplomatic missions in foreign countries. In some countries the New China News Agency performs a semidiplomatic service.

China's foreign affairs are also supervised by the Foreign Affairs Department of the Ministry of National Defense. This organization plays a special role in operating Chinese military missions abroad. It is independently responsible to the State Council for its foreign affairs activities. The State Council may also receive direction from the International Liaison Department of the Chinese Communist Party Central Committee, which has global responsibility for maintaining relations with communist parties in other countries, apparently independently of the foreign-affairs establishment (see Fig. 6.1).

An important part of Chinese diplomacy are people-to-people programs operated by the Chinese People's Institute of Foreign Affairs and the Chinese People's Association for Friendship with Foreign Countries. These two organizations also maintain contact with countries not having official relations with China, as well as with unofficial groups in other countries. This program is implemented by individual Friendship Asso-

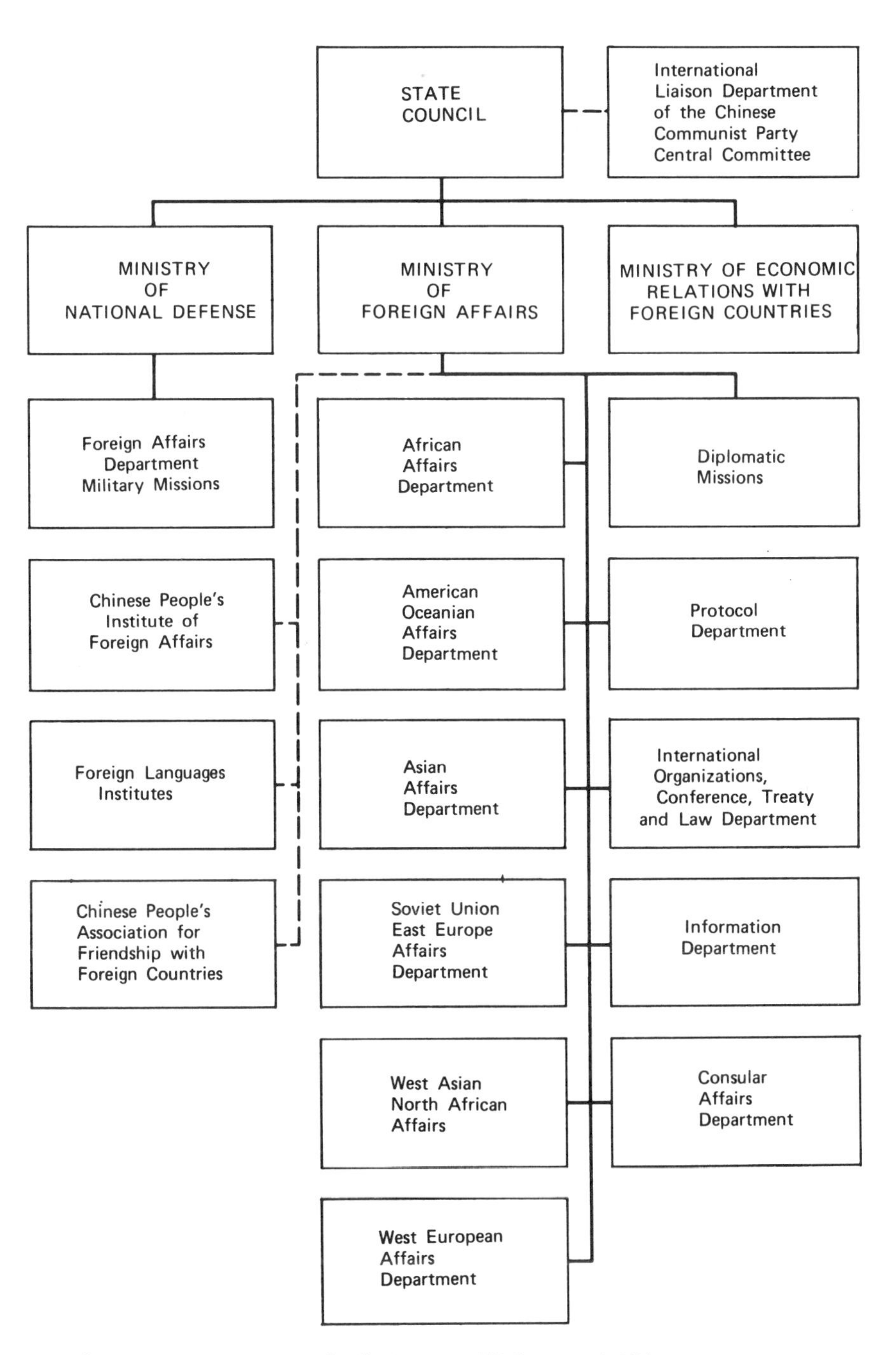

Figure 6.1 International relations establishment of China.

ciations; except for the Sino-African and Sino-Japanese, all are with communist countries.

The Foreign Languages Institutes are special schools that train personnel for various Chinese foreign-affairs organizations. The best known of those institutes is located in Peking.

The Ministry of Foreign Affairs, besides operating diplomatic missions abroad, has six geographical departments specializing in foreign affairs in particular regions. The West European Affairs Department is the most recent, formed in 1970 when considerable expansion of Chinese diplomacy in Europe was taking place. Western European affairs were previously handled by the American and Oceanian Affairs Department. The ministry also has several functional departments such as Protocol, Information, and Consular Affairs.

The Ministry of Economic Relations with Foreign Countries is not strictly a part of the Foreign Affairs Establishment of China because it does not engage directly in diplomacy. However, it is the ministry that is most likely responsible for the formulation and execution of Chinese foreign-aid programs. As such, it plays an important role complementary to and in support of China's foreign policy. This ministry was originally formed in 1961 as a bureau of the State Council and became a commission in 1964. The fact that it was raised to its present status in 1971 underscores the increasing importance of China's foreign-aid programs to its economy and foreign policy. The Ministry is headed by Minister Chen Mu-hua, who was appointed in 1977.

CHINA'S FOREIGN-AID PROGRAMS

China's foreign aid is an instrument of its foreign policy and as such is indicative of specific trade and political objectives pursued by China on the international scene. China is the only developing country in the world that extends both economic and military assistance to other developing countries within and without the communist sphere.

Since 1954 Chinese foreign aid was extended to about sixty different countries, and in 1976 those extensions were being drawn by fifty countries. Economic assistance predominates in Chinese foreign-aid programs but military assistance is of significance in the case of Egypt, Pakistan, and Tanzania. Previously North Korea and Albania were also major military-aid recipients. Some military assistance is provided for training and supplying guerrilla groups in several African countries, but the magnitude of this type of aid is generally considered to be insignificant. Contrary to some popular beliefs, anticolonialism and liberation movements are not the most important reasons for Chinese foreign-assistance programs.

China is believed to allocate its foreign aid primarily based on its perception of competing Soviet influences in LDCs. However, analysis of foreign-aid recipient countries also shows that most of those nations are existing or potential suppliers of strategic materials or other goods that China has to import from abroad due to inadequate domestic supplies. This highly political and commercial orientation of Chinese foreign aid is probably the reason why China does not channel any of its aid through international organizations such as the United Nations or World Bank.

The current program is regarded as a considerable foreign-policy success, particularly since 1970, when militant propaganda that accompanied previous foreign-aid programs was discontinued. Present programs are helpful to the recipient countries and appear to strengthen commercial ties, assuring China of future supplies of some of its most important import commodities.

In comparison with foreign-aid programs of the West and the Soviet Union, Chinese aid extensions are relatively modest. Since 1954 total extensions, including military aid to North Korea and North Vietnam, averaged about $330 million per year. In 1975 economic loans and grants to LDCs represented less than one tenth of a percent (0.08 percent) of the Chinese GNP. The comparable Soviet extensions in the same year represented 0.15 percent of the Soviet GNP, and the United States program amounted to 0.3 percent of the American GNP. However, Chinese extensions were not far behind the foreign aid programs of eastern European countries, which represented 0.1 percent of the combined GNP of the Eastern Bloc (see Table 6.1).

During the initial phase of China's foreign-aid program, at least 60 percent is estimated to have gone to North Korea, North Vietnam, and about 30 percent to other communist countries, including Albania, Cuba, Hungary, Mongolia, and Romania. Only 10 percent was extended to developing countries in the noncommunist world.

Since 1970 the pattern of Chinese foreign aid changed significantly with considerable shift of emphasis to LDCs of Africa and Asia. In 1970 total Chinese foreign-aid extensions were estimated at over $1.1 billion, of which no more than 30 percent went to communist countries. Since then Chinese foreign-aid promises have been declining, but in late 1975 China was reported to have pledged $1,000 million to Cambodia. It is not known if this amount has been formally extended so far, but this may well be the reason why Chinese economic aid and military deliveries to the LDCs in the noncommunist world have been reduced to only $208 million in 1976.

Table 6.1 Comparison of foreign aid programs to Less Developed Countries instituted by major donor countries since 1974

Donor Country	Gross Official Bilateral Capital Flows to LDCs[a] 1954–1974	Gross Official Bilateral Capital Flows to LDCs[a] 1974	Economic Loans and Grants to LDCs in 1975	Military Deliveries to LDCs in 1975
United States	66,183	4,705	4,822	4,707
France	19,497	1,806		285
West Germany	10,532	1,533		240
Japan	8,910	1,220		—
United Kingdom	8,534	688		275
Soviet Union	5,485	665	1,299	2,190[c]
Eastern Europe	1,565	145	422	200[c]
China	1,550	240	273	80[c]
Other[b]	14,697	2,417	—	430

Source: *U.S. Government Handbook of Economic Statistics, 1976 and 1977.*

[a] Includes grants, gross loans with maturities of 5 years or more, loans repayable in recipient currencies, and sales for recipient currencies. [b] Includes Australia, Austria, Belgium, Canada, Denmark, Finland, Italy, Norway, Netherlands, Sweden, Switzerland, New Zealand, Portugal (1954–1972). [c] Data for 1976.

If all those possible or expected extensions are taken into account, China's total foreign aid would amount to about $8,000 million, over 55 percent of which has been extended after 1970. Almost half of all the foreign aid since 1954 has been given to communist countries, although Cuba, Hungary and Mongolia have not received any credits since the mid-1960s. Africa is the largest region in the noncommunist world to receive Chinese foreign aid, and its share came to 27 percent of the total. South Asia, including Pakistan, Sri Lanka, and Nepal, is the next largest recipient region, accounting for over 10 percent of the total (see Table 6.2).

Chinese foreign aid is mostly extended in the form of interest-free loans for construction of projects or the purchase of Chinese products. It provides for repayment over at least 10 years with a 10-year grace period. During the first decade up to 20 percent of Chinese foreign aid was extended in the form of grants. Terms provide for repayment in commodity exports, or, on mutual agreement, in domestic or convertible currency. One of the advantages of Chinese foreign aid is that it often provides long-term financing to cover local costs of projects.

In comparison with the Soviets, Chinese foreign-aid extensions to noncommunist LDCs are dominated by economic programs, although some military aid is also given. In 1976 China concluded agreements for $100 million of military assistance, of which $80 million was delivered that year. However, the number of Chinese military advisers in LDCs

Table 6.2 Distribution of total estimated Chinese economic and military aid to various regions of the world

Region	*Total Extensions in Millions of $U.S. in 1954–1976*	*Percent of Total*
Total	8,000	100.0
Communist countries[a]	3,924	49.0
Africa	2,200	27.5
Southern Asia	826	10.3
Middle East	402	5.0
Eastern Asia	307	3.8
Latin America	153	1.9
Europe (Malta)	45	0.5

Source: Developed from several sources listed in references to this chapter.

[a] Includes Albania, Cambodia, Cuba, Hungary, Laos, Mongolia, North Korea, Vietnam, Romania. Estimate includes Chinese pledge to extend $1,000 million to Cambodia made in late 1975 and estimated $350 million annual aid to Vietnam since 1975.

declined in 1976 by more than 30 percent, while the presence of Soviet and eastern European military advisers went up by 10 percent during the same period of time.

It is interesting to note that the leading Chinese foreign-aid recipients are also suppliers of specific commodities or raw materials that are limited or unavailable in China. These include cobalt, nickel, chromium, copper, natural rubber, and diamonds. Most of the other foreign-aid recipients produce such goods as aluminum, bauxite, sugar, cotton, sulfur, and phosphates. All of these are also available in China, but their production has been inadequate and such commodities as aluminum or raw cotton have been imported recently in great quantities (see Table 6.3).

The effect of Chinese foreign aid on global raw-materials supply markets in the future does not appear to be very significant. Sudden changes in prices of some commodities could create temporary disruptions in certain countries as a result of foreign-aid repayment requirements. This could become significant only in countries where a large percentage of local production is committed for export to China.

Assistance to Communist Countries

Vietnam is probably the largest single recipient of Chinese economic and military aid since 1955. At least $1,660 million has been extended to Vietnam since 1954 for modernization of transportation and in military assist-

Table 6.3 Major recipients of Chinese foreign aid and their potential as suppliers of import commodities to China

Country	*Total Extensions in Millions of $U.S. during 1954–1976*	*Estimated Extensions in Millions of $U.S. during 1970–1976*	*Number of Chinese Technicians in 1976*	*Actual or Potential Major Chinese Imports*
Total[a]	8,000	4,500	20,415	
Vietnam	1,660	860	yes	natural rubber
Cambodia	1,092	1,000	yes	natural rubber
Albania	359+	large	yes	chromium and nickel
North Korea	300+	large		zinc, lead
Pakistan	405	250	200	cotton suppliers
Tanzania	359	305	1,000	copper, diamonds,
Zambia	307	290	5,700	cobalt, copper
Romania	300	265	yes	petroleum equip., trucks
Nepal	179	117	250	sugar, timber, jute
Sri Lanka	158	117	100	natural rubber supplier
Egypt	134	28	25	phosphates and cotton
Somalia	133	111	2,000	was major ship registry
North Yemen	106	49	400	cotton producer
Zaire	100	—	—	cobalt, copper, diamonds
Algeria	92	40	200	iron ore, petroleum
Mauritania	85	59	200	iron ore, copper
Ethiopia	85	84	100	sugar, oilseeds
Burma	84	57	40	sugar, lead
Sudan	82	82	800	cotton supplier
Guinea	77	11	400	bauxite, alumina
Afghanistan	73	45	125	cotton and wool
Cameroon	71	71		natural rubber, aluminum
Mali	68	4	350	cotton producer
Madagascar[b]	66	59	—	chromium, sugar
Chile	65	65	—	major copper supplier
Syria	61	45	50	cotton supplier
Upper Volta	60	60		cotton producer
Mozambique	59	59		sugar, timber, cotton
South Yemen	56	43	400	cotton producer
Chad	50	50		cotton producer

Source: Compiled by 21st Century Research from several sources listed at the end of this chapter.

[a] Estimates includes all economic and military aid to about 60 countries. [b] Malagasy Republic during 1958–1975.

ance. It is estimated that since 1975 China is providing $350 million annually in economic aid to Vietnam.

Cambodia received a total of almost $3 billion in U.S. economic and military aid prior to the communist takeover of Indochina in 1975. Current Chinese aid may be in the order of $25 to 30 million, but there are reports that $1,000 million has been pledged by China for economic development. This is unconfirmed and if extended will probably take several years to accomplish. If granted, it will make Cambodia one of the largest recipients of Chinese aid. Since Cambodia is not known to receive any assistance from the Soviet Union or eastern Europe, China may be developing an exclusive area of influence with a massive aid program. Laos received a Chinese aid commitment of $42 million in 1976, but it is also receiving aid from the Soviet Union and Vietnam.

Albania received Chinese aid since 1955, and total amounts were estimated at $359 million by 1971. Most of that aid was used by Albania to cover its annual deficit in foreign trade. Additional credits of unknown magnitude were extended in 1970, when China undertook the construction of thirty industrial projects for exploiting mineral resources. Albanian navy and air force use Chinese-built motor torpedo boats and aircraft. Albania was very important to China during the 1960s as the major supplier of chromium. In 1976 China appears to have reduced its aid to Albania, although a ferronickel plant and a steel works built with Chinese assistance have been completed. Some uncertainty about Sino-Albanian relations in the future may have developed as a result of the reported Albanian government search for additional foreign aid donors in 1976.

Although North Korea has been one of the largest recipients of Chinese economic and military aid in the past, it is not receiving any aid at present. A total of $330 million was extended to this country during 1953–1960, and an offer of assistance was made for North Korea's Six Year Plan of 1971–1976, but the magnitude of this offer is not known. North Korea is now an economy in many instances more advanced than China and participates in economic assistance to Cambodia.

Cuba, Hungary, and Mongolia received Chinese aid extensions before 1960 but have not received any assistance since that time. Since 1961 China helped Cuba by purchasing 500,000 tons of Cuban sugar while Western embargo was taking shape; however, this purchase was made possible by a $46-million credit extended by the Soviet Union to China to finance this transaction.

China extended a total of $300 million in aid to Romania since 1970. It is believed to be primarily a development credit under which China offered to construct light industrial plants. Romania is the largest communist trading partner of China and an important supplier of petroleum exploration and processing equipment since 1960.

Foreign-aid Programs in Africa

Tanzania is the largest recipient of Chinese economic and military aid in Africa and since 1954 received a total of $359 million. About 1025 Tanzanian military personnel have been trained in China, and about 1000 Chinese technicians were in Tanzania in 1976. Completed projects include an airport, habor facilities, a military academy, a pharmaceutical plant, and several medical and agricultural projects.

In July 1976 China completed the TANZAM railway and agreed to absorb $55 million in cost overruns, which brings the total cost of the railway to $455 million. At the peak of construction activity as many as 16,000 Chinese technicians were employed on this project. Of those, up to 15,000 are believed to have been provided by China's PLA Railway Engineering and Signal Corps. Some of the construction crews in Tanzania are now building a spur railway line and working on development of iron and coal mining.

Zanzibar, which forms part of Tanzania, has been reported to be phasing out Chinese assistance, while Tanzania has been shifting its dependence on military supplies to the Soviet Union. At present the Tanzanian armed forces are equipped with Chinese-built tanks and guns; the Tanzanian navy uses Chinese naval units and hydrofoils, and the air force is equipped primarily with Chinese-built MIG-21, -17, and -19 fighter planes.

Zambia is also one of the largest recipients of Chinese foreign aid, much of it accounted for by the TANZAM railroad project, which it shares with Tanzania. A total of $307 million has been extended to Zambia since 1954, and in 1976 the number of Chinese technicians in Zambia reached 5700. This was the largest number of Chinese technicians working that year in any single country under the foreign-assistance program. With completion of the TANZAM railroad Chinese technicians are expected to work on the Serenje–Sanfua road. Zambia has been getting a lot of attention from China because it is also a producer of cobalt, as well as copper, zinc, and lead, all of which are imported by China. In 1974 about 30 percent of Zambian copper output was exported to China.

Zaire is the world's second largest producer of cobalt, a strategic metal that China must import, and a major producer of copper and diamonds. It already received $100 million in Chinese credits. China is assisting in construction of a sugar plantation and refinery, a hospital, a sports stadium, and several experimental farms. After President Mobutu visited China in 1974, Zaire instituted some nationalization programs patterned after the Chinese model, but the country's deteriorating economic situation forced the government to rescind some of those undertakings.

Somalia has been one of the larger Chinese aid recipients, getting a total of $133 million, most of it since 1970. In 1976 there were still an estimated 2000 Chinese technicians in Somalia. A hospital has been built, and work continues on a sports complex. The major project, under a 1971 credit, is a $67 million road connecting Belet Uen with Burat. The road extends 970 kilometers (km) and should be completed by 1980. Somalia was important to China as a flag of convenience for a large portion of the Chinese Merchant Marine. After Somalia began developing more friendly relations with the Soviet Union in 1975, China transferred most of its fleet in Somalian registry to that of Panama.

Democratic Republic of Madagascar received a total of $66 million in aid extensions, most of which was granted in 1975. China may give even more attention to Madagascar in the future because it is a significant producer of chromium and sugar, both of which are imported by China. Under the economic agreement China will build a sugar refinery, a plant to produce agricultural implements, a porcelain factory, a model farm, and will improve 150 kilometers (km) of the road between Tananarive and the Port of Tamatave.

Guinea is an exporter of bauxite and received a total of $77 million in Chinese aid. Relations have been cool since Guinea supported Soviet action in Angola, but Chinese aid projects have continued. About 400 Chinese technicians are in Guinea, and an estimated 250 military personnel have been trained in China. Technical assistance continues in port development, construction of a match factory, and agriculture. Two Chinese-built fishing trawlers were delivered to Guinea in 1976 under credits extended in 1972.

Morocco, which was a major phosphates supplier, received $32 million of Chinese aid for the first time in 1975. The major project under this credit is a sports stadium near Rabat; the preliminary work on the stadium was begun by Chinese technicians in 1976.

Mozambique received $59 million of foreign credits from China for the first time in 1975, and fifty military personnel have been trained in China. Chinese and North Korean technicians in Mozambique are working mainly in agriculture.

Benin, formerly known as Dahomey, received a $44-million credit in 1972. China is completing construction of a sports complex and an experimental farm. In 1973 China took over a $1-million rice project originally started by Taiwan, presumably in the aftermath of that country's expulsion from the United Nations.

In August 1976 the Central African Empire resumed diplomatic relations with China and signed an agreement to reinstate an economic program involving agricultural and medical technical projects previously operated by Taiwan. This country is an exporter of cotton and diamonds, which may be of interest to China's foreign-trade planners.

China also signed an economic cooperation agreement with Equatorial Guinea; under this agreement China will build a power station and a transmission system. The agreement for economic and technical cooperation with the Comoro Islands near Madagascar is also new; it is the first agreement signed with a communist government by this new country.

Mali, a cotton-, sugar-, and gold-producing country, received a total of $68 million in aid from China. There are about 350 Chinese technicians active in agriculture. China is also building a third sugar mill at Sikasso and expanding a rice mill previously completed at Sevaire and a textile mill in Segou. New Chinese equipment has been installed at Radio Mali.

In 1974 China signed a technical-assistance agreement with Nigeria, but its provisions are unknown. Since 1976 about 100 Chinese technicians provided assistance in rice projects, water-well drilling, and industrial training. Under a $22-million credit program Chinese technicians are building a $16-million Kigali–Rusomo road in Rwanda and have completed a rice project at Kigali. No activity is apparent in Sao Tome and Principe under an economic and technical agreement signed with China in 1976.

Senegal received a total of $49 million credits, under which the Chinese are building dams and an athletic stadium and are involved in rice projects. Sierra Leone, which has bauxite, diamonds and rutile, received $30 million in Chinese credits since 1971, under which China completed the Mangeh Bridge and eight agricultural stations.

Upper Volta, another cotton producer, has received a total of $60 million in Chinese credits since 1973. In 1976 China completed the 250-hectare rice project in Kou Valley, which was also taken over from Taiwan in 1973.

China's activity in Africa in recent years was less pronounced than in the past, although two new first economic agreements were signed; one with Botswana and the other with the Comoro Islands. Chinese technicians arrived in Botswana to assess development needs and recommend specific projects.

In Algeria, China agreed to construct a rice-processing plant and several small agricultural projects. In Ghana, China's current extensions amount to $42 million and are dominated by a $13-million irrigation project in Afife. Ghana is a potential supplier of bauxite, alumina, and diamonds, all of which are imported by China.

Chinese Programs in Asia

Pakistan depends on China for military assistance and is the largest noncommunist recipient of Chinese economic aid, which was estimated at $405 million by 1976. At least 200 Chinese technicians are working in Pakistan and 375 Pakistani military personnel have been trained in China, the second largest contingent after the Tanzanian military. Pakistan is also the only noncommunist country receiving Chinese assistance in the development of heavy industry, probably to counteract similar initiatives by the Soviet Union. Under this program China has built a munitions plant, a steel mill, and power-distribution network.

Nepal, a potential sugar and timber supplier, is also one of the largest recipients of Chinese credits, which by 1976 amounted to $174 million, $80 million of which was extended in 1975. An estimated 200 Chinese technicians were involved in a $4-million irrigation project at Pokhara and a 65-km road from Naranghar to Gurka.

Sri Lanka has been traditionally a large recipient of Chinese foreign aid, now estimated to total $58 million. The rice-for-rubber barter agreements have continued ever since Sri Lanka broke the embargo on sales of natural rubber to China during the Korean war. At least 100 technicians are presently in Sri Lanka, and their number is expected to increase with new flood control and textile-mill projects.

In Afghanistan, China completed an irrigation project in Parwan and is planning to expand the program. A textile mill is under construction at Bagram and a hospital in Kandahar; an agreement was signed to construct a paper mill. About 125 Chinese technicians were reported in Afghanistan in 1976, and the total credits granted are estimated at $73 million. Afghanistan is a possible supplier of cotton and wool for the Chinese textile industry.

In eastern Asia a new economic assistance agreement was signed with western Samoa, and a five-year trade agreement with the Philippines for the supply of Chinese oil went into effect in 1976. A Sino-Philippino trade committee was established to redress a trade balance that is in China's favor. A total of $37 million in credits was extended during 1975 and 1976. The Philippines is shaping up as a very promising trade partner because it can supply chromium, nickel, copper, and sugar and can absorb and process Chinese oil.

Chinese Foreign Aid in the Middle East

Egypt received a total of $134 million in Chinese credits and is among the ten largest recipients of Chinese foreign aid. Since the cutoff of Soviet military deliveries to Egypt in 1974, China signed an agreement to supply Egypt with engine replacements for Soviet-built tanks and aircraft after Egypt failed to obtain spare parts from India, which also uses similar Soviet equipment. Egypt is an important cotton supplier to China and is an alternate source of phosphates. A recent trade protocol reportedly called for expansion of Sino-Egyptian trade to $600 million annually, apparently in an attempt to make Egypt one of the top trading partners of China.

In North Yemen $27 million of new Chinese credits was granted in 1976, bringing the total to $106 million. China has at least 400 technicians in North Yemen working on industrial plant and road construction. China also has 400 technicians in South Yemen, but aid extensions are in the order of only $56 million.

In Sudan, Chinese economic assistance is apparently regarded as best among all communist donor countries. Eight hundred Chinese technicians were in Sudan in 1976 completing the Wad medani–Gedaref highway, a textile mill, and a conference hall in Khartoum. Total extensions to Sudan amount to $82 million, and about 200 Sudanese military personnel have been trained in China.

Foreign Aid in Latin America

In Peru, China is providing irrigation equipment under a 1971 credit of $42 million. Sino-Peruvian trade agreements also provide for purchases of copper, lead, zinc, and fishmeal for a total of $57 million. In 1974 China accounted for 10 to 15 percent of Peruvian exports of lead, and its imports of copper are also significant. China's economic-aid extensions to Peru have been overshadowed by Soviet military-assistance agreements estimated at $250 million that include 36 SU-22 supersonic fighter–bomber aircraft as well as helicopters and advanced weapons for the army.

Chile and Guyana received a total of $65 and $36 million of credits, respectively. Chile is a major supplier of copper to China, but because of the ouster of Allende government there was little activity in Chinese aid program to that country. Imports of copper, which account for a significant portion of Chilean output, are continuing, and in 1977 Sino-Chilean assistance terms have been renegotiated. Guyana is a major bauxite and alumina producer and its sugar, timber, and diamonds are also of interest to China.

Jamaica became the recipient of Chinese foreign aid in 1974, when $9 million was extended for the construction of a polyester and cotton textile plant and a $1-million six-year credit was granted in 1976 for imports of 5000 tons of Chinese rice. China's only communist aid competitor in Jamaica is Cuba, which had over 280 Cuban technicians working there in 1976. Aluminum and sugar are possible repayment commodities of interest to China.

REFERENCES

Central Intelligence Agency. Communist Aid to the Less Developed Countries of the Free World, 1976. ER77-10296, August 1977.

Cooper, John F. "China's Foreign Aid in 1976," *Current Scene* June–July 1977, Hong Kong.

Current Scene. "China's Foreign Affairs Establishment," July 1976, Hong Kong.

Current Scene. "China's Foreign Aid in 1972," December 1973, Hong Kong.

Current Scene. "China and the TAZARA Railroad Project," May–June 1975, Hong Kong.

Doyle, George A. "China, the Chinese, and the True Leap Forward; the Chinese Perspective with Implications for United States Foreign Policy," in *China and the Chinese,* Joint Economic Committee, U.S. Congress, November 1976, p. 38.

FBIS SOV 75-48. "PRC's Commentary on Trade Relations with Africa," Moscow Radio in English, March 9, 1975.

Fogarty, Carol H. "China's Economic Relations with the Third World," in *China: A Reassessment of the Economy,* Joint Economic Committee, U.S. Congress, July 1975, p. 730.

Forward. "Diary of Events," August 1973, Rangoon, Burma.

Forward. "Sino-Burmese Friendship Tames Turbulent River," May 1, 1974, Rangoon, Burma.

Kovner, Milton. "Communist China's Foreign Aid to Less-Developed Countries," in *An Economic Profile of Mainland China,* Joint Economic Committee, U.S. Congress, February 1967, p. 609.

Kux, Ernst. "China and Europe," *Current Scene,* May 1973, Hong Kong.

Mirza Quibal. "China backs £ 34 M Pakistan Plant," Karachi, October 8, 1974.

Peking Review. "People of Poor Countries Have Courage—Notes on Building the TANZAM Railway," No. 31, July 1976.

Rafferty, Kevin. "Support for the Third World," *Financial Times,* August 29, 1975, London.

Tansky, Leo. "China's Foreign Aid: the Record," *Current Scene,* September 1972, Hong Kong.

Tansky, Leo. "Chinese Foreign Aid," in *China: An Economic Assessment,* Joint Economic Committee, U.S. Congress, 1972, p. 609.

7

Energy and Fuels

China is the fourth largest energy-producing country in the world after the United States, the Soviet Union, and Saudi Arabia. It also ranks as the fourth largest energy consumer behind the United States, Soviet Union, and Japan, with a total consumption level comparable to that of West Germany. Per capita consumption, however, is extremely low although it is at least twice as high as that of India, Pakistan, Indonesia, or Bangladesh.

Total primary energy production, including coal, crude oil, natural gas, and hydroelectric and geothermal power was estimated at 497 million metric tons of coal equivalent in 1976. China ranks third in the world in production of coal, fifth in natural gas, and eleventh in crude oil. Coal accounts for two-thirds of all primary energy production, but oil is rapidly becoming a major source of energy and exports.

Discovery of large oil deposits in 1960 and rapidly growing oil production since then may have been a significant factor that allowed China to sever its close economic ties with the Soviet Union. Until about 1965 China was a net importer of energy in the form of crude oil and petroleum products from the Soviet Union, Romania, and Albania, which amounted to as much as 11 percent of total Chinese imports at one time. Since 1965 China has been self-sufficient in crude oil supply. Today it is a net exporter of energy in the form of oil and coal to Japan, North Korea, Vietnam, and other Southeast Asian countries.

Rapidly growing oil exports led to considerable speculation that China may become a major oil exporter in the near future. Experts believe that under present oil-production development policies China may be able to export up to 100 million metric tons in coal equivalent of crude oil per year by 1985. Rapidly growing domestic demand would limit larger exports of oil under existing investment and consumption conditions. On the other hand, China will derive increasing hard-currency revenues from its growing exports of oil. This will give her the necessary funds to purchase the most advanced oil exploration and production equipment in the West and considerably accelerate its oil production and export program if it chooses to do so.

To develop its oil industry and modernize its massive coal industry, China imports advanced mining equipment and technology from abroad. During the 1950s the Soviet Union and Poland were the major suppliers. Romania replaced the Soviet Union as a supplier of oil-exploration and production equipment during the 1960s. In the early 1970s China continued to import coal-mining equipment from Poland but also turned to West Germany, the United Kingdom, Japan, and the United States for more advanced mining equipment and off-shore exploration technology.

China also exploits its uranium resources and produces plutonium, but so far there is no evidence that nuclear technology is applied for generation of electric power. Geothermal energy is exploited for power generation on an experimental basis. Other forms of energy used in China include wood for household fuel, organic waste, and waterpower aside from electricity generation and wind.

ENERGY RESOURCES AND RESERVES

Chinese energy resources are comparable to those of the United States or the Soviet Union in order of magnitude and are considered to be sufficient for centuries to come. Unfortunately, in the absence of official Chinese statistics, estimates of energy resources and reserves vary widely from one source to another and are often revised.

Energy resources include known energy reserves as well as deposits of oil, natural gas, and coal as yet undiscovered. Resources also include identified deposits of fuels or hydroelectric potential that cannot be exploited immediately for various economic or political reasons. However, reserves are that portion of total resources that are assessed as exploitable under existing economic conditions and with available technology.

Estimates of coal resources in China range from 9000 billion metric tons made in 1958 to more sober levels of 1500 to 2000 billion metric tons. United Nations estimates put total Chinese coal resources at 1011 billion metric tons, of which 300 billion are considered known economic reserves in place. Until recently this figure was sometimes also quoted as proven Chinese coal reserves that are now estimated at about 80 billion metric tons. Even this lower figure is equivalent to about 100 years of coal supply at present consumption rates in China. But since the use of coal in total energy supply is constantly declining, the current proven coal reserves of China may last even longer.

About 40 percent of the territory of China is believed to possess oil and natural-gas potential. Total oil resources in China have been estimated as high as 10 billion metric tons in 1972. In addition, the oil content of Chinese oil shale deposits is estimated at 11 to 20 billion metric tons. So far China has not published any official data on its oil resources, but in 1973 Chinese officials told an American lawyer representing U.S. oil equipment manufacturing firms that they believed China to be the world's third richest oil region behind the Middle East and North America. Estimates of Chinese oil resources are undergoing constant reassessment as oil exploration of western China and offshore areas continues. Natural-gas resources in China have been estimated to be in the order of 30 trillion cubic meters, which is also the same order of magnitude as those of the United States or the Soviet Union. However, a relatively much smaller portion of Chinese natural-gas resources is at present regarded as natural gas reserves.

Hydroelectric resources of China are very significant and are estimated in the range of 0.5 to 1.0 billion kW, but only a very small percentage of the total is exploited. More details of hydroelectric resources are found in Chapter 8 under the section on hydroelectric power stations. The magnitude of geothermal and uranium resources is unknown at present.

Although estimates of Chinese energy resources are very tentative and are constantly revised, estimates of proven energy reserves are more precise and are expected to improve with increased application of modern exploration techniques. Table 7.1 presents a comparison of the most recent Chinese energy reserves estimates with those of several primary energy-producing countries.

China's proven energy reserves appear to be very similar to those of the United Kingdom or Mexico in oil and natural gas. Some industry observers suggest, however, that proven reserves of China's oil in particular may grow considerably from the present estimates with increasing exploration of western China and offshore oil regions. Oil and gas reserves of China excluding coal are also comparable to those of Nigeria and surpass those of Indonesia and Venezuela. Thus China is currently estimated to possess the tenth largest oil reserves in the world. This is not to be confused with China's crude-oil production, which is now about eleventh in the world, comparable to that of Indonesia or Canada.

World significance of Chinese energy resources lies in the fact that proven reserves are sufficient to maintain economic development of

Table 7.1 World significance of Chinese proven reserves of fossil fuels ranked by crude oil reserves at yearend 1976

Country	*Crude Oil in Millions of Barrels*	*Natural Gas in Billions of Cubic Feet*	*Coal in Millions of Metric Tons*
Saudi Arabia	158,000	105,940	—
Kuwait	71,000	35,310	—
Iran	60,000	600,350	190
Soviet Union	40,000	812,240	136,600[a]
United States	39,000	218,950	181,700
Iraq	36,000	35,310	—
United Arab Emirates	31,000	35,000	—
Libya	25,000	24,720	—
Mexico	25,000	42,380	630
China	20,000	24,720	80,000
United Kingdom	20,000	45,910	3,870
Nigeria	19,000	45,910	180
Indonesia	14,000	21,190	1,060
Venezuela	14,000	42,380	11

Source: U.S. Government Handbook of Economic Statistics, 1977, September 1977; U.S. Government National Foreign Assessment Center, *International Energy Biweekly Statistical Review,* 11 December 1977.

[a] Includes brown coal and lignite.

China well into the 21st century without dependence on foreign imports of energy. Future Chinese energy supplies will be governed primarily by capital investment levels in exploitation of known energy reserves, and if recent growth rates are maintained by 1990 China may reach energy-consumption levels comparable to those in the United States in about 1968. By the year 2000 China has the potential to become the third largest energy consumer and economic power in the world.

PRIMARY ENERGY PRODUCTION AND CONSUMPTION

During the last 27 years production of primary energy in China has grown remarkably. In 1975 China became the fourth largest energy-producing country behind the United States, the Soviet Union, and Saudi Arabia. China displaced Iran in the fourth place mainly because the latter country has reduced its oil production in recent years.

Energy production increased rapidly during the 1950s and averaged 14.9 percent per year during 1953–1957. This growth rate slowed to 9.3 percent during 1958–1965 and further dropped to 9.1 percent during

1966–1970, which includes the Cultural Revolution period. In the first half of the 1970s growth averaged about 8.7 percent annually. Part of this slowdown in growth is not inconsistent with a large expanding industry base, but it is also believed to have been affected by the changing composition of energy production sources during the last 20 years.

Coal accounted for 96 percent of all primary energy produced in 1957, but by 1976 its share decreased to 66 percent. During the same time period production of oil increased from 2 to 23 percent of total primary energy, and the share of natural gas went up from 1 percent to almost 10 percent of the total. Contribution of hydroelectric energy remained steady at about 1 percent of all energy produced. Recent primary energy supplies, which take into account energy exports and imports and define domestic energy availability (see Fig. 7.1), differ only very slightly from

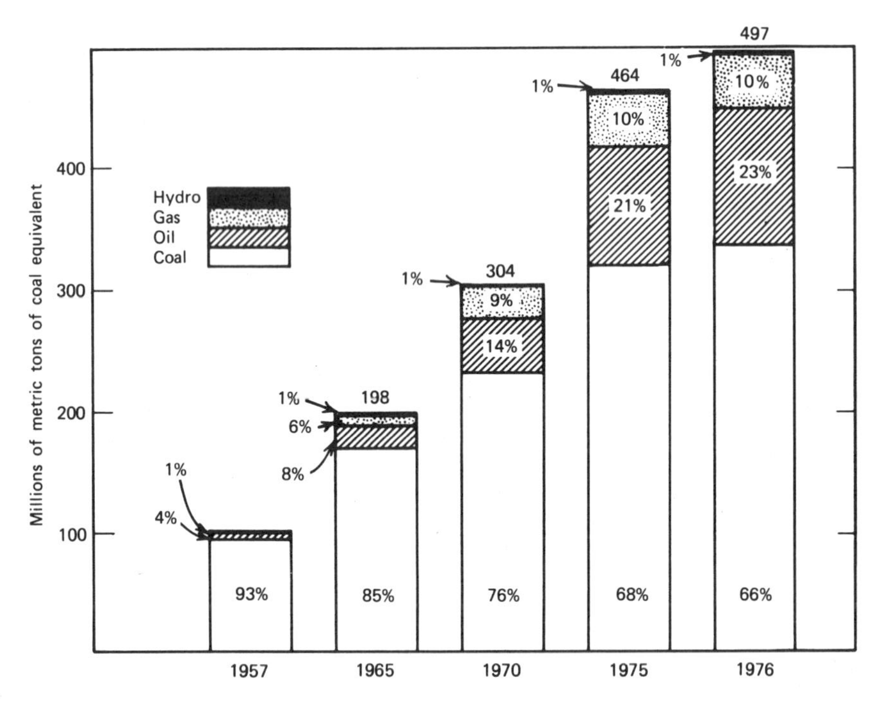

Figure 7.1 Growth of China's primary energy supply. (Source: *Adapted from data presented in* U.S. Government Research Aid, *A(ER)75-75, "China: Energy Balance Projections," November 1975; U.S. Government National Foreign Assessment Center, "China: Economic Indicators," ER77-10508, October 1977.*)

the pattern of energy production and are not expected to alter significantly until energy exports increase considerably.

Energy consumption grew from 97 million tons of coal equivalent in 1957 to 380 million tons in 1974 at an average 8.4-percent annual rate of increase. In 1975 total energy consumption reached 421 million tons and domestic energy supply was 464 million tons. This indicates handling and shipping losses and possible buildup of fuel inventories equivalent to almost 10 percent of energy production. By 1976 total energy supply was estimated at 497 million tons.

Per capita consumption in China is still very low, at 522 kg of coal equivalent per person per year. This figure differs according to population figures accepted for comparison. United Nations population figures for China are about 100 million lower than estimates of U.S. government agencies. Based on U.N. population figures of 838 million in 1975, the per capita energy consumption in China would be in the order of 632 kilograms of coal equivalent. This would be similar to per capita consumption of energy in countries such as Turkey, Syria, Peru, or Brazil at best (see Table 7.2 for comparisons). In either case China's per capita energy consumption is considerably higher than that in other populous countries such as India, Indonesia, Pakistan, or Nigeria.

Table 7.2 Primary energy production in 1976 and per capita consumption in selected countries in 1975

Country	*Total Energy in Millions of Metric Tons Coal Equivalent*		*Population in Millions*	*Per Capita Consumption in Kilograms of Coal Equivalent*
	Production	*Consumption*		
China	513	497[a]	951[a]	522
United States	2,494	2,681	213	12,586
Soviet Union	1,677	1,391	254	5,476
Japan	79	474	111	4,270
West Germany	183	347	62	5,597
Indonesia	103	24	131	183
India	114	132	614	214
Brazil	25	71	106	669
North Korea	24	44	16	2,750
Pakistan	8	12	72	166
Nigeria	131	6	65	92

Source: Figures derived from data in *United Nations Statistical Yearbook 1976* or *U.S. Government Handbok of Economic Statistics 1977.*

[a] Data for 1976.

Energy production and consumption are expected to grow rapidly through 1980, and the rate of this growth can be estimated using a recent energy–GNP elasticity coefficient that was 1.42 during the first half of the 1970s. For every 1-percent increase in GNP, energy consumption grew by 1.42 percent, and if China's GNP is expected to increase at an average 6 percent through 1980, its energy production would have to grow about 8.5 percent per year or even faster to provide surplus energy for export.

Industry and construction, including generation of electric power consume 62 percent of all primary energy supply in China, but agriculture, which uses only 6 percent of the total, has been by far the most rapidly growing energy consumer. While industrial energy consumption grew at an average annual rate of 11.6 percent in 1958–1974, energy use in agriculture was increasing at 26 percent annually. The (combined) residential and commercial sector is the second largest energy user in China and accounts for 27 percent of all consumption but has shown relatively modest growth of an average 4.2 percent during 1958–1974. Transportation consumes only 5 percent of all energy supply, and its energy use has grown at 4.7 percent in the same period. A more detailed energy supply by sources and consumption by sectors for 1975 is shown in Figure 7.2.

Uneven energy consumption growth rates underscore priorities assigned by Chinese leadership to industrialization and mechanization of agriculture. The share of energy consumed by industry increased from 37 percent in 1957 to 62 percent in 1976. During the same time the share of energy consumed by the transportation sector dropped from 9 percent to 5 percent, and that of the residential and commercial sector declined from 53 percent to 27 percent. Substitution of oil for coal has also primarily occurred in industry.

Growth of energy consumption is expected to continue at the fastest rate in agriculture, where it is estimated to be 15 to 20 percent annually until the early 1980s. Energy consumption in industry is estimated to grow at rates ranging from 10 percent to 13 percent per year with continuing substitution of oil for coal as a major industrial fuel.

Consumption of energy by the transportation sector is expected to continue growing at 4 to 6 percent annually, but the lowest increase in energy consumption is expected in the residential and commercial sector, ranging only between 3 percent and 5 percent per year.

Rapidly growing domestic demand for energy during the first half of 1980s will govern China's energy-production policies and influence its energy exports. As a result, by 1985 oil exports are unlikely to exceed 100 million tons of coal equivalent or about 1.3 million barrels of oil per day. China could increase its energy production and exports above that level

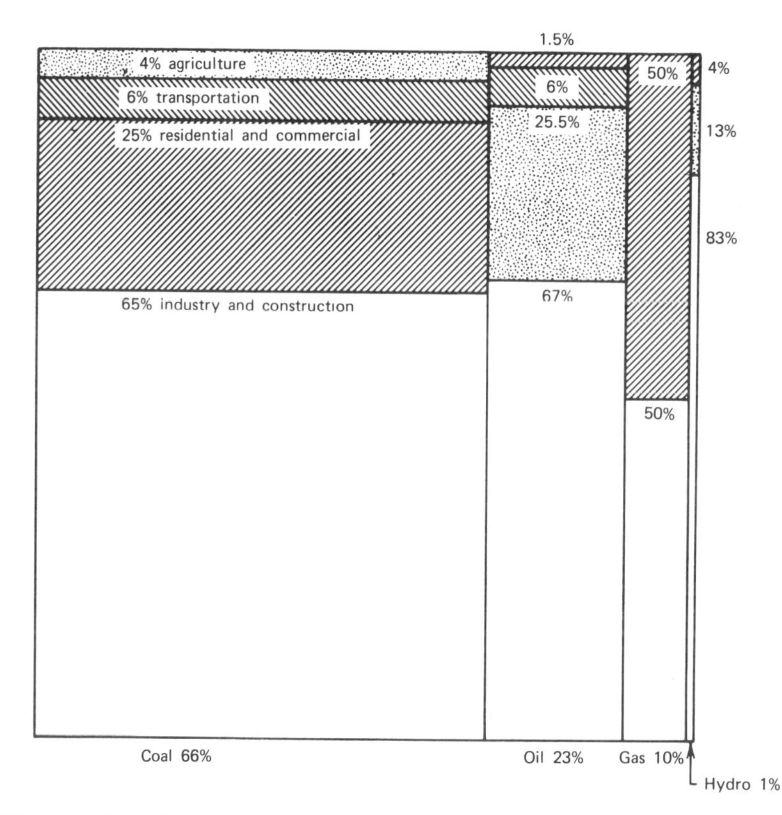

Figure 7.2 ***Primary energy supply and consumption in 1976. This double-bar chart is divided vertically according to the category of energy use and horizontally according to energy source. It shows quantitatively relative contribution of a particular fuel to a particular end use by area of the labeled item. For example, coal contributes 66 percent to the total energy supply (bottom left of figure), and coal used by transportation is 6 percent of 0.66 or 3.96 percent of the total energy supply.***

by increasing sharply the investment in oil and coal industries. Domestically this could only be done at the expense of other industries and overall economic growth and is unlikely to happen. The alternative is joint ventures with foreign countries and companies, which seems to be unlikely under present Chinese self-sufficiency policies. For this reason, further imports of advanced coal-mining and oil-exploration and production equipment appears to be the most likely course to follow for increasing the productivity of Chinese energy-producing industries.

Chinese energy resources are enormous, and annual production levels of 1000 million tons of coal and 1000 million tons of oil by the year 2000 are not unrealistic. Nevertheless, even at such high production levels Chinese per capita energy consumption would only reach the average world consumption level of today.

PETROLEUM PRODUCTION AND CONSUMPTION

China is a major oil producer, with its 1976 crude oil output estimated at 83.6 million tons, equivalent to 1.67 million barrels per day. This makes China the eleventh largest oil-producing country in the world, just ahead of Canada and Indonesia and not far behind Libya and the United Arab Emirates (see Table 7.3). During 1949–1975 oil production has grown at average annual growth rates in excess of 20 percent. In recent years oil production has slowed down and in 1976 was reported to have increased by only 13 percent. This unexpectedly low growth rate may reflect the effects of the severe Tangshan earthquake and recent political disruptions,

Table 7.3 World significance of Chinese proven oil reserves and production and depletion rates in 1976

Country	*Proven Reserves in Millions of Barrels*	*Percent of World Total*	*1976 Production Millions of Barrels*[a]	*Depletion Rate = Annual Production as Percent of Reserves*
World total	665,000	100.0	20,914	3.14
Saudi Arabia	158,000	23.7	3,129	1.98
Kuwait	71,000	10.7	783	1.10
Iran	60,000	9.0	2,148	3.58
Soviet Union	40,000	6.0	3,712	9.28
United States	39,000	5.9	2,963	7.59
Iraq	36,000	5.4	782	2.17
United Arab Emirates	31,000	4.7	706	2.27
Libya	25,000	3.7	706	2.82
Mexico	25,000	3.7	310	1.24
China	20,000	3.0	609	3.04
United Kingdom	20,000	3.0	292[b]	1.46
Nigeria	19,000	2.8	755	3.97
Indonesia	14,000	2.1	549	3.92
Venezuela	14,000	2.1	837	5.97

Source: Based on data from U.S. Government National Foreign Assessment Center, *International Energy Biweekly Statistical Review*, 11 December 1977.

[a] Data derived from daily production figures times 365 days. [b] Data for 1977.

and future growth is expected to average about 15 percent annually until about 1985.

If growth continues at such a rate, China could produce close to 300 million tons of crude oil in 1985, comparable to the output of Iran in 1976. This could establish China as the fifth or even fourth largest oil-producing country behind the Soviet Union, Saudi Arabia, the United States, and possibly Iran. Massive capital requirements for oil exploration, production, and transportation equipment are the main constraints on the growth of Chinese oil industry.

Chinese proven oil reserves are now estimated at about 20,000 million barrels, comparable to those of Mexico or Nigeria, and constitute about 3 percent of the total proven reserves of the world. Chinese production is also equal to about 3 percent of the world's total (see Table 7.3), and its current rate of depletion of its proved reserves is only 3.04 percent. This is less than half of the 7.59-percent rate for the United States and only a third of the Soviet depletion rate.

Beyond the year 1985 Chinese oil production is expected to grow at relatively slower rates, but few growth estimates have been made because there is a lack of authoritative data about the probable growth of associated supportive industries and industrial infrastructure. One Hungarian sinologist using Soviet and Western sources recently developed a forecast suggesting that Chinese oil production will grow at an annual average rate of 10 percent from 1985 to 1990. According to this projection, Chinese oil production could reach 1000 million tons by the year 2000. This is an enormous quantity of oil, comparable to the current combined output of Saudi Arabia and the Soviet Union, and its realization would require correspondingly massive investments in oil exploration and production facilities that only foreign credits and technology can supply.

This poses the question as to what extent China will be able to expand its oil production in view of its continued insistence on self-reliance as well as limited capital and technical resources. Because of its vast population China must be aware that its per capita energy consumption will continue to lag behind the world average probably well into the next century, even under the best circumstances that are economically feasible. Realization of these statistical facts may also limit Chinese oil exports only to the extent needed to finance critical technology and plant imports on an intermittent basis. As a result, China may never become a stable long-term oil supplier to other countries.

Petroleum consumption in China has grown from 3.3 million metric tons in 1957 to 74.5 million tons in 1976 and accounted for 23 percent of all the energy supply in 1976 (see Fig. 7.2). In 1957 industry and construction consumed over 40 percent of all petroleum supply, and transportation used almost 30 percent, but only 12.5 percent was consumed by agriculture. By 1976 petroleum-consumption patterns changed considerably, and industry and construction now account for 67 percent of all the supplies. In absolute terms this means a tremendous increase in mechanization of agriculture, which used only 480,000 tons of oil in 1957 and somewhere in the order of 18,000,000 tons of oil in 1976. Figure 7.3 demonstrates this rapid growth in petroleum consumption and priorities that have been clearly given to industry and agriculture.

The share consumed by industry is expected to grow further as modern power plants and petrochemical plants are coming on stream. Sub-

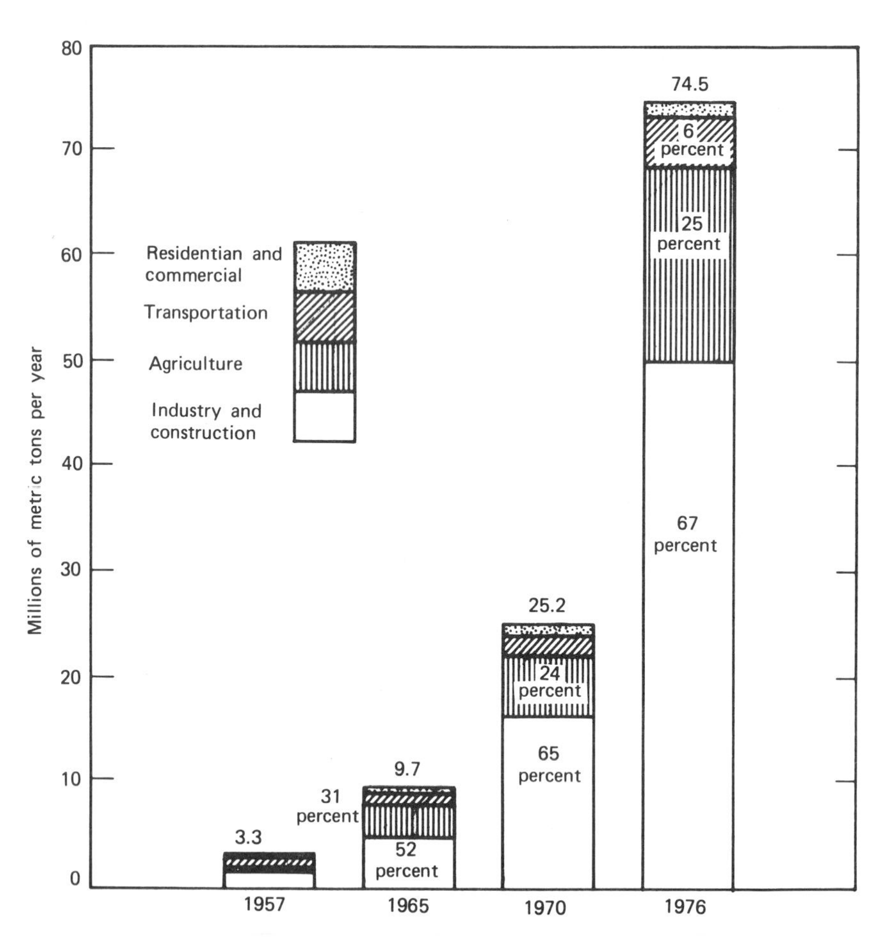

Figure 7.3 Growth of oil consumption. (Source: *Adapted from data presented in* U.S. Government Research Aid, *A(ER)*75-75, *"China: Energy Balance Projections," Washington, D.C.,* November 1975.)

stitution of oil for coal-burning plants has also occurred in efforts to modernize older industrial plants and increase their productivity with minimum new investment. Additional industrial petroleum consumption will come from the expansion of the Chinese petrochemical industry that is now taking place as production of fertilizers, synthetic textiles, synthetic rubber and other petroleum products increases.

The residential and commercial sector clearly receives the least priority for oil allocation and mostly relies on coal and natural gas as its basic fuel and power supply.

OILFIELDS, PIPELINES, AND REFINERIES

China's oil comes from about 100 commercial oilfields in fourteen known oil-bearing basins. Until the early 1960s, when the giant northeastern oilfields of Taching and Shengli began their production, most domestic crude came from the long-established oilfields of Karamai and Yumen in the relatively distant and isolated areas of western China (see Fig. 7.4). Because transportation continues to be a bottleneck the eastern oilfields, which are located close to large industrial areas, have been receiving most new investment in recent years and now produce at least two-thirds of the total Chinese oil supply (see Table 7.4). Nevertheless, in the long run the western fields are believed to hold the largest oil reserves in China.

Major producing oilfields of eastern and northeastern China are found in the Sung-Liao Basin in Manchuria and the North China Basin adjacent to Peking and Tientsin, which includes the offshore Pohai Bay oil deposits. In western and northwestern China the Dzungarian, Chiuchuan, North Shensi, Tsaidam, and Szechwan Basins also include major producing oilfields. Other oil-bearing basins that are not yet major oil-producing areas are the Tarim Basin in the remote Takla–Makan Desert, the adjoining Turfan Basin of western China, the Kwangsi–Kweichow Basin in South China; the Kiangsi Basin extending north of Shanghai well into the Yellow Sea, and the Liuchow–Taiwan Basin, which is a vast offshore area stretching from the Vietnamese border and the island of Hainan along the South China coast and beyond the Taiwan Straits (see Figure 7.3).

Taching is the largest oilfield of the Sung-Liao Basin and is also claimed to be the largest oilfield in the world. It accounts for at least 50 percent of total Chinese oil production. The latest Chinese releases indicate that Taching produced six times as much oil in 1976 as it did in

***Figure 7.4 Major oilfields and petroleum facilities.* (Source: *"China: A Reassessment of the Economy," Joint Economic Committee, U.S. Government Printing Office, Washington, D.C., July 1975.*)**

1965, which based on earlier estimates of various sinologists would put it at the 43.09-million-ton level for 1976.

Exploration of the Sung-Liao Basin first began in 1959 with the assistance of Soviet and Hungarian experts. The geological structure of the basin consists of deep, thick layers of sedimentary rock from the Mesozoic and Cenozoic eras. It is significant to the Chinese leadership because large parts of eastern China are of the same geological structure. In 1964 Taching was adopted as a model for industrial development and became an important symbol to the Chinese masses. Although Taching became a great success, it had been achieved in the face of extremely harsh working conditions with winter temperatures dropping down to as low as −30° centigrade and with very limited supplies of equipment and machinery.

Table 7.4 Output of major Chinese oilfields in millions of metric tons per year in 1975 (see also Fig. 7.4 for oilfields location)

Oilfield Name and Basin	*Estimated Output*
North and Northeast	
Taching, Sung-liao	43.09[a]
Shengli, North China	14.90
Takang, North China	4.34
Pan-shan, North China	4.05
Fu-yu, Sung-liao	3.25
I-tu, North China	1.50
Fushun, Synthetic oil from shale	1.72 to 2.80
Ankuang, Sung-liao	under 1.0
Yenchang, North Shensi	under 1.0
Far West China	
Karamai, Dzungarian	1.07
Yumen, Chiuchuan	0.78
Tushantzu, Dzungarian	under 1.0
Lenghu, Tsaidam	0.58
Tasharik, Tarim	under 1.0
Southern China	
Chienchiang, Hupeh	4.10
Nanchung, Szechwan	under 1.0
Maoming, Syntheic oil from shale	0.2 to 1.8

Source: Based on data in U.S. Government study *China Oil Production and Prospects,* ER77-10030U, June 1977.

[a] Data for 1976.

Unofficial estimates of exploitable reserves at Taching range from 400 to 900 million tons. It is believed that there are over 2000 producing wells extracting oil by water injection from an average depth of 1000 meters (m), although wells as deep as 4600 m have been drilled as far back as 1966. In 1976 China also announced that profit realized from the Taching oilfield was already equal to twice the amount of the total investment made during the 17 years of its operation.

Shengli and Takang in the North China Basin are the next two largest oilfields of great importance because they are located relatively close to populous and industrial areas of Tientsin and Peking. At the Shengli Oilfield, which is also close to the mouth of the Yellow River, oil was struck in 1962 and full operations were achieved in 1974. The Chinese believe that the Shengli oil reserves are comparable to those of Taching, and latest estimates suggest that its production was over 14 million tons in 1976. Takang, the third largest oilfield, began production in 1967 and extends out into the Pohai Bay. It is also believed that Takang reserves are equal to those of Taching or Shengli.

There are other less-known producing oilfields in northeastern China. These include Tengluku, Chinshankou, and Kungchuling in the Sung-liao Basin. The North China Basin also includes: (1) the Fu-yu Oilfield in Kirin Province, which lies at the southern edge of the Sung-liao Basin, (2) Pan-shan Oilfield in Lianoning Province with an estimated 4.05 million tons' production in 1975, (3) the I-tu Oilfield in Shantung, and (4) the Chienchiang Oilfield in Hupeh Province, with an annual output estimated at 4.1 million tons.

An oilfield identified as Tungan has also been reported between the Sung-liao Basin and the North China Basin north of the city of Shenyang. However, no oilfields have been identified so far in the Kiangsu Basin, which includes vast offshore areas in the Yellow Sea east and north of Shanghai.

The Karamai Oilfield in the Dzungaria Basin in the Sinkiang Uighur Autonomous Region of Northwest China appears to be the next largest oil-producing area. It was discovered in 1955, and by 1973 its output was estimated at 1.07 million tons. The Tushantzu Oilfield, also located in the Dzungaria Basin, dates back to 1938, when it was first developed by Soviet engineers and geologists. In 1950 it was operated by the Joint Sino-Soviet Petroleum Company, which was dissolved in 1954. Other oilfields identified in the basin are at Urho and Chiigu.

Lenghu in the Tsaidam Basin in the Tsinghai Province of western China is another large oilfield in operation believed to hold reserves upward of 150 million tons. Because of the remoteness and desert environment of the area, exploration of the Tsaidam Basin slowed down while northeastern China oilfields were developed. It is now reported to have picked up again. Other oilfields in this basin have been identified at Mangyai, Yuka, and Hsiaolienshan.

Yumen in the small Chiuchuan Basin in Kansu Province of western China is an old area, beginning operation in 1939. It was reported that output for the Yumen District rose to 3.01 million tons by 1973. Initial training of oil workers for other oilfields and production of domestic oil equipment had their start in the Yumen area. Yaerhsia, Laochunmiao, and Shihyoukou oilfields have been identified in this basin.

The Szechwan Basin is probably the next largest known oil-producing area, although it is best known for its output of natural gas. Oilfields are known to exist at Lungnussu, Penglaichen, Hochuan, and Loutuhsi, with Nanchung as the central producing and refining center.

The North Shensi Basin is the region where China's oldest oilfield at Yenchiang was first developed and the first oil well drilled as far back as

1907. It is a small oilfield, although reports suggest that attempts were made to expand its output in the early 1970s. Other oilfields identified in the North Shensi Basin include Yungping, Tsaoyuan, Shatintzi, and Machiatan. There is also an oilfield at Chingtuching, located between the North Shensi and the eastern end of the Chiuchuan Basin.

The Tarim Basin in extreme western China has been surveyed and oil and gas seepages have been reported over a distance of 800 km in the Aksu–Kashgar region. An exploratory oilfield is known to exist at Ichkelik in northeast area of the Tarim Basin, but remote location and lack of transportation preclude large-scale operations. In other areas one oilfield has been identified at Shengtingkou in the Turfan Basin, another at Paise in the Kwangsi-Kweichow Basin in South China, and another at Hopu near the Vietnamese border. There have also been persistent reports since 1974 that a new oilfield sometimes referred to as the "70" field is being developed in Kwangtung Province, but its location remains unknown.

Synthetic oil from shale is produced mostly at Fushan in Liaoning Province, where there are several plants with a joint output of 2.8 million tons. Some synthetic oil is also produced at Maoming in Kwangtung Province in South China. Shale-oil production commenced in 1929 and contributed significantly to total Chinese oil production until late 1950s. Estimates of shale-oil reserves vary widely, from 700 million tons to as much as 11,800 million, and represent a significant portion of energy resources. Soviet geologists who were in China up to 1960 estimated up to 21,900 million tons of shale reserves.

About 3500 km of pipelines have been identified in China in 1976 with another 2000 km under construction. These pipelines are 20, 25, 30, and 61 cm in diameter and are welded together electrically from 8-, 10-, and 12-m sections. The large 61-cm-diameter pipe is mostly imported from Japan because Chinese pipe larger than 30 cm in diameter is formed spirally and cannot withstand high pressures. Pipelines are normally wrapped in six alternative layers of plastic cloth and bitumen and buried.

The first two pipelines were built during the 1950s to connect the Karamai oilfield with the refinery at Tushantzu. These consisted of dual lines of 61 cm and 40 cm in diameter and 300 km long. Karamai is also connected by pipe with the railroad at Urumchi which can carry crude to the refining center in Lanchou. The longest pipeline so far appears to be the 1200-km-long line from Taching and Fu-yu to Chinghuangtao oil port and the Fangshan refinery in Peking. This line is 61 cm in diameter and includes nineteen pumping stations, which also supply heat during the winter operations to keep the oil from freezing. A second line, which is also about 1200 km long connects Taching with Tiehling near Anshan and the port of Dairen. The oil port of Hunagtao near the port of Tsingtao is connected with the Shenglo oilfield by a 200-km-long pipeline, and another 300-km pipeline connects the Takang oilfield with the Fangshan refinery in Peking.

In Szechwan the Lungnussu oilfield, the Nanchung refinery, and the industrial city of Chungking are connected by a 40-cm pipeline. The largest pipeline under construction is from Golmo in the Tsaidam Basin to Lhassa in Tibet and is estimated to be at least 1100 km long. Because Tibetan oil consumption is only a few hundred thousand tons a year, this pipeline is difficult to justify on economic grounds alone. It is probably being built for strategic purposes to guarantee oil supplies for military operations that may develop on the unsettled borders of Tibet. Several pipelines of lesser significance are also in operation.

The petroleum refining industry in China during the 1970s was compared to that of the United States or western Europe in the 1950s. However, despite a relatively backward oil technology, China already has 44 refineries which operate 13 thermal and 27 catalytic crackers an unusually large number comparable to those in the United States or the Soviet Union. The majority of refineries are also equipped with platforming and delayed coking units and produce over 400 different products, including a variety of feedstocks for the petrochemical industry.

By 1972 China's refining capacity fell behind its oil production and storage of crude oil became a significant problem. Marine tankers, railway tank cars, underground caves, and abandoned mines were all used to store oil. By 1975 total refining capacity was estimated at 58 million tons per year, which was well below the rapidly growing oil production reaching about 73 million tons that year. As a result, oil-export markets were pursued aggressively and many orders were placed in Japan and western Europe for the construction of the large petrochemical plants coming on stream during the late 1970s. If oil production continues to grow rapidly, it is clear that China will also need to modernize its existing refineries, which will become obsolete in the early 1980s, as well as provide for at least 200 percent larger refining capacity by 1985.

The first of the large modern refineries in China (see Table 7.5) was built at Lanchou in Kansu Province and was designed to process crude from Yumen and Karamai Oilfields. It was constructed during 1956–1959 with Soviet assistance and was expanded during the 1960s. In 1973 the Lanchou Refinery produced 160 different products, including aviation fuel, gasoline, diesel oil, lubricants, synthetic rubber, greases, cable oil, and plastics.

The largest refining complex at present is believed to be at Fushan in Lianoning Province, with an estimated total capacity of about 6.5 million tons per year and consists of four different plants. It processes oil

Table 7.5 Major oil refineries and their estimated crude processing capacities in millions of metric tons per year in mid-1970s

Refinery			*Estimated Annual Capacity*
Shengli Region			7.49
	Wangchuchuang	4.68	
	Chinan	1.77	
	Chinan Licheng	1.04	
Fushun			6.50
Taching Region			6.13
	Saerhtu	4.36	
	Lamatien	0.73	
	Chingchiaweitzu	1.04	
Dairen			5.0
Lanchou			5.0
Shanghai			5.0
Maoming			5.0
Takang Region			4.37
	Tsanchou	1.04	
	Taku	0.73	
	Tienching	2.60	
Peking General	Fangshan		4.00
Chinhsi			3.50
Nanching			3.00
Linhsiang			2.60
Anshan			2.50
Fuyu			2.08
Chingmen			2.00
Wuhan			1.77
Paoting			1.56
Tushantzu			1.56
Yumen			1.00
Hangchou			1.00
Yangliuching			0.83
Lenghu			0.31

Source: Adapted from data in *China: A Reassessment of the Economy,* Joint Economic Committee, U.S. Congress, 1975; *China's Oil Industry,* Sino-British Trade Council, 1975; and *China Oil Production Prospects,* U.S. Government Study ER77-10030U, June 1977.

from shale and coal as well as natural crude. Maoming in Kwangtung Province is a similar complex of shale and natural-oil refining plants. The oldest is said to be the Shanghai refinery, which processes crude from Taching and Shengli but is being expanded into a major petrochemical complex.

The Taching refinery, completed in 1966, and the Peking General Petrochemical Works (Fangshan), which began operations in 1969, are considered the two exemplary plants in all of China. The Taching refinery is believed to use some equipment imported from Italy. The Peking plant is widely publicized and is being expanded into a major petrochemical complex. It uses crude from Taching and Takang and produces jet fuel, kerosene, diesel oil, benzene, lubricants, wax, and asphalt and supplies twenty provinces and municipalities.

There may be more than one refinery in Dairen that receives crude by pipeline from Taching. Its No. 7 refinery plant processes about 2 million tons per year and produces lubricants, but total refinery capacity in Dairen is estimated at 5 million tons. The Shengli refinery in Shantung Province is apparently located near the huge oilfield, and most recent reports suggest that its capacity is being doubled. Chinhsi in Liaoning Province is a relatively new refinery using Taching crude from the pipeline also. Besides refineries listed in Table 7.5, other plants have been identified at Anching, Canton, Chinshawei, Haitzuching, Harbin, Kirin, Karamai, Loyang, Linhsian, Nanchung, Panshan, Urumchi, Yangchou and Tantung.

SIGNIFICANCE OF PETROLEUM EXPORTS

China derives immediate economic and political advantages from exports of crude-oil and petroleum products. Revenues from oil are needed to pay for imports of equipment and machinery to modernize several industries and the military forces during the fifth Five Year Plan of 1976–1980. Politically, China is believed to have already engaged in "oil diplomacy." By offering large quantities of oil for export to Japan it is reducing the possibility of significant joint Soviet–Japanese development schemes of the Siberian Tyumen and Sakhalin offshore oilfields. In the third world China is boosting its influence, particularly in nearby Asian countries, by providing oil at a "friendship price" to oil-importing developing countries that are all only too familiar with the hard-currency deficit problems.

China first began to export oil to North Korea in 1964 and to North Vietnam in 1965. By 1975 exports reached 10 million tons, of which at least 8 million went to Japan, but new oil clients such as Hong Kong, the Philippines, Laos, and Thailand were added to the list. Brazil, Australia, West Germany, and the United States already indicated an interest in Chinese oil imports. There is no question now that China is in the oil-export business, but it is not clear so far what quantities of oil will be available for how long at what price and to whom.

Speculation that China desires to become a major oil exporter in the future has been reinforced by its discussions of possible exports to Japan of 1 million barrels per day—equivalent to 50 million tons per year—by

1980. Such export levels are comparable to total Soviet exports to the West and are only 25 percent below total Soviet exports to eastern Europe. In view of recent disclosures of declining Soviet oil production and a possible need to import oil in the mid-1980s, Chinese export ambitions exhibit strong political overtones probably designed to impress eastern European countries, which rely almost entirely on Soviet oil imports.

In fact, to achieve exports of 50 million tons in 1980 China would have to cut down on its own domestic growth rate. Recent Western estimates suggest that exports of about 25 million tons in 1980 are much more realistic because these are possible with China's GNP present growth rate, maintained at an average 6 percent per year. Such export levels would represent 10 to 15 percent of Chinese oil production and would earn about $2.5 billion per year.

Oil exports during 1981–1985 will be governed strongly by domestic demand and could become negligible if China decides to increase its rate of development significantly. Although this remains a possibility, Peking will more likely adjust its oil production and consumption levels to provide an exportable surplus of oil of about 60 to 75 million tons by 1985. This would be well above the current Soviet oil exports to eastern Europe, comparable to oil exports of Canada, and would bring in from $5 billion to $8 billion in foreign currency each year. Such oil-export earnings alone would be sufficient for any purchases of grains and machinery similar in quantity to those imported by China in recent years.

The effect of large Chinese oil exports would be to reduce global dependence on Middle East producers. But the principal beneficiaries appear to be Japan, North Korea, Vietnam, Thailand, and the Philippines. American oil companies seeking to obtain Chinese crude in return for offshore oil exploration and production technology have been consistently rebuffed so far. Hence it is believed that China is acquiring and building its own offshore exploration and production equipment in an attempt to establish itself as an oil power in Asia, which it can only accomplish if it develops and maintains a significant and flexible oil-export potential. Economically, development of offshore oil for export is attractive because it reduces the demand on an already inadequate Chinese transportation system. Politically, it is also removed from immediate Soviet threat and establishes Chinese claims to the continental shelf, where some promising areas leased by Taiwan and South Korea are already in dispute.

Despite Sino-Soviet political and ideological differences there are also strong economic reasons for development of oil trade between China and the Soviet Union. The latest Soviet pipeline to Vladivostok is under construction along the Chinese border and passes relatively close to the Taching Oilfields in northeastern China. Similarly, western Chinese oilfields in Sinkiang are closer to Soviet Central Asian and Siberian markets than the Chinese industrial areas of the Northeast. In addition, trade between communist countries is conducted on a bilateral basis by annual trade agreements and does not require the expenditure of hard currencies.

The Soviet Union is also in a position to provide oil exploration and production equipment to China in return for oil and may offer other trade incentives that China may find difficult to resist and Western countries impossible to match. These could include, for example, resumption of shipments of chrome ores and nickel, which China has to import from Albania and from the West, though ample supplies of both metals are available in the Soviet Union. Viewed in the light of a possible reduction of the Sino-Soviet tensions, even temporary trade based on oil for strategic-metals exchanges may be appealing to both communist superpowers. In addition, such trade would give both China and the Soviet Union additional political capital in their respective dealings with the West without risking strategic material starvation on either side.

China may in the future use the prospects of oil exploration and production within its territorial waters as a means of applying political pressure more openly. This type of approach has been only recently demonstrated by Vietnam when informal proposals were reported to favor immediate approval of oil-drilling leases off the Vietnamese coast on the condition that diplomatic recognition and acceptance at the United Nations are not opposed. Clearly, the predicted Soviet oil shortages and future Sino-Soviet relations will be the dominant factor in China's oil diplomacy.

MARKETS FOR PETROLEUM EQUIPMENT

During 1953 to 1974 China invested an estimated $9 billion in fixed-capital equipment to develop oil production of about 65 million tons per year. Of this investment about $1.2 billion or 13 percent of the total was spent for imports of petroleum equipment from abroad. On the average the capital to output ratio was thus about $100 per ton of oil, and these factors can be used to project the possible markets for petroleum equipment through 1985.

Chinese oil production is expected to increase at about 15 percent per year until 1985. If production is about 100 million tons at the end of 1977 it may well reach close to 300 million tons by 1985. This means an increase in productive capacity of 200 million tons, which would require additional fixed-capital investment of about $20 billion before 1985. Because offshore exploration and production will play a much larger role than before, China may have to import a larger percentage of petroleum equipment from abroad until it develops a more significant offshore equipment manufacturing capability of its own. Therefore, if equipment imports are assumed to be 15 to 20 percent of the total fixed capital investment in coming years, this could amount to about $4 billion before 1985. There is no question that China can easily pay that amount in hard currency, even if its oil exports do not increase at all beyond its 1975 level of 10 million tons per year. It is clearly a buyer's market, with China as the buyer.

Western oil interests would like China to turn over offshore exploration to a consortium of leading oil companies working on a management contract basis and compensated in crude oil. This is perceived as the fastest method to provide large quantities of oil for domestic consumption as well as for export. So far China expressed no interest in such an arrangement, and the issue has been further complicated by oil-exploration leases issued to American companies by Taiwan and South Korea and even Japanese claims to some disputed areas. Until these issues are resolved it is unlikely that any joint ventures for offshore oil exploration will develop, and China will continue to import oil exploration and drilling equipment to build up its own capabilities. On the other hand, a deterioration of the Sino-Soviet relations and any increased threat to major Chinese oilfields in Sinkiang and Manchuria may prompt China to invite Western participation to help develop rapidly alternate oil sources off the southern coast of China, which are well removed from the Soviet borders. This is clearly in the interest of China's national security.

During 1949–1959 China accepted extensive Soviet and eastern European assistance to develop its oil industry and many aerial magnetic surveys, gravity-surface geological surveys, and seismic surveys were conducted by Soviet and eastern European technicians. Until about 1962 China imported equipment for geological survey, engineering, and petroleum and gas production valued at about $70 million from the Soviet Union. A very small amount came from Hungary, Romania, and East Germany. Onshore oil rigs, turbodrills, seismic equipment, and core sampling plants made up most of these imports. In addition, the Soviet Union supplied wide-diameter steel pipe and provided special assistance in construction of the Lanchou Petrochemical Machinery Plant, the Lanchou Refinery, and the Lanchou Petroleum Machinery Plant, all of which became the backbone plants of the Chinese petroleum-equipment industry.

China's petroleum-equipment manufacturing industry now operates an estimated 100 plants and supplies about 80 percent of demand. This "self-sufficiency" factor, however, may decline in the future as large-scale offshore oil exploration and production develop. China not only lacks the more sophisticated offshore technology, but rapidly growing demand for materials such as high-quality steel by competing industries like mining, automotive, and shipbuilding may limit the growth of petroleum-equipment manufacture. Iron and steel products already are China's largest import, and a sharp increase in imports of equipment and machinery may be a more expedient way to meet the growing demand, particularly when availability of hard currency should not present a problem in the future.

Since 1963 China turned to Japan and western Europe for import of refinery plants and seamless steel pipe plants and purchased some oil-drilling equipment from Berliet in France, which also became a large supplier of heavy trucks to China. It also continued to expand its trade with Romania, which already had an established petroleum industry and manufactured petroleum equipment to supply its own demand.

Romania became an important supplier of oil-drilling rigs and oil-well operation equipment, which became that country's most important export to China. Since it first demonstrated the 5000-m-deep drilling rig at the 1965 National Exhibition in Peking, Romania supplied eighty-six units up to 1973. Romania is also the largest supplier of railway tank cars to China and in recent years exported about 500 cars per year, which represented about 10 percent of Romania's annual tank-car production. The average price of a Romanian oil rig derived from foreign-trade statistics is in the range from $172,000 to $250,000 per unit based on official exchange-rates conversion. Romanian tank cars cost about $12,000 to $14,000 each. During 1970–1973 the total value of Romanian oil rigs exported to China was about $40 million, and tank cars accounted for another $21 million. Romania is also believed to have supplied a few drilling barges for early offshore exploration efforts and small quantities of petroleum products and steel pipe while importing some Chinese crude.

Since 1969 China began purchasing offshore drilling rigs, supply and maintenance vessels, and computerized seismic exploration equipment from Japan, western Europe, and the United States. It also continues to buy large quantities of seamless steel pipe primarily from Japan and West Germany (see Table 7.6). During 1972–1975 it is estimated that $127 million was spent on petroleum exploration and extraction equipment and at least 900 million on petrochemical and synthetic fiber plants (see Chapter 10).

Analysis of known imports during 1963–1977 suggests that China has not decided on one or two major suppliers of offshore petroleum equipment and technology but continues to purchase equipment from ten different countries (see Fig. 7.5). Japan has been the leading supplier, but if discussions with the French–German consortium about construction of six submersible sea rigs valued at about $310 million bear

Table 7.6 **Major imports of petroleum equipment by China during 1963–1977**

Year	Type of Equipment or Plant	Supplier and Country	Contract Value in Millions of $U.S.
1963	Gasoline refinery	Snam Progetti, Italy	5.00
1964	Oil cracking, olefin plants	Lurgi, West Germany	12.50
1965	Gasoline refinery	Snam Progetti, Italy	5.60
	Shale-oil refinery	ENI, Italy	9.00
	Oil refinery 2–3 MTA[a]	France	
	Oil-rig equipment	Berliet, France	4.86
	Seamless steel pipe plant	Innocenti, Italy	3.20
1966	Two drilling rigs	Romania	
	Petroleum test instruments	Stanhope-Seta, U.K.	
1967	Seamless steel pipe plant	Mannesman, West Germany	11.00
	Petroleum equipment	Ploesti, Romania	
1969	Sea bed drilling rig (canceled)	IHH, Japan	(10.00)
1970	Offshore drilling rig "Fuji" and 500-ton workship	Japan Offshore Drilling	9.80
1973	Offshore drilling rig	Asia Offshore Drilling, Japan	
	Seismic surveying equipment	Geospace, U.S.A.	5.50
	Suction Hopper Dredges	NV Industrieele, Netherlands	39.30
	Mining and petroleum drill bits	Hughes Tool Co., U.S.A.	
	Twenty blowout preventer stacks	Rucker, U.S.A.	2.00
	Five 600-ton offshore supply ships	Hitachi, Japan	10.00
	No. 2 Hakuryu offshore rig	Mitsubishi H.I., Japan	21.40
	Eight oil-rig supply-tow vessels	Weco Shipping, Denmark	20.00
1974	Unspecified drilling equipment	U.S.A.	
	Geostore seismic recorders	Racal-Thermionic, U.K.	
	Two offshore drilling rigs	Mitsubishi, Japan	12.50
	Catalyst for reforming	Haldor Topsoe, Denmark	
	One 33,822-ton tanker "Vesthav"	Norway	14.50
	Seismic surveying system	CGG, France/U.S.A.	7.00
	Offshore drilling rig (canceled)	Far East Levington, Singapore	36.00
	Twelve pipelayers	Komatsu, Japan	
	One 16,000 ton tanker "Pinghu"	Sweden	
1975	Five semisubmersible rigs talks	Hamburg, Germany	
	Four life-saving system/oil extract.	Nichimen, Japan	
	Petroleum equipment	Baker Trading, U.S.A.	
	Oil-exploration equipment	U.S.A. (unidentified supplier)	23.00
	Two ocean survey vessels	Japan Ocean Industry	
	Two rigs, 7000-m and 10,000-m, shown in Peking	Romania	
	Aromatics complex	Japan Gasoline Company	36.00
	Pipeline under-sea talks (120 cm)	Japan/Italy	
	Two jack-up offshore rigs	Robin Shipyard, Singapore	57.50
1976	Six submersible rigs deep-sea talks	German/French consortium	310.00
	Oil-well servicing equipment	Stewart & Stevenson, U.S.A.	5.80
	Eight oilfield fracturing units	Dowell-Schlumberger, U.S.A.	3.00
	Three mobile seismic monitors	U.S.A.	1.50
	Heavy oil-refinery interest	Mobil/Caltex, U.S.A.	
1977	Seismic prospecting equipment	Digital Resources, U.S.A.	1.70
	Two offshore platform helicopters	MBB, West Germany	0.60
	Drilling-rig anchoring chains	K Teufelelsberger, Austria	

Source: Compiled by 21st Century Research from various sources.

[a] MTA = metric tons per annum

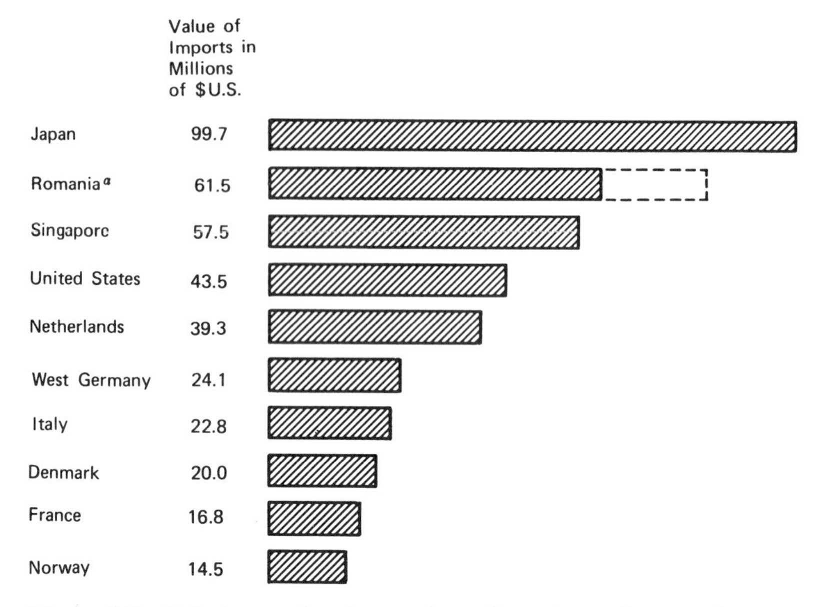

***Figure 7.5** Relative market shares of suppliers of petroleum exploration, production, refining, and transportation equipment identified during 1963–1977, excluding steel pipe and petrochemical plants.* **(Source: *Based on data compiled in Table 7.6.*) (*[a] Data for 1970–1973 only.*)**

fruit, both France and Germany may move into the lead as major off-shore petroleum-equipment suppliers. China clearly looks to the United States as a supplier of the most sophisticated computerized oil-exploration equipment. But here the pattern of purchases emerging is similar to that in the import of electronic equipment in general where China engages in "prototype purchasing" programs from several different suppliers and continues to develop domestic equipment incorporating the latest foreign innovations.

Major markets for petroleum equipment and technology imports are the oilfields, refineries, petroleum-equipment manufacturing plants, and specialized institutes engaged in petroleum exploration and production research. Most of those enterprises are subordinated to the Ministry of Petroleum and Chemical Industries (MPCI), which is headed by Minister Kang-Shih-en. The equipment-manufacturing department of this ministry is responsible for the production of specialized equipment for petroleum and chemical industries, although several categories of equipment are manufactured jointly with the appropriate production bureaus of the First Ministry of Machine Building (MMBI) and the Ministry of Light

Industry (MLI). Institutes engaged in petroleum-related research are operated by the Chinese Academy of Sciences (CAS) or by the Ministry of Petroleum and Chemical Industries, and sometimes jointly with the Ministry of Education.

The Chinese National Technical Import Corporation (TECHIMPORT) specializes in import of complete plants and manufacturing technology and has a special department to handle petroleum and petrochemical industry contracts. It is the Foreign Trade Corporation most likely to sign contracts for foreign petroleum equipment, although China National Machinery Import and Export Corporation (MACHIMPEX) may also be the pertinent trading organization involved with imports of specific machinery of equipment for use in the petroleum industry.

Among the sixteen research institutes identified in Table 7.7, those affiliated with the CAS are the most prestigious and may be considered as petroleum-technology gatekeepers in China. The Changchun Institute of Geology appears to concentrate on prospecting techniques and geological mapping, the Dairen and Lanchou Institutes of Petroleum Research are mainly involved in shale and natural oil-refining technology, and the

Table 7.7 Major research institutes engaged in petroleum exploration and production research in China

Name of Institute	*Year Established*	*Probable Affiliation*[a]
Changchun Institute of Geology		CAS
Chengtu College of Geology		(MG–MHE)
Dairen Institute of Petroleum Research	1952	CAS
Fushun Institute of Petroleum Engineering	1956	MCPI
Lanchou Institute of Petroleum Research	1958	CAS
Peking College of Petroleum Engineering	1953	
Peking Academy of Geology	1956	(MG)
Peking Academy of Petroleum Research		CAS
Peking Institute of Exploration Techniques	1956	(MG)
Peking Institute of Petroleum Refining	1956	MCPI
Shantung College of Geology		(MHE)
Shenyang N.E. College of Petroleum Engineering		(MHE–MP)
Sian College of Petroleum Engineering		(MHE–MP)
Sinkiang College of Petroleum Engineering	1958	MPCI
Szechwan College of Petroleum Engineering		(MHE–MP)
Taching Petroleum Production Technology Research Institute	1962	MPCI

Source: Compiled by 21st Century Research from various sources.

[a] CAS = Chinese Academy of Sciences; MG = Ministry of Geology; MHE = Ministry of Higher Education; MPCI = Ministry of Petroleum and Chemical Industry; MP = Ministry of Petroleum; abbreviations in parenthesis indicate that these ministries originally controlled the institute but are no longer in existence.

Peking Academy of Petroleum Research appears to specialize in automation in oil industry and application of geochemical and geophysical methods to geological exploration.

The remaining petroleum institutes are more comparable to petroleum research and development departments or specialized educational and training facilities. The Fu-shun Institute of Petroleum Engineering was formed in 1956 by amalgamation of No. 1 and No. 2 factories of the Ministry of Petroleum Industry (MPI) then in existence and specializes in oil-shale refining research and hydrogenation. The Peking College of Petroleum Engineering is more of an academic research and educational institution and operates departments of petrology, petroleum, drilling, oil refining, petroleum industrial economy, and petroleum machinery.

The Peking Academy of Geology seems to be the top administrative organization with authority over all geology-related institutes and originally operated under the Ministry of Geology, which is no longer in existence. This academy was involved in the first and second Five Year Plans and operates seven regional centers, each with laboratories on petrology and mineralogy. Specialized institutes of the academy include the Institute of Exploration Techniques, which investigates automation of petroleum equipment, and the Institutes of Geology and Geophysical Prospecting. This academy maintains a library of at least 340,000 volumes and subscribes to over 700 foreign periodicals.

Petroleum-research institutes present a limited market for various instruments and computers and specialized software but should not be overlooked as points of distribution to the Chinese petroleum industry as a whole regarding the latest information on the developments and advances in Western petroleum technology. This information also reaches the petroleum-equipment manufacturing enterprises that are trying to produce the equipment necessary for petroleum exploration, production, and refining.

Chinese petroleum-equipment manufacturing enterprises will continue to supply as much of the equipment required by the industry as possible. These plants and works are not an immediate market for Western petroleum equipment but rather for petroleum equipment technology and modern production and testing equipment such as specialized machine tools and instruments. Very little is known about these plants, but some information about petroleum equipment they produce is available and is useful in evaluating this market because Peking is unlikely to import equipment that is already produced or can be put into production in China.

Five major petroleum-equipment manufacturing regions have been defined in China. These are the Lanchou–Paochi region of northwestern

China, the Shanghai region, the Peking–Tientsin region, the southern central China region concentrating on cities like Canton and Wuhan, and the heavy industry plants region of Manchuria (see Table 7.8).

Table 7.8 Major plants engaged in petroleum-equipment manufacture and their basic petroleum-equipment products

Name of Plant	*Type of Equipment Manufactured*
Changchiakou Mining Machinery	mobile drilling rigs and derricks
Dairen Hungchi Shipyard	offshore drilling platform, tankers
Harbin Boiler Plant	asphalt-separation tower
Lanchou Petroleum & Chemical Equip.	75-ton mobile-oil-drilling rig
Lanchou Petrochemical Machinery	medium-depth drilling rigs
Lanchou General Machinery Works	mobile oil-fracturing equipment
Paochi No. 1 Machinery Plant	oil derricks and drilling tools
Paochi Oil Drilling Machinery	instruments and materials for rigs
Shanghai Boiler Plant	fractionating towers
Shanghai Hsi-chien Machinery	refining installations
Shanghai Hutung Shipyard	floating drilling rig (1974)
Shanghai Li-shen Machinery	shallow- to medium-depth drilling rigs
Shanghai Petrochemical Machinery	distillation towers
Shanghai Petroleum Equipment Parts	drilling tools and drilling bits
Shenyang Water Pump Works	oil-cracking pumps
Shihchianchuang Coal Mine Machinery	shallow-depth oil rigs
Sian Geophysical Instruments	logging and measuring equipment
Taiyuan Mining Machinery	oil-well equipment
Wuhan General Machinery	rotary drilling rig
Yumen Machinery Plant	drilling rigs and deep-well pumps

Source: Compiled by 21st Century Research from various sources listed at the end of this chapter.

The Hutung Shipyard in Shanghai and the Hungchi Shipyard in Dairen in Liaoning Province have already produced the first Chinese floating drilling rig and an offshore drilling platform, respectively, both of which have been given considerable publicity in Chinese media, which suggests that these are the state-of-the-art achievements in this industry. The floating drilling rig, known as "Kantan No. 1" (Prospector), is a large catamaran vessel constructed by joining together two hulls of Chinese cargo ships. In 1975 this rig was reported to be engaged in deep-sea oil prospecting in the south part of the Yellow Sea, probably in the Kiangsu Basin. The Hungchi Shipyard in Dairen, which specializes in oil-tanker construction, built a jack-up offshore drilling rig known as the "Pohai No. 1," which has been operating in the Pohai Gulf since 1972. This rig has four legs, each 73 m long and 2.5 m in diameter and supports a drilling platform weighing between 4000 and 5000 tons. In contrast, the "Fuji" offshore drilling rig purchased from Japan in 1972, renamed "Pohai No. 2," also has four legs but is only 52 m long.

Already in 1972 it was estimated that five to eight offshore drilling rigs were deployed in the shallow Pohai Gulf, some of which incorporated Romanian and other imported elements and operated down to depths of 200 feet. China and Vietnam are also believed to have conducted joint seismic surveys of the Tonkin Gulf, and China began onshore drilling on the Paracel Islands off the Vietnam coast in 1975. In that year the French Compagnie Generale de Geophysique was also reported to have sold China a "seismic ship" known as "Lady Isabel" and apparently also received service contracts for conducting surveys and on-the-job training. Another French firm, E.F.F.I., was also reported to have been awarded a $4.7-million contract to conduct aerial surveys over the Yellow Sea in 1975. These reports suggest that China does not as yet produce sufficient sophisticated offshore exploration equipment to undertake a more intensive oil-exploration program, and additional opportunities may exist for Western exploration services, particularly in the South China offshore regions.

Although offshore oil exploration and production appear to be the most promising markets for Western imports in the near future, China is also expected to continue importing deep-well drilling rigs for drilling below 3000-m levels. Exhibition by Romania of the 7000-m and the 10,000-m drilling rigs confirm the suspicion that so far China did not develop such equipment and will probably continue to rely on Romania as an established supplier.

In other petroleum equipment China is believed to produce steel cylinders for pressures up to 2000 atmospheres used in pumps and compressors, which means that deeper wells will require imports of equipment capable to withstand higher pressures. In oil transportation the largest-diameter steel pipe known to be made in China is 30 inches, and the largest tanker reported under construction at the Dairen Hungchi Shipyard in recent years was reported to be 50,000 tons, although the shipyard is believed to be developing the capacity to build 100,000-ton tankers.

NATURAL-GAS INDUSTRY

China is the fifth largest natural-gas producer in the world, and its gas production in 1975 was estimated at 34.6 billion cubic meters (m^3) or 46.0 million tons of coal equivalent. This gas production level is comparable to that of the United Kingdom or Romania but is only one eighth that of the Soviet Union and only one sixteenth the output of the United States (see Table 7.9). In 1976 natural-gas production was reported to have grown by 11 percent over that of 1975 and probably reached a total of over 39.0 billion m^3.

Table 7.9 World significance of Chinese natural-gas production in billions of cubic meters

Producing Country	*Production in 1976*	*Producing Country*	*Production in 1976*
United States	568.00	United Kingdom	36.00
Soviet Union	323.00	Romania	31.20[a]
Netherlands	97.00	Iran	24.00[a]
Canada	76.00	Mexico	14.00[b]
China	39.00	West Germany	18.00

Source: Handbook of Economic Statistics 1977, U.S. Government publication, September 1977, Washington, D.C.

[a] Data for 1975. [b] Data for 1974.

Major natural-gas resources are located in the Szechwan Basin in Central China (see Fig. 7.4) and some in the Tsaidam Basin. Exploitation of the Szechwan gas deposits is proceeding rapidly, and over 200 gas structures have been identified in that region. The Tsaidam gas fields are believed to be inoperative at present. In 1970 estimates of proved and probable natural-gas reserves by Meyerhoff rated the Szechwan Basin at 18,500 billion cubic feet and the Tsaidam Basin at 2850 billion cubic feet (1 m^3 = 35 cubic feet). At the World Petroleum Conference Chinese proven recoverable gas reserves were estimated at 30,000 billion cubic feet, equivalent to about 850 billion m^3. The U.S. government sources estimate proved reserves at 700 billion m^3.

Natural gas production increased at a rapid rate of about 36 percent per year during 1958–1965 but slowed down during 1966–1970 to 18 percent, and the growth further dropped to 8.6 percent during the early 1970s. Projections for growth in natural-gas supply up to 1980 range from 10 percent to a high of 20 percent annually, but actual growth will depend primarily on investment in pipelines, treatment facilities, and gas field equipment to distribute the gas to more distant regions.

In 1975 natural gas contributed 10 percent of the total primary energy production in China (see Fig. 7.2), and by 1980 its share may go as high as 13 percent of the total. Consumption of natural gas is believed to be divided evenly between the residential/commercial sector and the industrial users and this ratio appears to have remained steady for the

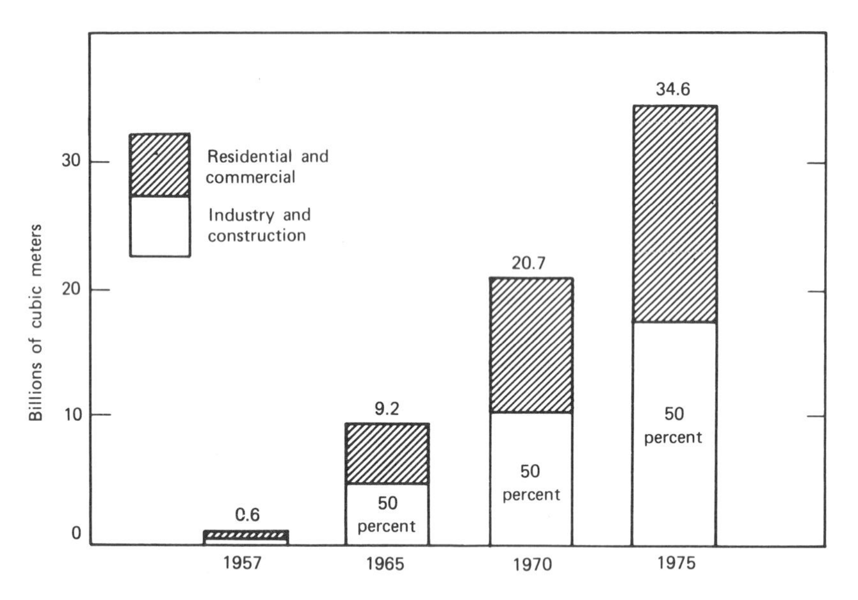

Figure 7.6 Growth of natural-gas supply and consumption. (**Source:** *Adapted from data presented in* **U.S. Government Research Aid,** *A(ER)75-75, "China: Energy Balance Projections," Washington, D.C., November 1975.*)

last 20 years (see Fig. 7.6). This is probably so because most of China's natural gas is produced and consumed in Szechwan Province.

There are over 1000 k of natural-gas pipelines in the province, but so far no long distance gas pipeline has been identified. Production of gas is well over the demand of the local industry and thus it is consumed by many households in Szechwan, which accounts for 11 percent of the total population in China. Small local pipelines distribute the gas to cities, towns, and villages. In industry two thirds of the iron and steel plants use natural gas, and more than 70 percent of nitrogen fertilizer produced in Szechwan is made from natural gas. It is also used in power generation, gas engines, and in production of carbon black, sulfur, medicines, and film materials. Sulfur in Chinese natural gas is apparently a problem, and several provincial institutes in Szechwan are reported to be working on finding a solution.

China was one of the first countries in the world to use natural gas, and the Tzuliuching (or Tzukung) gasfield has been known for more than 1000 years. The first known natural-gas transportation system was built in China over 3000 years ago and depended on natural reservoir pressure to move gas short distances in bamboo pipes.

Major known gasfields in the Szechwan Basin include Shihyukou, Luchow, Lungchang, Luchuan, Huanchianchan, Shentengshan, Tengchingkuan, Nahsi, and Chanyuanpa. In the Tsaidam Basin gasfields have been identified at Hsiaolienshan, Mahai, and Yenhu. There is also a gasfield near Shanghai in the Kiangsu Basin. Most of the natural gas produced outside the Szechwan Basin is oil-associated gas whose production has been growing very rapidly. Most likely the bulk of oil-associated gas is produced at oilfields such as Takang, Panshan, and Shengli and is piped to large nearly cities such as Peking, Tientsin, Shenyang, Liaoyang, Tsinan, and Tsingtao. Oilfields such as Yumen, Karamai, and Fu-yu may also supply natural gas to nearby cities.

China may be interested in exporting liquefied natural gas (LNG) to Japan in the near future, and officials of the Bridgestone Liquefaction Gas Company of Tokyo already discussed the scale of a small 150,000-ton/year liquefaction plant. China apparently is interested in a much larger LNG facility. In 1977 plans were announced to build the largest LNG plant in the world near Canton and Hong Kong, and the Japanese companies Kawasaki, Bridgestone, IHI, Mitsuit, and C. Itoh are believed to have been contacted about undertaking this project.

China also promotes production and use of marsh gas by individual families, and in 1975 it was announced that two-thirds of its provinces, municipalities, and autonomous regions were using marsh gas for rural cooking and lighting. This gas is produced from fermentation of night soil, grass, garbage, and other organic matter.

IMPORTANCE OF CHINESE COAL INDUSTRY

China is the third largest coal-producing country in the world, and in 1976 coal output was estimated at 448 million metric tons of raw coal, representing about 12 percent of global coal production. Coal reserves of China are comparable to those of the United States and the Soviet Union and have been estimated as high as 1500 billion tons. Proven reserves exploitable with available technology are at least 80 to 90 billion tons, which is sufficient for 90 years of supply at present depletion rates (see Table 7.10).

Although the use of oil has grown, coal still accounts for about two-thirds of all the primary energy produced in China. In contrast, coal produces one-third of the Soviet Union's primary energy and only one-fourth of that of the United States. Although coal's share in primary energy supply declined from about 85 percent in 1965 to 66 percent in 1976, it still remains and will continue to be for some time to come by far the most

Table 7.10 World significance of Chinese coal industry in 1976 (all data in millions of metric tons)

Country	Hard Coal Production	Brown Coal and Lignite Production	Proven Reserves[a]
China	448.00	negligible	80,000
United States	586.23	22.86	181,700
Soviet Union	492.00	163.00	136,600[a]
Poland	179.30	39.31	22,640[a]
United Kingdom	123.80	—	3,870
India	100.99	3.90	11,580
West Germany	92.84	134.53	na
South Africa	75.73	—	10,585
Australia	74.86	30.94	na
North Korea	31.20	8.30	na
Czechoslovakia	23.27	89.62	6,360[a]
East Germany	0.46	246.30	25,300[a]

Source: *U.S. Government Handbook of Economic Statistics 1977,* published in September 1977.

[a] Proven reserves are resources which have been assessed exploitable under local economic conditions and available technology. Data for communist countries include brown coal and lignite.

important source of fuel in China. This may be particularly true if China decides to become a large exporter of oil in the short term (see Fig. 7.1).

The industrial sector is the largest consumer of coal and accounts for an estimated 56 percent of total consumption or 65 percent if coal used to generate electricity is included (see Fig. 7.7). About 25 percent of the coal supply is used by the household sector, but recent trends suggest that increasing share of coal supply will continue to be allocated to the rapidly growing industrial sector.

China exports less than 1 percent of its coal production primarily to North Korea, which buys about 2 million tons of coking coal annually. Japan is the next largest importer, purchasing 330,000 tons of anthracite and 120,000 tons of steam coal in 1975, but it is seeking to import at least 1 million of coking coal from China by 1980. Coking-coal exports are also made to Vietnam, Pakistan, Hong Kong, and a few other Southeast Asian countries. Romania imports Chinese coke. Expansion of coal-loading facilities at the main coal port of Lienyunkiang in Kiangsu Province suggests that China intends to expand its coal exports in the future, but these will probably increase slowly because despite potential export markets, domestic needs are growing more rapidly and better returns may be obtained from exports of oil instead.

The richest coal deposits occur in northern China, and about three-

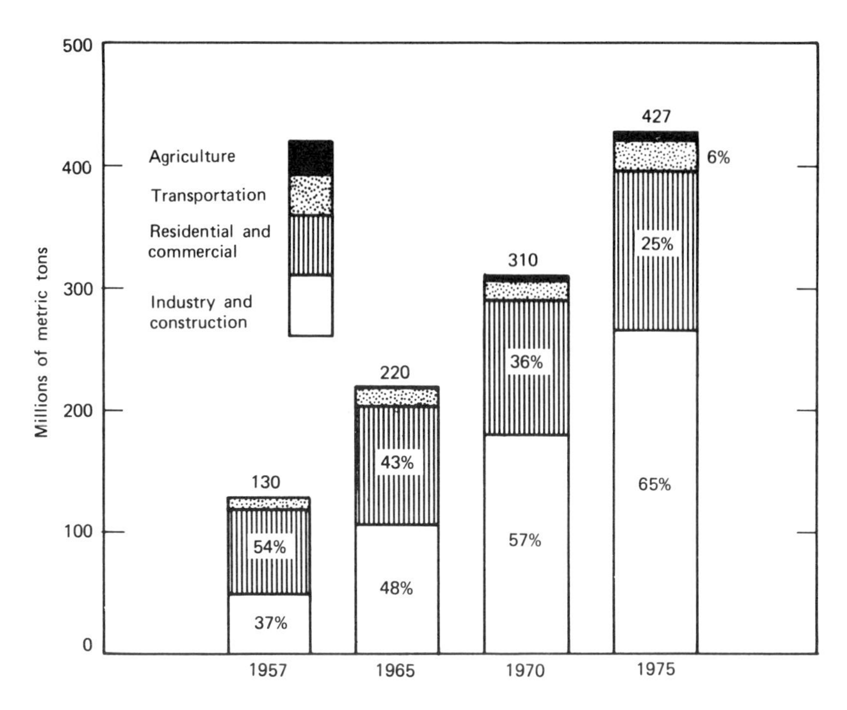

***Figure 7.7** Growth of coal supply and consumption.* (Source: *Adapted from data in* **U.S. Government Research Aid,** *A(ER)75-75, "China: Energy Balance Projections," Washington, D.C., November 1975;* **People's Republic of China Handbook of Economic Indicators,** *August 1976.*)

fourths of total reserves are believed to be in the mountainous Shansi Province and adjoining provinces of Hopeh, Honan, and Shensi. Large reserves are also found in the provinces of Helungkiang, Kirin, and Liaoning and in the Szechwan Basin of central China. Although coal reserves in northwestern China are considerable, they are not exploited because of their remote location. Substantial coal reserves also occur in the southern provinces of China but are poorer in quality and widely scattered in small deposits. An estimated 77 percent of Chinese output is bituminous coal, and 19 percent is anthracite, and the remaining 4 percent is lignite. Reserves of coking coal are adequate, but much of coking coal is considered of poor quality and requires considerable processing.

Growth in coal production averaged about 7 percent in 1966–1976, but production levels varied widely from year to year. Erratic and uneven growth resulted from recurring political disruptions as well. Coal mining is a labor-intensive industry and tends to be affected more than other industries when workers are required to attend meetings and political classes, which results in lower output and sluggish growth in the industry.

The Chinese coal industry appears to have improved its performance in 1975, when production increased 9.8 percent over 1974. In 1976 severe earthquakes affected the largest Kailuan coal mines, and output in that year is not expected to increase by more than 3 to 4 percent over 1975. Other major coal mines were urged to intensify their production during 1976 to compensate the losses. Coal output is projected to grow at rates between 6 percent and 7 percent annually until 1980, and higher growth rates are unlikely without significantly additional investment in modern coal-mining equipment and technology.

A basic and chronic problem of the Chinese coal industry is a low level of mechanization, which is the result of an abundant labor supply. Even the most recent industry expansion plans seem to concentrate on increasing production by opening new mines and intensifying operations rather than introduction of more advanced technology. The 10-year coal-industry development plan presented at a National Coal Conference in 1975 calls for general mechanization of major coal mines during the 1976–1980 Five Year Plan, but comprehensive mechanization is planned only in certain major coal mining areas. During the second stage of the program up to 1985, comprehensive mechanization is planned for all major mines, and the level of mechanization is to be raised in other mines. These objectives were reasserted at a more recent National Coal Conference in January 1977 attended by Chairman Hua Kuo-feng.

It is possible that China at present is sacrificing the modernization of its coal industry to develop its oil. When oil exports increase sufficiently to provide larger amounts of hard currency, China will almost certainly undertake extensive mechanization and modernization of its coal industry.

Two-thirds of the coal produced in China comes from Shansi and the six provinces of Shantung, Liaoning, Hopei, Anhwei, Honan, and Heilungkiang. These areas also receive most of new investment and account for 50 percent of new mine capacity and will probably continue to be primary development areas in the conceivable future. The Shansi Province alone produced a total of 70 million tons of coal in 1975, accounting for one-sixth to one-seventh of the total national output.

The Ministry of Coal Industry is responsible for the construction and operation of major coal mines, which are administered by at least eighty-two known Coal Mining Bureaus, also known as "Administrations," each of which operates up to a dozen major coal mines. By world standards Chinese coal mines are large, and at least five of the Coal Mining

Bureaus are believed to produce over 20 million tons of coal per year (see Table 7.11). The large Coal Mining Bureaus account for about 70 percent of all coal production in China. The remaining 30 percent of output comes from an estimated 100,000 small coal mines operated by local units.

Table 7.11 Major coal-mining administrations and estimated coal production in millions of tons per year about 1975

Coal Mine Administration	*Estimated Production*	*Coal Mine Administration*	*Estimated Production*
Kailuan, Hopei	25.2	Nantung, Szechwan	5.0 to 10.0
Tatung, Shansi	20.2	Mentoukou, Peking	5.0 to 10.0
Fushin, Liaoning	over 20.0	Poshan, Shantung	5.0 to 10.0
Fushun, Liaoning	over 20.0	Shuangyashan, Heilungkiang	5.0 to 10.0
Huainan, Anhwei	over 20.0		
Hokang, Heilungkiang	10.0 to 20.0	Shihkuaikou, Inner Mongolia	5.0 to 10.0
Pingtingshan, Honan	over 10.0		
Chihsi, Heilungkiang	11.0	Suchou, Kiangsu	5.0 to 10.0
Chinghsi, Peking	5.0 to 10.0	Tzupo, Shantung	5.0 to 10.0
Fengfang, Hopeh	5.0 to 10.0	Yangchuan, Shansi	5.0 to 10.0

Source: Compiled by 21st Century Research from various reports listed at the end of this chapter.

Kailuan Coal Mines in Hopeh Province are the largest coal-mining complex in China, reporting a total output of 25.2 million tons in 1974. It is often looked on as a model coal mine in the country. Its Chaokechuang mine established a daily record of 25,000 tons in 1974, and a workface at its Fankechuang mine is credited with a national record of 100,000 tons of coal for medium-thick seams. The Kailuan complex employs a total of 100,000 miners, which includes 10,000 women. Tatung in Shansi Province is believed to be the second largest coal-mining administration, producing at least 20.2 million tons in 1976. Current expansion plans are to triple coal production in Tatung by 1985. Yangchuan, also in Shansi Province, produces up to 10 million tons per year and is the largest anthracite supplier.

At Fushun in Liaoning Province the largest open-pit coal mines are worked in China, and Fushun and Fushin Coal Mining Administrations are each believed to produce over 20 million tons of coal annually. Huainan and Huaipei Administrations in Anhwei Province together produce over 20 million tons of coal per year, and at Huaipei there is a modern coal center with ten shaft mines and a large coal beneficiation plant. Other large coal-mining administrations include Chihsi, Hokang, and

Shuangyashan in Heilungkiang Province, which together are rated at about 30 million tons annually. Pingtingshan in Honan Province is another large administration, producing over 10 million tons, and is also well known for its coking coals.

There are as many as 100,000 estimated small coal mines in China operated by counties, communes, production brigades, or PLA units. Such mines produce less than 1000 tons of coal annually, but a few are known to have developed production levels up to 100,000 tons per year. Small mines play an important role in the Chinese economy, and their main purpose is to provide fuel for Chinese households and raw material for small-scale local industries. The ability to supply coal locally without straining the overburdened transportation system of China is an important factor justifying the existence of the small coal mines. These mines produced about 28 percent of total coal output in China in 1973 and have since appeared to maintain that level (25 to 30 percent). In relatively inaccessible areas such as Tibet small mines account for 100 percent of coal production. In Kwangtung and Yunnan Provinces of southern China about 50 percent of coal comes from small mines, and even in leading producing areas such as Shansi up to 40 percent of coal is produced by local mines.

Most small coal mines are only an open pit or shaft, and some are operated only on a part-time basis. These mines depend on large amounts of rural labor for operation and generally produce coal of poor quality. However, they have the advantage of very low capital requirements and can be started up rapidly to absorb even temporarily local labor that may have become available on completion of seasonal agricultural tasks. Thus these mines have also other purposes and probably help to regulate the local labor supply and demand.

When small-mine coal deposits run deeper or when underground water becomes a problem, small mines become very difficult to operate. Shallow coal deposits may be worked out after only two or three years, and small mines must be shut down or mechanized to continue in operation. As capital investment increases, many small mines become uneconomical to operate and considerably less safe to work.

MARKETS FOR COAL-MINING EQUIPMENT

Although China's coal industry is huge and its output may reach 700 million tons by 1985 and over 1 billion tons by the year 2000, it has not been receiving investment priority comparable to that of the oil industry. Hence it presents only a limited market for specialized coal-mining equipment at present. However, this situation may change during 1980–1985, when complete mechanization of all major coalmines is to be undertaken. Some idea of possible Chinese demand can be gained from analysis of the coal-mining technology in use, domestic equipment production, and recent imports.

In general, Chinese coal mines are characterized by a low level of mechanization, relatively poor coal-processing facilities, and obsolete technology. Much of the mining equipment dates back to the 1950s, when China relied heavily on Soviet and eastern European assistance. During the 1960s China operated its coal mines without the benefit of more advanced technology inputs from abroad and manufactured most of its own mining equipment in relative isolation. The present overall Chinese production methods and technology are considered comparable to those of the United States during 1909 to 1912 and of the Soviet Union during 1953 to 1955.

Chinese coal-mining specialists are believed to have considerable experience in mine exploration and planning, sinking mine shafts, and developing new mines. Their preferred technique is the longwall mining method, although they are familiar with the room-and-pillar method. Since 1965 hydraulic mining has been in use in China, and it is estimated to produce about 3 million tons of coal per year. There has also been some speculation in recent years that the Chinese are anxious to develop modern open-pit mining methods for small and large coal-mining operations.

There are several specialized research institutes in China that are engaged in development programs of new and better equipment for the coal industry; most of those are operated by the Ministry of Coal Industry. The Peking Academy of Coal Mine Design is known to have developed hydraulic coal-mining systems and researches horizontal mining techniques through different strata. The Academy of Coal Research, also located in Peking, has worked on concrete supports for tunnels, gasification of coal, automation of conveyors, and hydraulic mining. The Academy of Coal Dressing Design in Peking was involved in design of coal washing and dressing plants in 1970. Two Shanghai institutes conduct research on mine transportation. The Coal Mines Machines Institute was reported in 1975 to have developed China's first hydraulic propelled coal-mine tunneling machine. There are also other institutes specializing in coal-mine research in Fushun, Tsinan, and Tangshan.

Only a few plants specialize in manufacture of coal-mining equipment, and most equipment is produced in plants that manufacture small machinery and a variety of tools for use in different mining and construction operations. Only in recent years did the Chinese begin manufacturing magnetic separating and filtering equipment and hydraulic coal-lifting systems. Most commonly produced equipment consists mainly of drills, carrying cars, loaders, small powered shovels, scrapers, winches, and hoists. A few models of air compressors, water pumps, and excavators are also made by various plants of the First Ministry of Machine Building Industries (see Table 7.12).

Table 7.12 Major plants manufacturing coal-mining equipment in China

Name of Plant	*Type of Equipment Manufactured*
Chihsi Mining Machinery Plant	Donbass-type coal combines based on Soviet designs; ventilating and transporting equipment for coal mines
Chiangchiakou Coal Mining Machinery	large supplier of coal conveyors
Huainan Coal Mine Machinery Works	hoists and winches and compact mine car for deep shaft operations
Harbin Heavy Machinery Works	bucket-type material handling equipment for coal and ores including 160-ton and 260-ton units
Loyang Mining Machinery Plant	single-layer coal-sorting machine for grading and separating; centrifugal dehydrators, rock drills, excavators, hoists, conveyors, boring machines
Shanghai Mining Machinery Plant	3-m tunneling combine; loaders, crushers and rock drills
Shenyang Mining Machinery Works	drum-type magnetic concentrators; double drum hoist with 5-m drum diameter
Shihchiachuang Coal Mining Machinery	geological prospecting equipment for coal
Taiyuan Mining Machinery Works	large bucket excavators
Tientsin Coal Mining Equipment Plant	automatic control systems for hydraulic conveyors

Source: Compiled by 21st Century Research from various sources listed at the end of this chapter.

China first began importing Western mining equipment in 1965, when purchases were made from ASEA in Sweden, followed by sales by Atlas-Copco in 1967 of additional mining gear. A major upturn in imports took place in 1969, when China imported a total of $15-million worth of mining and construction equipment from ten different countries. In 1974 total imports of mining and construction equipment were valued at $84 million, including about $30 million from communist countries of East

Germany, Czechoslovakia, Poland, and Romania. Between 1969 and 1974 China purchased about $140-million worth of specific coal-mining equipment from Poland, the United Kingdom, West Germany, and the United States. Although no major sales have been noted since then, it is believed that considerable additional purchases will be made in the future as China exports more petroleum and earns additional hard currency. Poland was probably the largest supplier of coal-mining machinery to China in recent years. Starting in 1969, Poland supplied a total of $50 million worth of unspecified coal mining equipment with the last sale taking place in March 1974 (see Fig. 7.8). The United Kingdom was the largest Western supplier of coal-mining equipment in that period with total sales amounting to $48 million. British suppliers have the advantage in selling fully mechanized coal face equipment to China because of their own extensive use of the longwall mining method, which is also preferred in China.

In 1972 John Davis & Sons sold face communications and signaling equipment for an unspecified amount. This was soon followed by a sale in June 1973 by Gullick Dobson (Export) Ltd., who led a group of British firms that supplied the first fully mechanized mining installation to China valued at $13.36 million, including the manufacture and installation of over 1000 longwall roof supports. Other members of this group included Anderson Mavor, who provided double-ended ranging boom shearers and conveying equipment, Baldwin and Francis Ltd. with electrical panels for face machinery operation, and BICC Ltd. with cabling

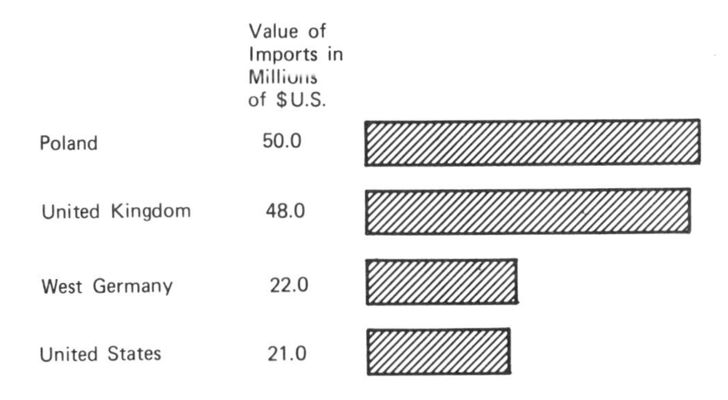

***Figure 7.8** Relative market shares of major coal-mining equipment suppliers during 1969–1974.* (**Source:** *Based on data in* **U.S. Government Research Aid,** *ER 76-10691 "China: The Coal Industry," November 1976 and on 21st Century Research compilation of Chinese foreign-trade statistics.*)

and power supply and Brush Transformers Ltd., who supplied the transformers for the installation.

In July 1973 another British firm, Dowty Mining Equipment Ltd., obtained an order for $29.52 million to provide complete mechanization for thirteen longwall faces in existing Chinese coal mines. This contract included coal-face equipment, powered roof supports with forty-one legs, conveyors, and power-loading machinery, face lighting, signaling equipment, and switchgear. In December 1973 Dowty Mining received an additional contract valued at $5 million to supply four sets of equipment for working coal seams less than one meter thick.

West Germany and the United States supplied smaller amounts of coal-mining equipment. In 1973 Hermann Hemscheidt of West Germany supplied ten pit support units valued at $12.5 million, and in 1974 GHH-Sterkrade sold twenty-three fully equipped steam-shovel extracting units valued at $7.49 million. The German company AEG-Telefunken provided electronic equipment for mechanized coal-face installations. The American firm Bucyrus-Erie sold electric shovels and blasting drills valued at $20 million in December 1973, and Reed Tool Inc. supplied rock bits for a total of $1 million in January 1974. A small sale was also recorded by McNally of Pittsburgh, who sold coal samplers to China for a total of $25,000 in 1974.

In addition to those clearly coal-mining sales to China, a variety of construction and mining equipment was shipped that could be used in coal mining, minerals extraction, or industrial construction in general. Because negotiations are conducted with Foreign Trade Organizations and not the end users, it is often impossible to ascertain for what specific end-user market more general equipment is destined. Some sales to China in this category include 120-ton, 75-ton and 35-ton mine trucks sold by WABCO, 20-ton tractors from Clark Equipment, 200-ton steam shovels from H. Ernault-Somua of France, and large quantities of ordinary trucks that China imports from many countries, including Japan, Italy, France, Romania, Czechoslovakia, Hungary, and the Soviet Union. Japan is not a major coal-mining equipment supplier to China, but it provides small quantities of equipment such as excavating, leveling, tamping, and boring machinery, as well as steel tubing, which are used in various industries, and it is very difficult to separate such sales from other markets in China.

REFERENCES

Ashton, John. "Development of Electric Energy Resources," in *An Economic Profile of Mainland China,* Joint Economic Committee, U.S. Congress, February 1967, pp. 297–317.

Business International, Hong Kong. "Industry Profile: Coal Mining," *Business China,* October 11, 1974.

Business Week. "China: a Bid for U.S. Help in Unlocking Its Oil," September 23, 1976.

Centre Francais Commerce Exterieure DIMEX, Paris. "Le Petrole en Chine," *Nouvelles Commerciales,* November–December, 1974.

Cheng, Chu-yuan. *China's Petroleum Industry,* Praeger, New York, 1976.

Cheng, Chu-yuan. "China's Energy Resources," *Current History,* pp. 73–84, September 1976.

China Reconstructs. "Report on Kailuan Mines," February–March 1977, Peking.

China Trade Report. "Big Hopes for China's Energy Resources," December 1974, Hong Kong.

China Trade Report. "Oil Guesswork," July 1975, Hong Kong.

CIA Economic Research (A[ER]75-75). *China: Energy Balance Projections,* November 1975.

CIA Economic Research (ER76-10540). "Production, Supply and Consumption of Primary Energy," *PRC Handbook of Economic Indicators,* August 1976, p. 27.

CIA Economic Research (ER-76/10691). *China: the Coal Industry,* November 1976.

CIA Economic Research (ER 10D SS 77-007). *International Oil Developments: Statistical Survey,* April 21, 1977.

Citibank. "Oil Exports—the Wild Card in China's Strategy," *Money International* **2**(9): 2–3, September 30, 1974, New York.

Current Scene. "PRC Coal Industry, Performance and Prospects," May 1976, Hong Kong.

Department of the Army. "Fuels and Power," in *Communist China: Bibliographic Survey,* Pamphlet No. 550-9, p. 248, Washington, D.C., 1971.

Dziennik Polski, "Rurociag Naftowy Wyrowna Chinskie Deficyty," 1975, London.

The Economist. "China's Looking for Oil Too," June 28, 1975.

Heymann, Hans, Jr. "Acquisition and Diffusion of Technology in China," in *China: a Reassessment of the Economy,* Joint Economic Committee, U.S. Congress, July 10, 1975, pp. 678–729.

Hidasi, Gabor. *Ekonomika i Doktryna Maoistowskich Chin, Panstwowe* Wydawnictwo Ekonomiczne, Warsaw, 1976.

Howard, John. "China's in the Market for North Sea Know-How," *Industrial Management* (U.K.), July 1975.

Ikonnikov, A. B. "The Capacity of China's Coal Industry," *Current Scene,* April 1973, Hong Kong.

Kambara, Titsu. "The Petroleum Industry in China," *The China Quarterly,* No. 60, December 1974, London.

Klinghoffer, Arthur Jay. "Sino-Soviet Relations and the Politics of Oil," *Asian Survey* **16**, 540–552, June 1976.

Le Monde, Paris. "Pekin Pourrait Ressearer ses Liens avec du Sud-Est," July 10, 1975.

Mainichi Daily News. "10 Million Tons of China Oil Next Year?," October 23, 1974, Tokyo.

Moscow Radio Peace and Progress, in English to Asia. "Peking Using Oil as Weapon in Asian Strategy," in *Foreign Broadcast Information Service,* SOV 75-70, III USSR International Affairs, page C1, April 10, 1975.

National Council for U.S.–China Trade. *U.S. China Business Review* (now *China Business Review*), bimonthly, Washington, D.C., several issues.

O'Connor, D'Arcy. "Chinese Market for Oil Exploration Gear is Tapped Cautiously by U.S. Companies," *Wall Street Journal,* September 6, 1974.

Perspektywy. "Chiny: Polityka Naftowa Pekinu," February 20, 1976, Warsaw.

Pravda. "PRC Military Presence in Paracels," in *Foreign Broadcast Information Service,* SOV 75-34, III U.S.S.R. International Affairs, page C1. February 19, 1975, Moscow.

Selig, Harrison S. "China: the Next Oil Giant," *Foreign Policy,* Fall 1975.

Sino British Trade Council. *A Survey of Chinese Petroleum Industry,* 1973, London.

Sino British Trade Council. *China's Oil Industry Survey,* 1975, London.

Sino British Trade Council. *Sino British Trade Review,* January 1977, London.

Smil, Vaclav. "Energy in the PRC," *Current Scene,* **13**(2): 1–15, February 1975, Hong Kong.

Smil, Vaclav. "Energy in China: Achievements and Prospects," *The China Quarterly,* pp. 54–81, March 1976, London.

Stanford Research Institute. "Trade with People's Republic of China," *Executive Summary Report No. 512,* April 1974.

Taching Oilfield. PRC Brochure, 1974.

Time Break (Geospace House Magazine). "The Expanding Oil Industry of the PRC," Winter 1974–1975, Houston.

21st Century Research. *China Industries, Markets, Imports and Competition 1975–1985,* February 1977, North Berger, New Jersey.

United Nations. *Statistical Yearbook, 1975,* New York.

U.S. China Business Review. "China Economic Notes from Chinese Media Reports," **3**, 59 (Table 4), November–December 1976.

U.S. Department of the Interior. "The Mineral Industry of Mainland China," *Bureau of Mines Yearbooks, 1971, 1973, 1975.*

U.S. Government, *China Oil Production Prospects,* ER77-10030U, June 1977.

U.S. Government, National Foreign Assessment Center, *International Energy Biweekly Statistical Review,* ER 10D SS 77-025, 14 December 1977.

U.S. Government Printing Office. *People's Republic of China Atlas,* Stock No. 415-0001, November 1971, Washington, D.C.

Wang, K. P. *PRC, New Industrial Power with Strong Mineral Base,* U.S. Bureau of Mines, Department of the Interior, Washington, D.C., 1975.

Williams, Bobby A. "The Chinese Petroleum Industry: Growth and Prospects," in *China a Reassessment of the Economy,* Joint Economic Committee, U.S. Congress, July 10, 1975, pp. 225–236.

8

Electric Power Generation and Distribution

Electric power production in China was estimated at 121 billion kW-h in 1975, which makes China the ninth largest electric power-producing country in the world. China also has the world's ninth largest generating capacity, equal to an estimated 34,000,000 kW. However, the per capita electricity production is extremely low, it is below that of India and Pakistan, even though China's total energy production is several times the level of those countries. Only 3.5 percent of all the electricity produced is available to the residential and commercial consumers.

The recent electric power-generation level in China is comparable to that in the United States in about 1930 and in the Soviet Union during the early 1950s. Power production has been growing unevenly during the last 10 years, at an average annual rate of over 15 percent, but is slowing down to about 12.5 percent at present and may average only 10 percent during the early 1980s.

Electric power supply is inadequate to meet all industrial and consumer demands. In some areas factories are known to shut down for one day per week in rotation to avoid voltage drops. Household consumption is often limited to a single naked bulb, and authorization is required to purchase a bulb more powerful than 40 W. Power loads are often adapted to suit power-plant output on a shift-allocation basis.

Overall, the electric power industry in China lags behind industrialized countries. This is particularly noticeable in comparing capacities of power plants, sizes of generating units, transmission-line voltages, and research and development activity. The average power-plant capacities in China are relatively smaller, with the largest thermal power plants having

650-MW (megawatt) capacity and the largest hydroelectric plants being rated at 1225-MW and 900-MW capacity. There are at least fifty hydroelectric power plants in other countries with generating capacity larger than the largest power plant in China. A total of only about twenty Chinese power plants are believed to exceed 250-MW generating capacity.

Chinese leaders are aware of their technological gap and are developing the manufacture of larger generating units at 300 MW to meet the demand. Some immediate needs for increases in generating capacity are being met by imports from abroad of large generating units and technology, and further imports of electrical power equipment are expected to continue during the 1980s.

ELECTRIC POWER PRODUCTION AND CONSUMPTION

Total electric power production in China in 1975 was estimated to be 121 billion kW-h, about 20 percent less than power production of Italy and about 25 percent higher than that of Poland. It was equal to only about 5 percent of the electric power produced in the United States and 12 percent of that in the Soviet Union. Annual per capita consumption is extremely low at 128 kW, lower even than that of India and Pakistan and at least ten times lower than per capita consumption in some neighboring countries such as North Korea, Taiwan, and Hong Kong. On the other hand, it is two to three times higher than per capita consumption in Vietnam and Burma and other populous countries like Indonesia and Nigeria, which have abundant energy resources. Per capita power consumption in large Latin American countries like Brazil and Mexico is several times above the Chinese levels (see Table 8.1).

Industry is by far the largest consumer of electric power in China, using an estimated 70 percent of the total supply. The chemical industry appeared to be the largest single consuming sector, accounting for almost 13 percent of all production, and iron and steel production consumed another 10.4 percent. The nonferrous metals industry used another 11 percent of all electric power produced. Most of it is probably used in aluminum production based on Soviet refining technology, which is a heavy consumer of electricity.

Government policies to increase mechanization of agriculture mark this sector as another rapidly growing consumer, and electric power is available for agricultural use at lower unit rates. Transportation is a minor user, primarily because only a small section of Chinese railways is electrified and urban subway systems are still in their infancy. Residential and

Table 8.1 Production of electric power, installed power-generating capacity, and per capita electricity consumption in 1975

Country	*Electricity Production in Billion kW-h*	*Installed Generating Capacity in kW*	*Annual Per Capita Production in kW-h*
United States	2,122	524,270,000	8,912
Soviet Union	1,038	218,000,000	4,048
Japan	445	110,400,000	4,077
West Germany	301	75,766,000	4,800
Canada	277	59,576,000	10,000
United Kingdom	272	82,000,000	5,100
France	178	46,300,000	3,300
Italy	149	43,586,000	2,750
China	121	34,000,000	128
Poland	97	20,600,000	2,840
East Germany	84	17,230,000	5,010
Brazil	80	20,120,000	780
India	82	21,360,000	137
Mexico	43	13,300,000	830
North Korea	23	3,700,000	1,337
Taiwan	22	5,500,000	1,300
South Korea	20	5,100,000	555
Pakistan	11	2,730,000	158
Hong Kong	7	2,600,000	1,600
Indonesia	5	1,700,000	39
Nigeria	3	1,127,000	47
Vietnam[a]	2.65	1,133,000	58
Burma	0.76	450,000	24
Nepal	0.13	57,000	10

Source: *U.S. Government Handbook of Economic Statistics, 1976.*

[a] Derived 1974 estimates obtained from separate data for both North and South Vietnam for that year.

commercial use is only 3.5 percent of the total (see Fig. 8.1).

Rapidly industrializing countries exhibit a typical ratio of growth in electric power production to overall industrial growth of about 1.4 : 1.0. Assuming that industrial growth will average 9 percent during 1975–1980 and slow down to about 7.5 percent during 1980–1985, electric power generation should grow by 12.5 percent and 10 percent, respectively. Using these growth factors China should be producing about 220 billion kW-h of electricity in 1980 and about 350 billion kW-h in 1985 (see Table 8.2).

Although these projections are considered reasonable by Western analysts, there are forecasts made by COMECON analysts who generally assign lower growth rates for Chinese electric power-production capabil-

Economic Sector or Industry	Percent of Total
Chemicals	12.9
Nonferrous metals	10.9
Iron and steel	10.4
Agriculture	8.0
Machine building	7.2
Petroleum industry	5.7
Coal industry	5.7
Residential/commercial	3.5
Construction industry	3.1
Textiles manufacture	3.1
Foodstuffs precessing	3.1
Building materials	2.7
Paper industry	2.2
Transportation	0.7
Other uses	5.0
Losses	16.0
Total	100.0

***Figure 8.1** Relative electric power consumption by major economic sectors and industries in China in 1973.* (**Source:** *Adapted from analysis of electric power consumption in China during 1973 by sectors of the economy from U.S. Department of Commerce, "Electric Power Equipment," a market assessment for the People's Republic of China, Bureau of East–West Trade, Washington, D.C., February 1975.*)

ities. They envisage an average growth of only 8.7 percent during 1976–1980 and 8.3 percent during 1981–1985. Thus eastern European economists estimate that Chinese electricity production may be 175 billion kW-h in 1980 and only 260 billion kW-h in 1985, or lagging in projected production levels by about 5 years behind Western estimates.

There is some evidence that the Chinese would like to increase their electric power production by 10 to 15 percent per year, and reports of

Table 8.2 Growth in electric power generation, installed generating capacity, and change in average utilization factor

Year	*Electricity Generation Billion kW-h*	*Installed Generating Capacity in kW*	*Average Utilization Factor in Hours/Year*
1960	47	10,200,000	4608
1965	42	11,800,000	3559
1970	72	19,400,000	3711
1971	86	21,100,000	4075
1972	93	23,600,000	3940
1973	101	26,800,000	3768
1974	108	30,000,000	3600
1975	121	34,000,000	3558
1980	220[a]	58,000,000[a]	3754[a]
1985	350[a]	90,700,000[a]	3858[a]

Source: Based on *U.S. Government Handbook of Economic Statistics, 1976* and U.S. Department of Commerce projections.

[a] Figures are estimates based on growth rates of electricity production discussed in the text.

even higher production increases for individual cities or provinces sometimes appear in the Chinese media. In the highly industrialized Shantung Province, for example, electric power generation was reported to have increased by 49.2 percent in 1975 over the previous year. The city of Shanghai claimed a 10-percent increase and Peking a 13.7-percent increase for the same period. Perhaps more revealing was one report indicating that the province of Hupeh actually planned a 10-percent increase in power generation during 1975, although there is no indication whether this target was met.

Whether China will be able to maintain the higher electric power-production growth rates will depend on its ability to add to generating capacity, which should almost triple from the 1975 levels to meet that demand. Increased output also depends on power-plant utilization rates, which in China have been in the order of 3600 h annually in recent years. In contrast, American power stations obtain utilization factors ranging from 4500 to 4800 h per year.

It is interesting to note that in 1960, when most electricity was generated by a few large power stations, China's utilization factor was about 4600 h per year. A subsequent drop in utilization factor and its fluctuation since are probably due to the increasing role of the estimated 60,000 small rural power plants and hydroelectric units, many of which operate only 1000 to 1500 h per year (see Table 8.2).

China's ability to add significant generating capacity depends even more on whether its electric power-equipment manufacturing industry can increase its output at 9 to 10 percent per year and produce large and efficient turbogenerator units. There is evidence that China is expanding this industry, but recent imports of power-generating equipment suggest that production shortfalls will be met by further imports of large power-generating units from Western and COMECON suppliers.

ELECTRIC POWER PLANNING AND DEVELOPMENT

All the large and important power stations and power-transmission networks are controlled by the Ministry of Water Conservancy and Power, headed by Minister Chien Cheng-ying. This ministry established in 1975 replaces the former ministry of Water Conservancy and Electric Power created in 1968 by a merger of the Ministry of Water Conservation, existing since 1949, and the Ministry of Electric Power. It is responsible for the construction of thermal and hydroelectric power plants, transmission lines, and water-conservation projects of regional importance. The ministry also operates major power-generating stations and distribution systems and as such is the most important end user of imported power-generating equipment as well as construction, earthmoving, and transportation machinery.

In addition to the large power stations under direct control of the Ministry of Water Conservancy and Power, there are many small and medium-sized power plants constructed and operated by other ministries serving specific industrial enterprises. Other small municipal and rural power plants are controlled by local authorities, but these do not present opportunities for imports and are almost all equipped with Chinese-built machinery.

It is believed that China follows a policy of maintaining a four to one ratio between thermal and hydroelectric generating capacity. During the last decade new construction appears to have been concentrated on fossil-fueled power plants, resulting in a relative reduction of hydroelectric resources development. This may have been a result of numerous problems and delays experienced with construction of dams and hydroelectric power plants and relatively poor long-distance power-transmission facilities.

The Ministry of Water Conservancy and Power and the State Planning Commission are the ultimate decision-making authorities with regard to large-scale electric power-plant development. However, at least sixteen institutes have been identified that specialize in electric power research and probably influence final decisions in matters of technology and equipment availability. Leading among those is the Institute of Electrical Engineering of the Chinese Academy of Sciences in Peking, which has been in operation since 1959. It has been involved in the design of the largest power plants in China, such as the hydroelectric scheme of the San Men Gorge, and is known to have done research on application of computers to control power fluctuations in networks (see Table 8.3).

Table 8.3 Major institutes identified in China that are engaged in electric power-industry research and development

Name of Institute	*Year Established*	*Probable Affiliation*[a]
Peking Institute of Electrical Engineering	1959	CAS
Canton Institute of Hydroelectric Engineering	1955	MWCP
Changchun Academy of Electrical Power Engineering Design	1957	MWCP
Chengchou Institute of Water Conservation	1950	YRWCC
Chekiang Water Conservation and Power Planning Institute	—	—
Nanching Institute of Hydrotechnology	1958	MWCP–MC
Peking Academy of Hydrotechnical Engineering Design	—	MWCP
Peking Academy of Hydrotechnology	—	CAS
Peking Institute of Electric Power Construction	—	MWCP
Shanghai Academy of Hydroelectric Engineering Design	1958	MWCP
Sian Academy of Electric Power Design	—	MWCP
Sian Northwestern Institute of Hydrotechnology	—	MWCP
Talien Northeastern Academy of Water Conservation and Power	1959	MWCP
Tientsin Electric Power Transmission Design Institute	1958	—
Wuhan Academy of Electric Power Engineering Design	—	MWCP
Wuhan College of Water Conservation and Electric Power	1955	—

Source: Compiled by 21st Century Research from various publications.

[a] CAS = Chinese Academy of Sciences, MWCP = Ministry of Water Conservation and Power, MC = Ministry of Communications, YRWCC = Yellow River Water Conservation Committee.

Primary power generation and consumption centers are the major industrial areas. These include the Shenyang and Anshan regions in northeastern China, Peking and Tientsin in the north, and Shanghai and Nanking in the east. Lanchou, Wuhan, Chengtu, Chungking, and Canton are also major power-consumption areas.

Power distribution is regionalized in several major networks, of which the northeastern grid is the largest, with a total estimated capacity of at least 4,000,000 kW, which is still relatively small by Western standards. Shanghai is the center of the second largest power grid in China, estimated at over 3,000,000 kW, and Peking is third with over 2,000,000 kW capacity. Expansion of generating capacity in those areas probably proceeds at rates higher than the national average, and at least one report in 1977 indicated that Peking Electricity Administration was to add 800,000 kW of capacity in that year alone.

THERMAL POWER STATIONS

Thermal power stations constitute about 70 percent of all generating capacity and produce over 75 percent of all electric power. Most are coal-fired steam turbine units, although oil-fired power stations are being introduced. The trend to build more fossil-fueled power stations is likely to continue because of an abundance of coal and increasing supplies of fuel oil. New thermal stations are likely to be oil-fired depending on their location. Although there is abundant natural gas supply in certain regions of China, there is no evidence that natural gas is used as fuel in large power plants.

Thermal power stations are fired with coal ranging from anthracite and bituminous to lignite. Most subbituminous coal used has a heating value in the range of 5400 to 9000 BTU (British thermal units) per pound. Since 1956, when consumption of coal was 0.594 kg per kW-h power-generating efficiency has improved and is estimated to range from 0.5 to 0.4 kg/kW-h for the major power plants. The average heat rate reported in 1964 for thermal power plants was 17,360 BTU, which is comparable to that achieved by the United States in the 1930s.

A British delegation from the British Electrical and Allied Manufacturers Association, which visited China in 1974, was impressed by Chinese achievements in developing their electric power industry. They have also reported on Chinese ability to expand existing power-plant capacity by combining generating equipment from various suppliers. One Shanghai power station was reported to operate an original Brown Boveri generator dating back to 1928, alongside Soviet and recently made Chinese equipment.

About thirty thermal power plants have been identified in China which are believed to have generating capacities larger than 100 MW. Of those, half a dozen have a generating capacity larger than 500 MW, and the largest thermal plant is believed to have 650-MW capacity (see Table 8.4).

The largest turbogenerator units in regular use are the 125-MW sets that were first installed at the Wuching Power Plant in Shanghai in 1969. Domestic generating sets of 200-MW capacity are also believed to be in trial operation, and a Chinese 300-MW unit is probably being tested. In contrast, the largest units operating in the United States are in the 1000 to 1200-MW range.

The capacity of basic generating units in China is 50 MW in large

Table 8.4 Location and generating capacity of the largest thermal power plants identified in China up to 1977

Location of Plant	*Province*	*Generating Capacity in kW*
Hsintien	Shantung	600,000
Shihchingshan	Peking Municipality	600,000+
Wangting	Shanghai Municipality	600,000
Unknown; under construction	Liaoning	600,000
Kaoching	Peking Municipality	500,000
Chaoyang	Liaoning	400,000
Wuching	Shanghai Municipality	350,000
Chingling	Shensi	250,000
Chinghsi	Peking (West)	200,000
Tanho	Honan	200,000
Tanshan	Hopeh	125,000
Chapei	Shanghai Municipality	125,000+
Huainan	Anhwei	125,000

Source: Compiled by 21st Century Research from various reports and publications.

new thermal power plants compared with 72.5 to 100 MW in new hydroelectric power plants. In addition, there are generating sets serially manufactured in China in the 25-MW, 75-MW, and 100-MW categories, as well as many smaller units. Large power plants are connected into power grid systems by 110- or 220-kV (kilovolt) transmission lines, and a 330 kV line is in operation on an experimental basis. High-voltage direct current (HVDC) transmission is under study for bringing power from remote hydroelectric power plants.

Thermal power plants driven by gasoline and diesel engines are usually small units operating in remote areas and are believed to contribute a very small proportion of the total output. Although China shows considerable interest in gas-turbine generating plants, only a small number of imported units are in operation, probably used during power peak periods of demand. Those and smaller Chinese-built gas turbine units do not contribute significantly to overall power production. Power plants where gas-turbine exhausts are combined to drive steam turbines are not known to exist in China at present.

HYDROELECTRIC POWER RESOURCES AND PRODUCTION

China has an extensive river system whose total water power potential has been estimated at 535,000,000 kW, which is comparable to the total installed generating capacity of the United States, but probably no more than 10,000,000 kW or less than 2 percent of the potential is being exploited. Over 70 percent of all water-power resources are found in southwestern China, primarily in the upper Yangtze River and the Brahmaputra River in Tibet.

The total Yangtze River basin is estimated to possess water-power resources of 217,000,000 kW, or 40.5 percent of all hydroelectric power potential, and over 20,000 various hydroelectric power stations have been reported in operation in the Yangtze Basin in 1977. The potential of Tibetan rivers, primarily the Brahmaputra, is estimated at almost 22 percent of the total, while other southwestern rivers are estimated to possess another 17 percent of the potential. Other major river systems with hydroelectric potential are the Yellow River, the Hsi-kiang Basin, the southeastern coastal rivers, northeastern rivers, and northern rivers, which together make up the remaining hydroelectric potential.

Hydroelectric power production in 1975 was estimated at about 27 billion kW-h, which represents about 22.5 percent of all electric power generated in China. If this amount of electricity was generated by estimated 10,000,000 kW installed hydroelectric generating capacity, this suggests a hydroelectric power plant average utilization factor of only 2600 h per year. One reason for this relatively low utilization factor is the large number of small and inefficient hydroelectric power plants scattered throughout rural China. Nevertheless, most of the hydroelectric generating capacity growth is expected to continue in the small hydroelectric plants used for agriculture; these plants have been expanding rapidly, growing at an average 14 percent annually during the early 1970s.

Although Chinese rivers have relatively abundant water and flow through natural environments suitable for construction of considerable hydroelectric generating capacity and require relatively small investment, irregular river flows present numerous problems in hydroelectric power generation.

Flood-water flows of many rivers may be ten to 1000 times the normal flow and thus present a threat to hydraulic installations. Low water levels during winter are often inadequate to operate hydroelectric stations at capacity; therefore, efficient use of Chinese rivers for electricity generation requires extensive river-flow regulation by means of dams, reservoirs, and canals. All this is reflected in the fact that water management and power production are the responsibility of a single Ministry of Water Conservancy and Power.

Another problem in hydroelectrical generation is a high level of silting of the Chinese rivers; this is particularly acute on the Yellow River, which carries an average of 40 billion cubic feet of suspended material per year. The Yangtze averages over 10 billion cubic feet of silt per year, but its greater flow makes this problem less severe.

The third major obstacle to rapid development of hydroelectric power potential on China's rivers is their importance as a major transportation network of China. Most rivers have a deep navigable channel and do not freeze in winter and hence are suitable for year-round navigation. Total navigable waterways are estimated to be at least 55,000 miles long, which is about twice the mileage of China's railway system. Importance of rivers for irrigation of agricultural land is also a factor in hydroelectric potential development. Because Chinese authorities regard large-scale hydroelectric power projects as capital-intensive, difficult, and time-consuming, they normally consider them as part of irrigation and flood-control programs, which probably further complicates their implementation and delays their execution.

In 1976 a Chinese broadcast indicated that there were 100 large and medium-sized hydroelectric power stations in China, with the largest identified at Liuchiahsia, Yengkuo, Papan, Chintung, and San Men Gorge. The largest of them all in terms of generating capacity is Liuchiahsia on the upper course of the Yellow River, which has a generating capacity of 1,225,000 kW. Several hydroelectric power plants in China have a capacity of 200,000 to 700,000 kW and at least sixteen hydroelectric power plants have been identified with generating capacity over 100,000 kW (see Table 8.5).

Table 8.5 Location and generating capacity of the largest hydroelectric power plants identified in China up to 1977

Location of Plant	*Province*	*Generating Capacity in kW*
Liuchiahsia	Kansu	1,225,000
Tanchiangkou	Hupeh	900,000
Hsinankiang	Chekiang	625,500
Yenkuo	Tsinghai	design 595,000
Chenchi	Hunan	435,000
Chingtung	Ningsia Hiu AR	260,000
San Men Gorge	Hunan	200,000
Ansha	Fukien	150,000
Chentsun	Anhwei	150,000
Huangungtan	`Hupeh	150,000
Shihchien	Shensi	135,000
Papanhsia	Kansu	108,000

Source: Compiled by 21st Century Research from Chinese news releases and various reports and publications.

The dam and plant construction at Liuchiahsia began in 1964, and electricity was first produced in April 1969. The plant has announced 5.7 billion kW-h of annual production capacity, which appears to give it a relatively high utilization factor in the 4600-h/year range. It operates five generator sets, including one 300-MW turbine, which is the largest built so far in China and has water cooled rotor and stator. The station also has four units of 225-MW capacity built by the Harbin Electric Machinery Plant.

The San Men Gorge power station on the Yellow River was the first large water conservation and hydroelectric project and was originally designed with Soviet assistance with a planned capacity of 1,200,000 kW, using eight 150-MW generating sets. Construction began in 1957, but according to Soviet sources inadequate attention given to silting problems led to delays and flooding and eventual decision to redesign the project. Chinese specialists blame the Soviet planners for improper design of the project; the power plant was finally rebuilt, but its present capacity is only 200,000 kW, using four 50-MW turbogenerator sets.

GEOTHERMAL, TIDAL, AND NUCLEAR POWER

Geothermal power production is experimental in China, and its contribution to the total electricity production is negligible. One experimental plant of unknown capacity is reported to operate in Hualai County within the Peking Municipality. In 1976 a geothermal power station of 300-kW capacity was announced in operation at Huichang in Hunan Province. Considering the fact that there exists considerable seismic activity throughout China more geothermal power stations may be in operation or being built, but these are probably low capacity units developed to take advantage of specific local conditions.

No nuclear power plants have been identified in China, although there are many nuclear reactors in operation, and a nuclear industry exists primarily dedicated to nuclear materials and weapons production. In fact, the largest Chinese hydroelectric power plant at Liuchiahsia is believed to provide much of its output to the Lanchou uranium gaseous diffusion plant. On the other hand, Chinese visitors to nuclear facilities in Japan and western Europe are reported to have shown definite interest in acquisition of nuclear power plants in the future.

At least one visitor to China reported a small tidal power station in operation in Kanchutan, which was claimed to generate a total of 12 million kW-h of electricity per year. It was described as a typical local project consisting of a concrete gravity dam about 12.5 m wide exploiting low heads of tide ranging only from 0.4 to 1.0 m in height. Power was generated by low-speed horizontal propeller units operating ten generators of 200-kW and twelve units of 250-kW capacity. In 1973 a pumping and energy storage system of 11,000-kW capacity was constructed for a reservoir near Peking specifically to regulate the peak loads of electricity supply in that region.

ROLE OF SMALL GENERATING STATIONS

Massive construction of small rural power plants was first undertaken at the outset of the Great Leap Forward, and by the end of 1960 rural hydroelectric capacity was claimed to have reached 520,000 kW, compared with only 20,000 kW, which existed at the beginning of 1958. In subsequent years many of those plants were abandoned, and it is estimated that by 1966 no more than 200,000 kW of capacity remained in service. Plans were laid to distribute electric power from large power stations then under construction in the Yangtze River delta, the Pearl River delta near Canton, and in the Tungting Lake region. That policy did not result in meeting the demand of vast agricultural areas for electric power, and after the Cultural Revolution China returned to widespread construction of small power plants in the rural areas, at the same time pressing on with the construction and expansion of large power plants in major industrial regions. The speed with which small power plants can be put into operation and their independence of extensive and sophisticated power distribution systems must have been significant factors in this decision, particularly if mechanization of agriculture is considered as one of the priorities. Abundant hydroelectric potential, local supplies of coal, and more recently, availability of fuel oil were also important factors.

By 1975 about 60,000 rural hydroelectric plants had been built, with an estimated capacity of about 3,000,000 kW. There are also several thousand small steam and diesel power plants with a generating capacity comparable to that of the rural hydroelectric units. Although small power plants are believed to represent nearly 15 percent of the national electric power-generating capacity, they actually produce only about 5 percent of the total electricity output primarily because they are seldom used to capacity and operate at low utilization factors. Most small power plants in China operate for only 1000 to 2000 h per year, compared to 4000 to 6000 h for large modern power plants.

Small plants range in size from 1-kW units to about 1000-kW stations, with the average unit believed to be in the order of 30 kW. About 80 percent of their output is used for irrigation and drainage, and in southern China small power plants are an important source of power for flooding of rice paddies. Some output is consumed by small local enterprises, radio and loudspeaker diffusion systems, and by peasant households who use low 10 to 15-watt bulbs for lighting. Small power plants are constructed employing local labor and construction materials and often use second-hand generating equipment and require little copper for transmission systems. Sometimes small plants are built as a result of the construction of dams and canals for water-conservation projects. In other cases they exploit small streams that would not otherwise be put to work and are often inadequate during dry seasons.

Although small plant construction costs are low, these units are considered inefficient due to low outputs, and the actual investment cost per kilowatt of output is believed to be about double that of a larger power station. Savings in power-transmission costs and low opportunity costs of local resources are the major justifications for construction of small generating units, particularly in remote areas. These small plants do not represent a market for imported equipment and use always locally produced parts probably manufactured to designs obtained from central authorities or research institutes. Visitors to such small power plants reported seeing items such as conduits, scroll cases, draft tubes, housings, switching gear, transformers, and shafts, all manufactured in local workshops and machine repair plants.

ELECTRIC POWER-EQUIPMENT INDUSTRY

Practically all the equipment required by the electric power industry is manufactured in China under the supervision of at least two production bureaus of the First Ministry of Machine Building. Over seventy plants manufacturing steam and hydraulic turbines, generators, boilers, transformers, switchgear, wire and cable, and all types of components for small rural stations have been identified. Production of small gas turbine-generating units has begun recently, but manufacture of nuclear power reactors and more sophisticated process-control instrumentation and communications equipment is probably in its infancy. Major manufacturing centers are located in Harbin, Shenyang, Shanghai, Peking, and Sian (see Table 8.6).

China claimed in 1973 to have the capability to build "several thousand megawatts of generating capacity," and the latest estimates are that the industry produced about 5500-MW electric power generators during 1975.

However, in the early 1970s the net additions to generating capacity

Table 8.6 Major plants engaged in electric power-equipment manufacture identified in China in recent years

Name and Location of Plant	*Type of Equipment Manufactured*
Harbin Turbine Plant	steam and hydraulic turbines
Harbin Electrical Machinery Plant	serial production of generators
Shanghai Turbine Plant	steam and gas turbines manufacture
Shanghai Electrical Machinery Plant	serial production of generators
Peking Heavy Electrical Machinery	steam turbines and generators
Peking Steam Turbine and Generator	turbine and generator development
Szechwan Tungfang Plant	steam and hydraulic turbogenerators
Harbin Boiler Plant	large boilers for 200-MW turbines
Shanghai Boiler Plant	boilers for turbines up to 125 MW
Peking Boiler Plant	boilers for Peking turbine plants
Shenyang Transformer Plant	largest transformer plant in China
Sian Transformer Plant	produced first 330-MVA[a] transformer
Paoting Transformer Plant	produced a 100-MVA transformer
Shenyang Switchgear Plant	circuit breakers and other switchgear
Sian High Voltage Switchgear Plant	produces circuit breakers
Shanghai Switchgear Plant	oil circuit breakers
Shenyang Wire and Cable Plant	aluminum and copper wire cable
Shanghai Wire and Cable Plant	oil-filled 220-kV conductors

Source: Adapted from "Electric Power Equipment," A Market Survey for the PRC, U.S. Department of Commerce, Bureau of East–West Trade, February 1975, Washington, D.C.

[a] MVA = megavolt amperes.

averaged only 1200 MW annually when manufacturing capacity appeared to be in the order of 3500 MW. This led some observers to believe that total announced manufacturing capacity may include double counting, combining rated megawattage of turbines and generators or even boilers.

The largest steam turbine in serial production is the 125-MW unit, although a 200-MW unit has been constructed at the Harbin Turbine Plant and the first 300-MW turbine was reported developed at the Shanghai Turbine Plant. Larger turbogenerator units in the 600-MW category may not be produced until more operating experience with 300-MW units is gained. Some analysts think, however, that 300-MW units may be optimal because Chinese power grids lack high-capacity interconnections.

The largest hydroturbines in serial production are 72.5-MW units manufactured by the Harbin Turbine Plant, which also developed a 225-MW unit and the first 300-MW hydraulic unit. The new Tungfang Plant in Szechwan Province also produced a hydrogenerator of 210-MW capacity.

China appears to have built its first 1.5-MW gas turbine in 1964 at Nanching, and it is believed to have been the prototype for the 6-MW gas turbine produced in 1965–1966 at the Shanghai Turbine Plant. There have been rumors that a 25-MW gas turbine has been built, but other reports suggest that the 6-MW gas turbine was only adapted for power use in 1974. A portable 1-MW gas turbine was also made at the Nanching plant in 1970, and a 0.5-MW gas turbine was reported in production at the Canton Power Machinery Works in 1975.

The Harbin Electrical Machinery Plant and the Shanghai Electrical Machinery Plant are the largest manufacturers of generators to match the turbines produced in other plants. The Shanghai plant produced the 300-MW generator with water-cooled rotor, which is believed to be a significant Chinese design development. China first began production of 100-MW generators in the 1960s, and analysis for known prices for several units in 1973 suggests that 1 kW of thermal generating capacity had a cost of about $35 to the Chinese end user. Several models of Chinese generators have been identified with 0.75-kW, 1.5-kW, 10.0-kW, 120-kW, 130-kW, 160-kW, 6000-kW, 12,500-kW, 25,000-kW, 50,000-kW, 100,000-kW, and 200,000-kW generating capacities.

Power boiler plants in Harbin, Shanghai, and Peking manufacture boilers to match turbines made in other plants in those cities. The Harbin Boiler Plant has made steam boilers to 670 metric tons per hour capacity for large 200-MW turbines. Largest boilers in serial production are 400 metric tons per hour units, which support the 100-MW and 125-MW turbines. Besides the three major plants there are about twelve other boiler factories in Wuhan, Hangchou, and other cities producing small drum-type and industrial boilers.

The most important transformer plant is the one in Shenyang, which produces a large variety of transformer models, the largest of which is rated at 260 MVA. The Sian Transformer Plant produced the first 330-MVA power transformer in China, and the Paoting Transformer Plant is known to have manufactured a 100-MVA unit in 1971.

Circuit breakers and switch gear are produced at the Shenyang Switchgear Plant, the Sian High Voltage and Mercury Rectifier Plant, and the Shanghai Switchgear Plant. The largest circuit breakers in serial production are the minimum-oil type with 220-kV, 1000-A, 700-MVA rating for outdoor use; however, air-blast breakers with 15,000-MVA interrupting capacity at 330 kV have also been made. Metal-clad and metal-enclosed types of switchgear for up to the highest transformer and circuit breaker ratings are also made.

Shenyang Wire and Cable Plant is believed to be the largest such manufacturer in China and produces both aluminum and copper wire and cable, including the 330-kV power cables. This plant is a major producer of oil-filled 220-kV conductors and paper-insulated cable up to 300 kV. Several other plants exist in Chengchou, Chungking, Kunming,

Tientsin, and other cities. There are also numerous smaller plants manufacturing separate components for construction of small local hydroelectric and other rural power stations.

China developed its electric power industry on an original Japanese base with considerable assistance of the Soviet Union, Czechoslovakia, and East Germany during the 1950s. Several major plants were built or expanded with Soviet assistance, including: (1) the Steam Boiler Plant, Steam Turbine Plant, and Power Equipment Plant in Harbin, (2) the Voltage Equipment Plant, High Voltage Switch and Mercury Rectifier Plant, and High Voltage Ceramics Plant in Sian, (3) the Transformer Plant and the Wire and Cable Plant in Shenyang, and (4) the Wuhan Steam Boiler Plant. Extensive power-equipment installations based on original Soviet designs is probably one of the reasons why China continues to import power-generating equipment from the Soviet Union of eastern Europe.

Industry observers believe that many technological innovations are inevitable, if not imminent, in Chinese power-generating industry. These will probably include the construction of a 1,000,000-kW thermal power station and use of turbogenerators of 500 MW and even higher capacity not presently known to exist in China. In long-distance power-transmission voltages are expected to double, and a 330-kV line of 500-km length is already in existence on an experimental basis.

Because of the large requirements to increase electric power-generating capacity, China will increase the output of its plants, particularly by trying to put into serial production large turbine and generator units of the 200-MW, 300-MW range in steam turbines and at least 150-MW units in hydroelectric sets. But serialized production of 300-MW turbines and all the other associated equipment required to standardize China's major power-generating networks will probably take some years; hence additional imports of large power generating units in the 200 to 300-MW may continue. The large number of suppliers chosen so far may also be an indication of a research process to find the most suitable Western supplier who could provide the technology for the construction of manufacturing plants for the production of large power-generating equipment in China.

MARKETS FOR ELECTRIC POWER-GENERATING EQUIPMENT

To generate 350 billion kW-h of electric power by 1985 as previously projected, China would have to almost triple its present estimated 34,000,-000-kW capacity to about 90,000,000 kW and improve its utilization factor, at least to about 4000 h per year during the early 1980s. This would require additional generating capacity of 56,000,000 kW to be installed and all the associated transmission and distribution equipment to be put in place at an average rate of about 7,000,000 kW annually, which is probably beyond the capacity of the Chinese electric power industry at this time.

If the Chinese power-generating industry is able to supply about 5,500,000 kW of capacity per year, China may still have to import about 2,000,000 kW of capacity worth in the neighborhood of $200 million a year to keep the industrial growth at reasonable levels well into the 1980s. On the other hand, a 10-percent growth in domestic equipment production could reduce the levels of these imports appreciably because imports of such large units are extremely sensitive to changes in levels of domestic production.

Analysis of sales and deliveries of generating equipment and power plants to China in recent years indicate that a total of almost 5,000,000 kW, valued at over $550 million, was contracted for during 1972–1975. This suggests that China was willing to pay an average $110 per kilowatt of imported capacity in the form of large steam turbines and generator sets. Comparable prices for Chinese turbines and generators in the 100 to 200-MW range were estimated at only 100 Yuan per kilowatt, or about 50 percent lower than the imported unit price (see Table 8.7).

During the 1960s imports of power-generating and switchgear equipment from all countries to China ranged from only $5 to $15 million per year, with the Soviet Union, Czechoslovakia, East Germany, Poland, and to some extent Japan as the major suppliers. In 1972 imports doubled to over $30 million and although the Soviet Union still supplied almost 50% of the market, France and the United Kingdom also became significant Western suppliers. In 1973 over $43 million of equipment was shipped to China, signaling the start of deliveries of power-plant equipment, which was now also being ordered from Japan, West Germany, Italy, and Switzerland (see Fig. 8.2).

Analysis of Chinese orders for electrical power-generating equipment between 1972 and 1975 also suggests that progressively larger turbines and generating units predominantly for thermal power stations were being purchased. At least twelve different manufacturers from ten different countries have been involved, which suggests a desire to obtain additional generating capacity quickly but could also be interpreted as a practical research program to obtain and compare the most advanced large turbogenerator technologies. Such a program could lead to incorporation of innovations found on imported equipment into Chinese power-equipment manufacture.

Table 8.7 Deliveries and contracts for imported power plants and generating equipment identified since 1970

Year	Type of Equipment Imported	Generating Capacity in kW	Supplier and Country	Contract Value in Millions of $U.S.[a]
1972	Three 37-MW hydrogenerators	111,000	ASEA/KMV, Sweden	4.0
	Four 75-MW steam turbogenerators	300,000	U.S.S.R.	8.2
	One 100-MW steam turbine +	100,000	Mitsubishi/Hitachi/Toshiba	13.0
	One 125-MW thermal plant	125,000	GIE, Italy	8.5
	Three 25-MW gas turbines	75,000	John Brown, U.K.	8.4
	Two 125-MW steam turbogenerators	250,000	Hitachi, Japan	30.0
	Two 75-MW Kaplan hydroturbines	150,000	Alsthom/Creusot-Loire, France	10.0
1973	Seven 100-MW steam turbogenerators	700,000	U.S.S.R.	16.4
	Three 60-MW hydroturbines	180,000	Althom, France	7.5
	Three 7.5-MW gas turbines	22,500	ACEC/Westinghouse, Belgium	5.0
	Three 100-MW turboalternators	300,000	Skoda, Czechoslovakia	50.0
	Five 25-MW gas turbines	125,000	John Brown, U.K.	8.2
	Two 250-MW steam turbogenerators	500,000	Hitachi, Japan	72.0
	Two 325 MW turbogenerators	650,000	GIE, Italy	86.2
1974	One 300-MW turbogenerator	300,000	Kraftwerk Union, Germany	60.0+
	Electrical substations		Brown Boveri	5.0
	One 200-MW steam turbines	200,000	U.S.S.R.	
	Four 100-MW	400,000	U.S.S.R.	
	One 300-MW steam turbogenerator	300,000	CEM/Sulzer, France	40.0
1975	Two 25-MW gas turbogenerators	50,000	Hitachi, Japan	5.4
	One 150-MW steam turbine technol.	150,000	Siemens, Germany	6.0
1976	Two 200-MW steam turbines	400,000	U.S.S.R.	12.5

Source: Compiled by 21st Century Research from several publications listed in the references to this chapter.

[a] Contract values are in many cases reported estimates of value of equipment under negotiation and reflect the magnitude of transactions rather than final contracts signed, which may differ from figures above.

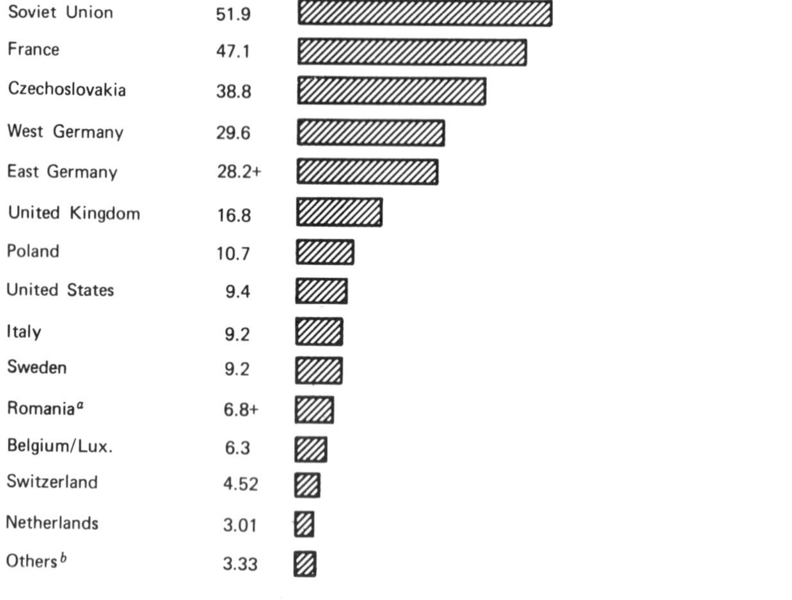

Figure 8.2 ***Relative market shares of suppliers of electric power generation, transmission, and distribution equipment identified during 1967–1975, excluding boilers and turbines for use in power plants.*** **(Source:** ***Based on export statistics for individual countries under SITC commodity codes 722 and 723 or equivalent for COMECON countries.*****)** **([a]** ***Data available for 1967–1973 only.*** **[b]** ***Other exporters of these products include Denmark, Finland, Austria, and Canada.*****)**

On the other hand, if equipment production cannot be increased to meet the expected demand, China may be interested in choosing a supplier to build a complete power-generating equipment-manufacturing facility on a turnkey basis. In this respect the technology to manufacture in China of large steam turbines should be of particular interest to Chinese authorities because they are believed to be more advanced in solution of problems attendant on manufacture of large generators rather than power-generating turbine equipment. The recent sale by Siemens of 150-MW steam turbine technology to China seems to support this argument.

According to an assessment of the Chinese market by U.S. Department of Commerce analysts, the Chinese are expected to make further significant purchases of power-generating equipment in the 200-MW or larger category. The 300 to 600-MW equipment range, in which China has very little experience, may have to be supplemented by equipment and technology imports from abroad.

In 1976 imports of power-generating equipment amounted to $189 million, over three times the average imports in this category in recent years. The Soviet Union was the leading supplier, with $58 million of steam turbines, followed by France with $46 million. United States, Japan, and Italy were the other major suppliers, with $24, $22, and $23 million, respectively. Most of these imports are probably deliveries of power equipment ordered in previous years, but these indicate the importance of imports of this equipment to China.

REFERENCES

Ashton, John. "Development of Electric Energy Resources in Communist China," in *An Economic Profile of Mainland China,* Joint Economic Committee, U.S. Congress, February 1967, pp. 279

Bouc, Alain. "La Production Chinoise d'Electricite atteindrait 110 a 120 milliard de KWH," *LeMonde,* March 13, 1975, Paris.

Bouc, Alain. "La Fleuve Jaune et le Developpement de la Chine," *Le Monde,* August 9, 1975, Paris.

CIA Economic Research. "China: Energy Balance Projections," November 1975.

Carin, Robert. "*Power Industry in Communist China,*" Union Research Institute, Honk Kong, 1969.

Foreign Broadcast Information Service, "PRC to blame for need to rebuild San Men Gorge Dam," Moscow broadcast in Mandarin to China, April 3, 1975.

Hidasi, Gabor. "*Ekonomika i Doktryna Maoistowskich Chin,*" PWE, Warsaw, Poland, 1976, p. 52.

McDougall, Colina. "China building up power capacity," *Financial Times,* April 9, 1974, London.

McDougall, Colina. "UK Electrical Team impressed by Chinese innovation at plants," *Financial Times,* November 6, 1974, London.

Peking Power Supply Company. "A Modern Transformer Station," October 1971, *translated by JPRS* # 57170, Washington, D.C.

U.S. Department of Commerce. "Electric Power Equipment; A Market Assessment for the People's Republic of China," Bureau of East–West Trade, February 1975.

U.S. Government Research Aid. "China: Role of Small Plants in Economic Development," A(ER)74-60, May 1974, p. 11.

9

Mining and Metallurgy

China has considerable reserves of most minerals and metals required to support a major industrial expansion program for many years to come. Reserves of antimony and tungsten are believed the largest in the world. Reserves of bauxites, coal, fluorspar, lead, magnesite, manganese, mercury, molybdenum, tin, and talc are very substantial. Some strategic metals such as chromium, cobalt, and nickel are in very short supply and must be imported (see Fig. 9.1).

The iron and steel industry in China is the fifth largest in the world. China is the second largest iron-ore producer, but the quality of ores is poor and increasing beneficiation is required. Growing amounts of higher-grade iron ores are being imported from Australia to provide input to an expanding steel-making industry.

It is estimated that China will require two to three times its present steel-making capacity to maintain the planned industrialization of the economy. Steel output is being expanded by imports of complete steel-finishing plants from Japan and West Germany. Demand for steel products is being met by imports of finished steel from Japan, West Germany, and several capitalist and communist countries. By value, imports of iron and steel account for 20 percent of all Chinese imports and are the largest single import category in China's foreign trade. Ferroalloys, except for chromium, cobalt, and nickel are in plentiful supply.

There are also extensive reserves of nonferrous metals such as tin, lead, zinc, mercury, aluminum, and copper. However, aluminum and copper ores are poor in quality, and large quantities of these metals are being imported while world prices are attractive. Imports of nonferrous metals accounted for over 6 percent of all China's imports in 1975. Some of those metals are imported from developing countries who are recipients of Chinese foreign aid or who purchase oil in return.

Surplus domestic capacity	Antimony	Fluorspar
	Manganese	Magnesite
	Mercury	Salt
	Molybdenum	Talc
	Tin	
	Tungsten	
Adequate domestic capacity	Aluminum	Anthracite
	Iron ore	Coal
	Lead	Cement
	Zinc	Pyrites
Deficient domestic supply	Chromium	Phosphates
	Cobalt	Sulfur
	Copper	
	Nickel	
	Steel	

***Figure 9.1** China's minerals and metals position.*

Despite substantial domestic minerals and metals resources, China remains a major importer of metals, although its trade is seldom a major factor in the metals trade of other countries. Except for steel imports from Japan and copper from Chile, such trade accounts for less than 5 percent of the trading country metals exports. Although this trade has little impact on world markets, minerals and metals represent 25 to 30 percent of total Chinese imports and about 5 percent of total exports. As a result, China is making efforts to reduce its dependence on such imports. This depends on considerable investment in development of domestic resources, which is not readily available because of petroleum-industry exploration and production priorities. In addition, exports of Chinese minerals and metals are similar to those of other developing countries and tend to increase world market supplies and weaken the prices. Relatively low demand for such products further contributes to a slow growth in exports. These factors may have driven China to voice support for Third

World exporters who are making efforts to raise export prices by limiting supplies through cartel arrangements.

THE IRON AND STEEL INDUSTRY

Chinese leaders have always been aware that iron and steel are the "key link" between agriculture and a modern industrialized economy, and China is well endowed with coal and iron ore, which are the basic inputs to a large steel industry. Consequently, the steel industry received considerable attention and investment priority, although Mao's political and social campaigns played their part in disrupting and slowing down its development.

China is the fifth largest crude and finished steel producer in the world after the Soviet Union, United States, Japan, and West Germany. Its crude-steel production was 26 million tons in 1975, about three times as large as that of India, Australia, or Brazil. It is comparable to the steel industries of Italy, France, and the United Kingdom, although those countries probably have more idle productive capacity (see Table 9.1).

According to a recent U.S. Bureau of Mines evaluation, China may need a steel industry of 50 to 100 million tons per year in the foreseeable future to continue planned industrialization of its economy. That magni-

Table 9.1 Significance of Chinese iron and steel industry relative to selected major steel-producing countries in 1975 ranked by total crude-steel output (Production figures in millions of metric tons)

Country	*Iron Ore*	*Coke*	*Pig Iron*	*Crude Steel*	*Rolled-steel Products*
Soviet Union	233.0	83.5	103.0	141.2	115.0
United States	81.6	51.9	72.5	105.9	72.5
Japan	—	47.1	87.09	102.3	86.35
West Germany	3.29	35.2	30.13	40.42	31.93
China	109.0	24.3	33.8	26.0	19.50
Italy	—	8.3	13.72	21.84	18.56
France	46.65	11.8	17.93	21.56	17.04
United Kingdom	4.37	15.8	12.25	20.16	15.93
Romania	3.3	2.0	6.6	9.55	6.81
Brazil	44.32[a]	2.3	6.92	8.36	—
Australia	99.42	5.19	7.6	8.06	—
India	40.27	8.84	8.54	7.81	—

Source: *U.S. Government Handbook of Economic Statistics, 1976; United Nations Statistical Yearbook, 1976.*

[a] Data for 1974.

tude of output is comparable to the steel industry of the United States or Japan today, and there is little doubt that China has no alternative but to expand its steel industry in the future to such production levels.

However, due to considerable overcapacity in the steel industries all over the world and readily available steel imports, China may be deferring faster growth of its own capacity at present. This policy would free additional resources to develop China's oil industry and transportation.

Production of Iron Ore

China has huge deposits of low-grade iron ore and a few deposits of high-grade ore. Some fantastic claims of iron-ore reserves, in the order of 100 billion tons, have been made by the Chinese. These are not given much credence by industry observers, but it is generally agreed that Chinese iron-ore reserves are more than adequate for continued exploitation well into the 21st century.

Chinese iron ores are generally poor in quality, containing only 25 to 35 percent iron, compared with over 50 percent iron content of the ores in Australia or Japan. The average grade of iron ore mined in China declined from about 50 percent during the 1950s to just over 30 percent in the early 1970s. Modernization of the iron and steel industry is now dependent on extensive introduction of coal and iron-ore beneficiation plants.

Production of iron ore was estimated at 109 million tons in 1975, second only to that of the Soviet Union and about 30 percent larger than that in the United States. Two-thirds of all the iron-ore production comes from large mines, and the rest is output of small mines for processing by local steel plants (see Table 9.1). China supplements its iron-ore production by importing about 3 percent of its total requirements mainly from Australia. This further underscores the deterioration of domestic iron-ore quality, because until 1969 China was itself an exporter of iron ore. Imports of iron ore are now continuing, possibly because they provide immediate input to a growing steel industry and are cheaper than the considerable investment required in time and resources for the modernization of the Chinese iron-ore mines.

Pig-iron Production and Scrap-iron Supply

Pig-iron production was estimated at 33.8 million tons in 1975, higher even than that of West Germany, which makes China the fourth largest pig-iron producer in the world. However, this level of pig-iron output was still insufficient for Chinese steel-making capacity, and about 3 percent or 1.0 million tons of pig iron was being imported annually.

Industry observers believe that Chinese blast furnaces are not as efficient as those of Western companies and that additional output can be achieved from improved operations. This may be one reason why construction of new blast furnaces appears to be lagging in China, and imports of pig iron are expected to continue or even increase in the future. Beneficiation of coal and iron ore used as inputs will probably provide the greatest blast-furnace production gains in the immediate future.

Scrap-iron input to the blast furnaces is also in short supply because of much longer working life of many steel products and recycling of materials for other purposes. As a result, China began importing scrap iron in the late 1960s. In 1974 scrap imports were 300,000 tons, and it is expected that these will continue and increase when scrap prices are depressed in world markets.

China uses relatively modern blast furnaces based on Soviet designs of the 1950s. Several standard units are built in China, ranging from small furnaces with 25,000 tons annual capacity to large units with over 1,000,000 tons of output. Over two-thirds of Chinese pig-iron capacity is believed to consist of blast furnaces under 500,000 tons capacity and smaller than 1000 m^3 in volume.

China also manufactures very small "native" blast furnaces for use in rural areas. Their capacities range from a few thousand to as high as 10,000 or 20,000 tons of pig iron per year, and they are often built from second-hand materials. Much of the iron produced in them is of low quality and is converted to cast iron for use in simple agricultural implements.

Crude-steel Production

Crude-steel production capacity is believed to be larger than domestic pig-iron and scrap-iron supplies. In 1975 crude steel output was estimated at 26.0 million tons but suffered a setback in 1976 dropping to 23.0 million tons that year. China is the world's fifth largest crude-steel producer, and its production is comparable to that of Italy, France, or The United Kingdom in magnitude (see Table 9.1).

Crowth in crude-steel production was extremely rapid during the late 1950s, increasing from less than 3.0 million tons in 1955 to 18.7 million tons in 1960. The introduction of a large number of small plants since 1958 is estimated to have contributed over 30 percent of low-quality steel to the total industry output. However, many production claims of the Great Leap Forward period have been overstated, and industry observers are uncertain about actual production levels in those years (see Fig. 9.2).

Crude steel production dropped to a low of only 8.0 million tons in 1963 and did not recover to its claimed 1960 level until 1970, when 17.8 million tons was produced. Since 1970 small and medium steel plants once again began playing an increasing role, although they do not at present contribute more than 12 percent of the total crude-steel output.

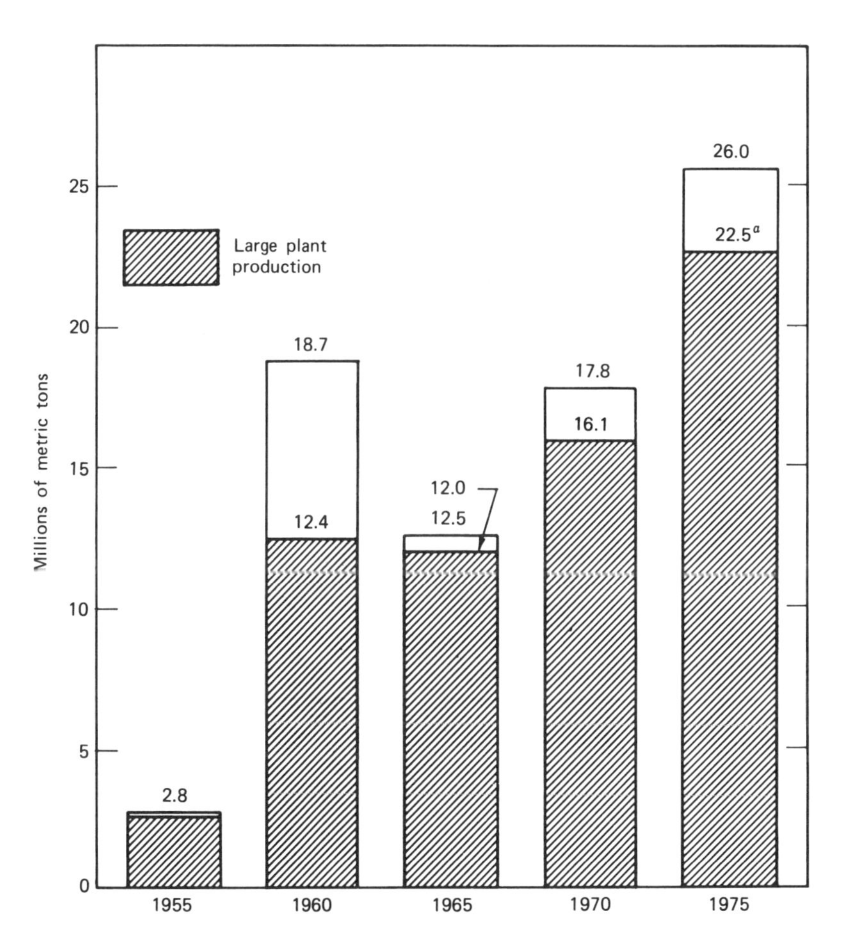

Figure 9.2 ***Growth trends in crude-steel production.*** (Source: ***Based on crude-steel production statistics in "China: A Reassessment of the Economy," Joint Economic Committee, U.S. Congress, July 1975.***) ([a] ***Authors' estimates based on data for 1974.***)

About 60 to 70 percent of total steel-making capacity consists of open-hearth furnaces (OHF). This is particularly suitable for China because this type of furnace readily processes a wide variety of hot metal, cold pig-iron, or scrap-iron charges. Basic oxygen furnaces (BOF) account for 15 to 25 percent of steel-making capacity, and another 10 to 15 percent is provided by side-blown converters (SBC). Only 5 to 10 percent of capacity is believed to consist of electric steel furnaces (ESF). The largest OHF units are 500-ton furnaces, comparable to the largest units in the United States or the Soviet Union. Smaller OHF furnaces of 200-, 50-, and 25-ton capacity are believed to be in general use in most Chinese steel-making plants. China first built its own 30-ton BOF furnace during the mid-1960s, and it has an annual output capacity of about 500,000 tons. Two 55-ton BOF units were also imported in 1965 from Austria. Imports of large air-separation plants, computers, and steel-analysis instruments suggested that China had to obtain its BOF technology from abroad before mastering their production. Many small BOF units of 3 to 10-ton capacity are also built for local use.

Large numbers of SBC units were also built when small plants were being put into operation during 1958–1960. These are mostly obsolete and inefficient and produce only mediocre steel. China also makes its own ESF units, which previously had to be imported. These are also small and inefficient and range from 10 to 20 tons in capacity. Major interest in more advanced steel-making technology now centers on vacuum and electron-beam furnaces.

In recent years China claimed to produce over 1200 kinds of steel. However, it is believed that its capacity to produce large quantities of stainless steels, electric steels, and superalloys is limited. Chromium, nickel, and cobalt must be imported as they do not occur in significant amounts in China. This is clearly shown by growing imports of these strategic metals as well as imports of high-quality steels. Industry observers feel that China will not attain self-sufficiency in specialty steels production without extensive foreign technical assistance.

Finished Steel Production

Although China is the fifth largest finished steel producer in the world, its production is inadequate to meet domestic consumption needs. Finished steel output was estimated at 19.5 million tons in 1975, up 9.5 percent over the 1974 output of 17.8 million tons, but only 2 percent above the high of 19.1 million tons reached in 1973.

China has very few continuous or semicontinuous sheet mills, tinplating, or galvanizing lines because its production of consumer goods that use the products of such mills is limited. Most Chinese steel mills are designed for hot rolling, but there are also extensive foundry and forging facilities. Only a few large blooming, rail, and structural mills are in operation. A considerable number of small and medium-sized mills produce rod, bar, plate, and sheet, as well as welded and seamless tube products. Continuous casting is one variety of steel-finishing technology developed entirely by the Chinese industry.

Some of China's deficiencies in finished-steel production technology are being met by imports of steel finishing turnkey plants from Japan and West Germany. The Wuhan finishing mills now being installed at one of the largest iron and steel plants in China will alleviate the shortage of flat-rolled steel products. The new plants may even eliminate some of the imports in this category. However, growth in demand for flat-rolled steel products is expected to continue to exceed the Chinese capacity to produce such products for some years to come.

Major Iron and Steel Plants

At least ten major iron and steel plants whose output is larger than 1.0 million tons of crude steel per year have been identified in China. At least fifteen more medium-sized plants have been also identified with annual production capacities ranging from 100,000 to 1.0 million tons of crude steel annually (see Table 9.2).

Table 9.2 Major iron and steel plants identified in China during the 1970s

Location and Province	*Production Capacity in Tons/Year*	*Blast Furnaces Installed*	*Number and Type of Steel Furnaces Installed*[a]
Anshan, Liaoning	7,000,000	11	25 OHF, 2 BOF
Shanghai	4,200,000	8	1 OHF, 3 BOF, 1 ESF
Wuhan, Hupeh	2,500,000	3	6 OHF, 1 BOF, 1 ESF
Paotou, Inner Mongolia	1,600,000	3	5 OHF, 2 BOF
Peking	1,600,000	4	1 OHF, 1 BOF
Chungking, Szechwan	1,200,000	3	4 OHF, 1 ESF
Maanshan, Anhwei	1,200,000	13	2 OHF, 21 SBC, 1 ESF
Taiyuan, Shansi	1,200,000	5	2 BOF, 3 OHF, 1 ESF
Canton, Kwangtung	1,000,000	3	1 OHF, 1 BOF
Penhsi, Liaoning	1,000,000	5	1 OHF, 1 ESF
Tangshan, Hopeh	707,000		
Anyang, Honan	500,000		1 BOF

Source: U.S. Congress, Joint Economic Committee, July 1975.

[a] OHF = open hearth furnace, BOF = basic oxygen furnace, ESF = electric steel furnace, SBC = side-blown converter.

Anshan, in Liaoning Province, is the largest integrated iron and steel complex. It operates eleven blast furnaces, and its capacity is estimated at about 7.0 million tons, which is approximately 25 percent of all the crude-steel output in China. This complex operates three steel-making plants as well as several dozen mining, smelting, rolling, power-generation, and transport enterprises. It also has its own oil refinery, an oxygen plant, and gas-transmission system. Anshan produces silicon sheet in coils, seamless and welded tubes, rails, bars, rods, wire, structural steel, and tinplate, as well as hot and cold rolled sheet.

Large iron and steel complexes are also operating in Shanghai, Wuhan, Paotou, Peking, Chingking, Maanshan, Taiyuan, Canton, and Penhsi. Medium-sized plants are located at Fulaerhchi, Tangshan, Tayeh, Anyang, Chianyu, Dairen, Hsiangtan, Hsuanhua, Kunming, Shenyang, and Tientsin, as well as in Fushun, Lienyuan, Sian, and Hainan. Seamless tube is also produced at Anyang, Canton, Chingking and Hopei. Special steels and alloys are reportedly produced at Peking, Sining, Yentai and Hsining, and ferrochrome is produced in Tangshan.

Role of Small Iron-producing Plants

About 20 percent of all the pig iron and 15 percent of crude steel in China is produced at small iron and steel plants. At the height of the Great Leap Forward movement there were an estimated 600,000 backyard furnaces in operation, and in the peak year of 1960 nearly 50 percent of all pig iron and almost 30 percent of crude steel was made by small local plants. The backyard furnaces were quickly abandoned, and small plant production was sharply reduced during the 1960s, but since 1968 some return of small iron and steel plants has been noticeable. By 1974 almost 9 million tons of pig iron was produced in small plants, which is more than the total pig-iron production of India or Australia (see Table 9.1). A typical small plant produces up to 10,000 tons of pig iron per year. It employs relatively primitive blowing equipment, skip hoists, and low-quality furnace linings. Earlier local plants made steel mainly in side-blown converters. Because of difficulties in removing impurities and great variations in grades of locally available iron ore, only low quality steel is produced. Small plants are less efficient and have been estimated to use 0.9 tons of coke per ton of pig iron produced, compared with 0.65 tons of coke required per ton of pig iron made in large modern plants.

Steel produced in small plants is mainly used for local construction and in production of simple farm implements by nearby communes and villages. By supplying such needs, small plant output allows large steel plants to provide higher-grade steel for building more complex machinery in industry. Because of an inadequate and overburdened transportation network, local steel production also plays a role in relieving some transportation delays and bottlenecks. Small iron- and steel-plant output is not efficient, but it is generally achieved at a low cost in labor and materials. It is often forgotten that 85 percent of the Chinese people reside in rural areas, and a great number of workers are always available for other work, particularly during the slack seasons.

For the foreseeable future small iron and steel plants will certainly retain their role as valuable supplements to the modern steel-producing complexes. Because of a need to develop beneficiation of iron ores a new role may be given to those plants that also operate their own mines. It is conceivable that beneficiation and preparation of iron ores could be performed by such local units. Their output could be higher-grade ores and concentrates that could be shipped for processing at larger regional steel-producing plants. Such activity, however, may also depend on a better transportation network than is at present available in many rural areas.

Some small plants situated near large and rich iron-ore deposits will undoubtedly be upgraded. It is estimated that at least forty converters similar to modern BOFs have already been installed in some small plants. At least two are also believed to operate continuous casting equipment. However, most use extremely simple finishing equipment and produce mainly small bars, sheets, light rails, wire, and tubes. As the industry expands and more large modern plants come on stream, many small plants will probably disappear altogether. For the next decade or so they will serve their supplementary function, but most of them do not represent any potential for imports of foreign steel-making equipment.

IMPORTS OF STEEL-MAKING TECHNOLOGY

The Chinese had only negligible modern steel-making experience at the time of communist takeover in 1949. Most facilities were damaged or destroyed, and much of the rolling-mills equipment had been removed by the Soviet Union, which occupied Manchuria after World War II. Most of the plants were of Japanese or western European origin and in 1943, their peak year, produced only 900,000 tons of steel. After 1949 China began rebuilding its steel industry with extensive Soviet assistance and during 1952–1957 the iron and steel industry grew at an annual average rate of 28.7 percent admittedly from a relatively limited base. Crucial to this growth was Soviet assistance in building three major plants for the construction of metallurgical equipment.

The largest plant for building complete sets of steel-rolling and smelting equipment was built at Fulaerhchi in Heilungkiang Province. It was completed in 1959 and produced very large blooming machines for steel rails, seamless tubes, and beams as well as blast furnaces. The Fulaerhchi Iron and Steel Plant was also built with Soviet assistance and became a major manufacturer of high-quality alloy steels using electric furnaces.

Complete sets of iron-smelting and steel-rolling equipment were also put into production at the Loyang Mining Machinery Plant in Honan Province, which was also built with Soviet aid. The Taiyuan Heavy Machinery Plant in Shansi Province was another "backbone" plant built with Soviet assistance to manufacture thin-plate rolling mills with 30,000 to 50,000-ton annual capacities.

The Wuhan and Paotou Iron and Steel Complexes were also originally developed with Soviet assistance. Soviet equipment planned for Wuhan was not all delivered, and this plant is now the site of considerable steel-finishing mill expansion under contracts with Japanese and West German steel-mill equipment suppliers.

By 1965 China began its first round of steel-making technology acquisition in the form of imports of Japanese and European steel-making and steel-finishing equipment. Two 55-ton BOFs were purchased from VOEST in Austria and were installed at the Taiyuan Iron and Steel Plant in 1969. In 1965 China also ordered several steel-rolling mills from West Germany, Austria, Italy, and Japan (see Table 9.3).

Negotiations for large rolling mills for the Wuhan Iron and Steel Complex also began in the mid-1960s, and final contracts were signed in 1974. Between 1972 and 1975 China made a second round of steel-making technology purchases, ordering a total of five rolling mills of different types and one iron works for an estimated total of $635 million. This was the second largest turnkey-plant expenditure after an expenditure of about $900 million on petrochemical and synthetic-fiber plant imports in the same period.

The largest steel-finishing plant ordered by China was the 3,000,000-ton/year capacity steel-sheet and strip mill. This is being provided by a Japanese consortium led by Nippon Steel and includes Mitsubishi Heavy Industries, Ishikawajima-Harima Heavy Industries, Toshiba, Hitachi, Mitsubishi Electric, and ten other Japanese suppliers. The contract also includes a 70,000-ton/year silicon steel-sheet mill and a 100,000-ton/year tinning plant.

About 7 percent of the total contract, or over $15 million, is estimated to be the value of technology transferred to China during this by China since 1965

Year	Type of Plant Imported	Capacity in Metric Tons/Year	Supplier and Country	Contract Value in Millions of $U.S.
1965	Cold-rolling mill		Schloemann, West Germany	17.0
	Two 55-ton BOF furnaces		VOEST, Austria	
	Linz-Donau steel mill	650,000	VOEST, Austria	12.0
	Iron-ore pelletizing plant		Japan	4.0
	Pipe and tube plant		Italy	3.2
	Seamless-tube plant		Innocenti, Italy	3.2
	Wire-drawing plant		Japan	5.0
1968	Seamless-tube plant	40,000	Mannesman AG, West Germany	11.0
1972	Iron works		Hitachi, Japan	
	Continuous-slab rolling mill		Sumitomo, Japan	5.24
	Small-rolling mill		DEMAG AG, West Germany	80.0
1974	One 3000-ton forging press		SHI, Japan	
	Hot-rolled continuous-sheet mill	3,000,000	Nippon Steel et al., Japan	228.50
	Cold-rolling mill	1,000,000	DEMAG AG, West Germany	198.0
	Water-recirculation plant		Kurita, Japan	65.0
	Continuous casting mill	1,500,000	DEMAG AG, West Germany	58.0
1975	Two ladle degassing units		ASEA, Sweden	1.6
	Oxygen plant	35,000[a]	Japan Oxygen Company, Japan	12.8
	Isostatic press powder metallurgy		ASEA, Sweden	0.85
1976	Lime-burning system		Imperial-Krauss-Maffei, West Germany	5.0
1977	Hot-transfer forging press, 6000-ton		SHI, Japan	
	Steel foundry		Otto Wolff, West Germany	10.0

Source: Compiled by 21st Century Research from various articles and publications listed in references to this chapter.

[a] In cubic meters of oxygen released per hour.

project. Under the terms of the contract, 10 percent was paid as deposit and 20 percent was due on shipment of equipment.

Another 10 percent was to be paid when trial operations began, and 60 percent was arranged to be paid on a deferred-payments basis. The contract also stipulated training for over 200 Chinese technicians at three major Nippon Steel plants in Japan and provided for 350 Japanese experts to supervise plant construction and start-up in China.

Contracts for 1,000,000-ton/year cold-rolling sheet-steel mill and a 1,500,000-ton/year continuous-casting mill were also awarded to a western European consortium in 1974. It is headed by the West German companies DEMAG AG and Schloemann-Siemag AG, who are jointly responsible for the project. Valued at a total $256 million, those projects are being fulfilled by at least twenty firms from West Germany, Belgium, Switzerland, and Austria. These include such well-known companies as ACEC, AEG, August Thyssen-Huette, Brown-Boveri, Mannesmann, MAN, Siemens, and Otto Wolff.

The advantage of this particular consortium apparently lies in successful completion of a similar project for Venezuela that is basically being reproduced for the Wuhan Iron and Steel Complex. A computerized project-design system, coupled with a photoassembly information concept, is being used to help in construction of this plant, which was scheduled to begin in February 1977. Under the terms of this contract, China was to transfer 90 percent of the purchase price in German marks to the Deutsche Bank in Duisburg, the headquarters town of the DEMAG AG company. Initially a down payment of 10 percent was released, with another 10 percent due after 10 months of work. A further 70 percent of contract value was staggered from the 13th month to the 33rd month after the contract was signed. A final 10 percent is to be paid in two installments after performance tests and materials guarantees in 1977, with the last payment due in March 1979.

The contract with the western European consortium also provided for training of about 300 Chinese technicians for up to 15 months in Germany, and up to 150 German specialists were allowed to travel, and live in China. The German workers in China were accommodated in a specially built air-conditioned hotel designed along Western lines, and schooling facilities were arranged for their families.

On the assumption that China intends to control its rapidly growing imports of steel products, additional purchases of steel-finishing plants and process-control automation may be expected in the future. As China's steel industry develops to depend on larger production units such turnkey plants, imports are again likely to come about whenever iron ore, crude-steel output, and steel-finishing capacity get out of balance.

IMPORTS OF STEEL PRODUCTS

Iron and steel products account for over 20 percent of all Chinese imports and are the largest single import category in Chinese foreign trade. Imports of steel increased almost tenfold from $153 million in 1965 to $1,445 million in 1976. Between 1970 and 1975 over $5,100 million-worth of steel products were imported into China from over twenty countries (see Fig. 9.3). In 1976 a total of 4.2 million metric tons of finished steel was imported, more than in any previous years so far.

Japan is by far the largest supplier of iron and steel products, accounting for about 60 percent of the total in 1975. Exports to China also

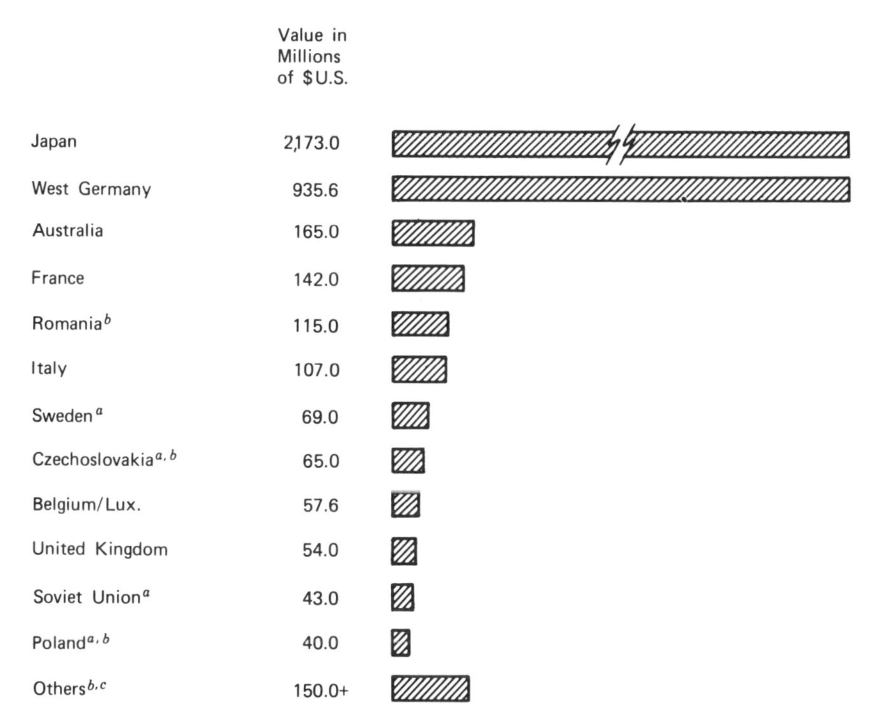

***Figure 9.3** Relative market shares of iron- and steel-product supplier countries identified during 1970–1975.* (Source: *OECD Commodity Trade Statistics Series B for SITC code 67 or equivalent and foreign-trade yearbooks of individual countries.*) [a] *Data for 1970 not included* [b] *Amounts estimated from volumes shipped and average prices.* [c] *Other suppliers include Brazil, United States (16.6), Finland (5.4), Hungary (2.3), Canada (1.1), Norway and Yugoslavia (35.0), Spain (31.7), Netherlands (29.0), Austria (26.5). Some of these values are estimates derived from volumes and prices known at the time.*

account for almost 10 percent of all Japanese steel exports. Over 40 percent of all the steel products imported into China during 1970–1975 came from Japan. But in 1976 China reduced its steel imports from Japan to 2.25 million tons, down considerably from 3 to 4-ton levels in previous years. However, by 1977 business reports indicated that China was again planning to buy 1.6 million tons of Japanese steel in the first half of that year, which would suggest a return to higher import levels.

Nippon Steel is the single largest Japanese steel supplier, and the Japan Iron and Steel Federation is active in promoting Japanese steel plants and technology sales to China. In a typical semiannual 1975 agreement, Nippon Steel and five other Japanese steel producers agreed to export 1.5 million tons of steel products during April to September of 1975, valued at $400 million at an average price of $267 per ton. For the first time, payments were to be settled on a $U.S. basis instead of the RMB-denominated yen-payment basis as before. The deferred-payment period was extended to 10 to 12 months from the previous 150 days for such transactions.

Because Japan accounts for almost two-thirds of all Chinese foreign steel purchases, an analysis of those imports is a good indication of Chinese demand for specific types of steel products. In terms of volume steel sheet, ingots, plates, tubes, and wire rods account for over 86 percent of all the steel imports.

In terms of cost during 1970–1975, China spent about 50 percent over $5.1 billion on universals, plates, and sheet, with at least half of this amount going to purchase plate and sheet less than 3 mm thick. Almost 20 percent of the total was spent for tubes and pipes, most of which are seamless tubes believed to be in great demand by the petroleum, petrochemical, and chemical industries. Bars, rods, angles, and shapes accounted for about 15 percent of the total import value.

West Germany is the second largest supplier of steel products, with its steel exports to China rising rapidly from just over $60 million a year in 1970 to over $300 million in 1975. However, in contrast to Japan, almost 70 percent of German steel imports are large-diameter seamless pipe, with some plate and sheet imports as well. Although German steel exports reached 455,000 metric tons in 1975, this volume represents only about 3 percent of all the German steel exports (see Table 9.4).

Until 1973, the last year for which Romanian foreign trade statistics are available, Romania was the third largest supplier of steel products to China. During 1970 to 1975 Romania was probably the fifth largest supplier after France and Australia when the total value of all iron and steel products are taken into account. In 1973 over 40 percent of Romanian steel products imported by China consisted of 44,200 tons of steel tubes,

Table 9.4 Relative volumes of imports of typical steel mill products from major supplier countries during 1975 in thousands of metric tons

Supplier Country	*Steel-products Exports to China*	*Ingots and Semis*	*Rail-road Steels*	*Sections*		*Wire Rods*	*Strip*	*Plates*	*Sheet*	*Steel Tubes*	*Wire*	*Tin Plate*	*Wheels, Tyres, Axles*
				Heavy	*Light*								
Japan	2830.2	539.7	6.6	36.6	154.0	407.8	75.1	489.2	675.7	334.8	11.1	79.9	19.6
West Germany	455.9	0.01	—	—	13.8	—	28.0	45.1	33.9	307.5	1.24	26.9	—
Romania[a]	101.7									44.2			
Czechoslovakia	94.2	—	—	—	—	32.0	—	15.9	31.2	15.1	—	—	—
Italy	88.1	16.1	—	0.0	1.3	0.0	6.1	2.7	36.5	25.4	0.1	—	—
France	85.9	—	—	—	3.3	36.0	—	1.4	23.6	11.7	0.2	9.7	—
Australia[b]	71.5	—	—	—	—	20.3	—	—	28.1	23.0	—	—	—
Belgium/ Luxembourg	65.0	—	—	—	—	16.0	9.0	1.0	19.0	—	2.0	18.0	—
Spain	59.4	1.8	—	—	32.5	0.5	2.4	—	13.1	9.1	—	—	—
Poland	41.4	—	—	0.5	9.6	5.7	—	25.6	—	—	—	—	—
Soviet Union	35.8	—	—	—	—	—	—	20.3	5.3	10.2	—	—	—
United Kingdom	19.7	—	—	—	1.1	10.6	0.4	—	2.1	4.6	0.8	0.1	—
Netherlands	15.6	—	0.29	—	—	—	0.01	2.9	0.0	3.1	0.0	9.7	—
Finland	10.3	—	—	—	—	—	—	7.1	3.2	0.0	—	—	—
Hungary	9.0	—	—	1.0	0.0	—	—	—	1.0	7.0	—	—	0.0
Sweden	7.3	—	—	—	1.3	—	0.06	3.8	0.0	1.45	0.67	—	—
United States	6.54	—	—	—	—	—	0.0	[c]	1.2	5.3	0.01	—	—
Yugoslavia	4.3	—	—	—	4.1	—	—	—	—	0.2	—	—	—
Austria	1.53	0.03	—	—	1.2	—	0.06	0.12	0.03	0.09	—	—	—
Denmark	0.11	—	—	0.04	—	—	—	—	—	0.07	—	—	—

Source: *United Nations Statistics of World Trade in Steel, 1976.*

[a] Data for 1973 available from *Romanian Foreign Trade Yearbook.*
[b] Data for Australia available to June 1974 only.
[c] Shipments of plates included in figure showing shipments of sheet.

which makes Romania the third largest supplier of steel tubes after Japan and Germany.

Australia is the third largest iron and steel supplier to China. Its exports were valued at $165 million for the 5-year period of 1970–1975, accounting for only 3 percent of all Chinese imports in this category. However, Australia exports a rather different product mix than Japan or West Germany. The Australian Broken Hill Proprietary (BHP) Company is the major supplier of iron ore, and pig iron, as well as some finished steel products to China. China imports about 3 percent of its total iron-ore supply, and over 90 percent of it comes from Australia. A large portion of Chinese pig-iron imports, estimated at 800,000 tons in 1974, also comes from Australia. In the early 1970s about 50 percent of the BHP Kwinana Plant of 800,000-ton annual capacity was exported to China, and in 1977 Australia again sold 360,000 tons of pig iron to China valued at $40 million. In terms of value, Australian exports to China represent about 4.5 percent of all Australian exports in this category.

France and Italy are the only other iron and steel suppliers who shipped over $100 million worth of steel products to China during 1970–1975. Most of their exports also consisted of steel sheet, tubes, and wire rods. Exports to China account for only 1.0 to 1.5 percent of all steel exports of those countries (see Table 9.4).

Sweden, Czechoslovakia, Belgium/Luxembourg, the United Kingdom, and the Soviet Union are the next five largest suppliers of steel to China, but their combined shipments during 1970–1975 were about equal to those of West Germany in 1975 alone. Those countries, as well as Poland, Yugoslavia, Spain, Netherlands, and Austria, are relatively minor steel exporters to China. Their sales to China represent only 1 to 3 percent of their total steel exports in most cases and ranged from $5 to $15 million in 1975. United States became a major steel-scrap supplier to China in 1975, accounting for 60 percent of all China's scrap imports in recent years, and shipped a total of $15 million in that year.

Despite large imports, China has also exported 300,000 to 400,000 tons of finished steel annually in recent years. These exports account for only 1.5 percent of Chinese output and have been primarily made to North Korea, Vietnam, Albania, and some developing countries probably within the framework of various Chinese foreign-aid programs.

Future imports of finished-steel products are contingent on China's ability to increase its domestic production, which suffered a setback in 1976. China's apparent consumption in 1975 was over 26 million tons, of which almost 4.0 million was imported. If this consumption level continues to increase at 8 percent until 1980 and even slows down to 7 percent during 1980–1985, China's demand for finished steel will be nearing 38 million metric tons in 1980 and about 53 million metric tons in 1985.

In 1977 China's total finished-steel production capacity was probably about 28 million tons, including the new 3 million tons of imported hot-sheet mill capacity being installed in Wuhan. If this capacity increases at growth rates comparable to those of demand for finished steel, China would have to continue importing at least 3 million tons in 1980 and 4 million tons in 1985. Failure of Chinese steel production to keep up with this growth rate could increase future steel imports to 5 or 6 million tons per year. Because of a very large investment in time and money required to increase finished-steel production to such levels, there is a strong probability that steel imports will both continue and increase in the future.

COPPER PRODUCTION AND IMPORTS

Chinese copper reserves are believed to be larger than those of Chile, which has the third largest mine copper output in the world after the United States and the Soviet Union. Domestic copper deposits are believed to be of poor quality, however, and are located in isolated areas; hence it is probably cheaper for China to import copper concentrates and copper metal than to expand domestic production.

Chinese production is estimated at 150,000 metric tons of refined copper per year, but actual demand is believed to be two or three times this amount and was estimated at 300,000 tons in 1975. Chinese smelter output and refined copper production is about one tenth that of the United States and is comparable to the copper production of Spain (see Table 9.5).

Chinese mine copper output is estimated at about 100,000 metric tons from low-grade ores, and together with an organized effort to collect scrap on a national basis, it provides input for Chinese smelters. Most copper mines in China are relatively small, with an estimated maximum output of 15,000 metric tons of mine copper per year. The Hungtoushan Copper Mine near Fushun in Liaoning Province is estimated to produce about 12,000 tons of mine copper per year using 1.2 percent copper ore. Another copper mine at Tungkuanshan in Anhwei Province was estimated to produce about 5000 to 10,000 tons of copper in a year. Over twenty copper mines and copper-ore deposits have been identified in several provinces, but details of their output are unknown.

The largest copper smelter known to be operating in China is in Shanghai and processes mainly imported blister copper and scrap. It is believed to have an output of over 10,000 tons of refined copper per year. Also in Shanghai are three copper mills manufacturing copper tubes, copper wire, and other copper products.

The second largest copper smelter is believed to operate in Shen-

Table 9.5 Production of mine copper, smelter, and refined copper output in selected countries during 1975 (All figures in thousands of metric tons of copper)

Country	*Mine Copper*	*Smelter Output*	*Refined Copper*
World	7300	7350	8220
United States	1282	1312	1620
Soviet Union	1100	1100	1420
Japan	85	821	818
Chile	931	724	535
Zambia	806	640	619
Canada	724	496	529
Zaire	499	462	304
Peru	173	165	53
Spain	19.4	116	133
China	100	100	150
Mexico	78	76	73
India	23.8	16.3	16.3
Brazil	3.0	2.5	1.7
United Kingdom	—	—	151
Belgium	—	—	357
Philippines	225	—	—
New Guinea	172	—	—

Source: *United Nations Statistical Yearbook, 1976.*

yang in Liaoning Province and has a capacity reported as several tens of thousands of tons, which may not be all copper-refining capacity. It uses copper from the Hungtoushan mine not too far away, as well as copper scrap. Another ten small copper refineries and scrap-processing plants have been identified, with a few thousand tons of output each. Furukawa Mining Company of Japan has been reported negotiating with China to build flash copper smelters in China, but no contracts have been announced to date.

Imports of Copper

Copper is the single most important nonferrous metal imported by China. Most of it is in the form of refined copper, but since 1975 small amounts of concentrates have been imported from the Philippines and New Guinea. During 1970 to 1975 almost $1.0 billion was spent by China on copper imports. Since 1970 copper accounted for over 50 percent of all nonferrous metal imports. However, although imports of copper are a major position in Chinese foreign trade, even during the peak year of 1973, when 170,000 tons were imported, this amounted to only 5 percent of the total world copper exports; thus the Chinese copper trade has relatively little effect on world's copper market. In 1976 China imported about 150,000 tons of copper, up from 120,000 tons in 1975.

China used to purchase copper through the London Metals Exchange (LME), but during the 1970s it entered into agreements with several copper-producing countries whose copper industries are government-controlled. Chile, Peru, and Zambia became major copper suppliers to China, accounting for 90,000 tons every year. Other countries like Japan, France, West Germany, and Yugoslavia are additional suppliers, but in 1973 China again purchased large quantities of copper through the LME (see Fig. 9.4).

If Chinese copper consumption continues to increase at a rate of 5 percent every year, it would be in the order of 500,000 tons per year in

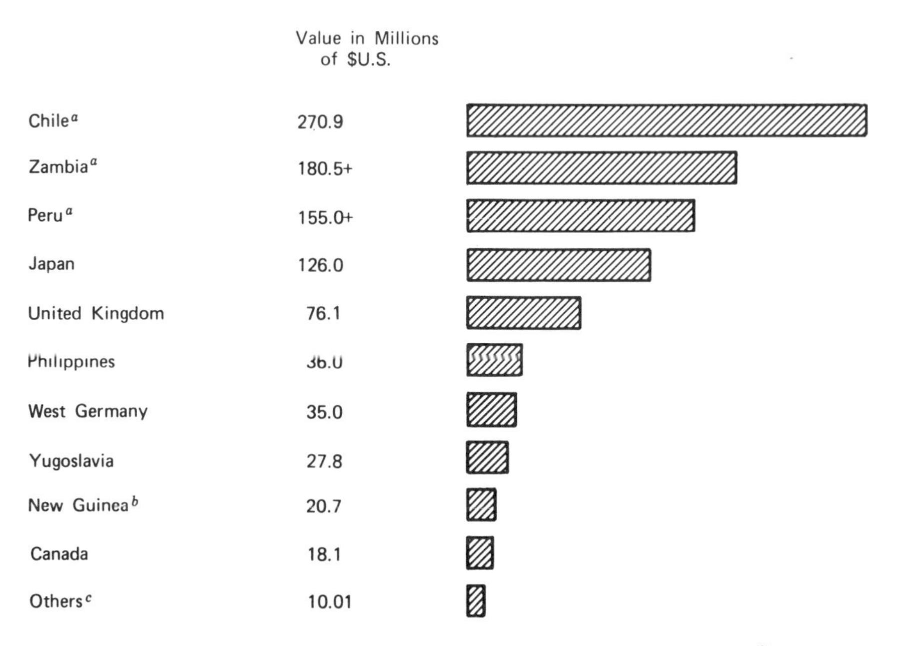

***Figure 9.4** Estimates of relative market shares of copper suppliers to China identified during 1970–1975.* (Source: *OECD Commodity Trade Statistics and individual country reports for SITC code 682.*) [a] *Data for 1970–1974 only.* [b] *Data derived from shipments and average prices.* [c] *Others include Australia (4.0), Belgium (0.1), Finland (0.4), France (1.9), Sweden (2.6), United States (0.01), Italy (1.0).*

1985, which is probably higher than the industry could absorb at that time. However, considering this as an upper limit of possible consumption, it is unlikely that Chinese copper production will more than double in the same period, which will give it an output of about 300,000 tons per year by 1985. This would leave a shortage equal in amount to recent annual imports. Some industry observers believe that China may have been stockpiling copper for future use not knowing to what degree it will be able to expand its domestic production.

It should, therefore, be expected that China will continue to develop its copper trade with a country like the Philippines because opportunities for barter agreements exist in exchange of oil for copper. In mid-1977 China agreed to purchase at least 40,000 tons of copper concentrates from the Philippines for about $12 million, and Chinese sales of oil to that country are expected to reach 900,000 tons during the same year.

Peru, Chile, and Zambia, though major copper exporters to China, are not in the same trade partner category because their copper industries are controlled by their governments and their economies are relatively small consumers of imported oil.

The Chilean government agreed to export 65,000 tons of copper per year in the early 1970s; however, although the agreement is believed to have lapsed since deposition of President Allende, Chile continues to be a major copper supplier though not exporting as much as originally planned.

Peru had an agreement with China to export about 40,000 tons of copper annually in return for Chinese assistance to develop the new Tintaya copper deposits. If China agreed to such a transaction, this clearly shows the political nature of its copper trade because it should have applied the same resources to development of its own copper deposits in line with the self-sufficiency policies it espouses. Now apparently China entered into a long-term political agreement with Peru, where the latter country is to export 18 percent of its refined copper as well as up to 25,000 tons of blister copper in return for rice, wheat, and other commodities from China.

Zambian copper imports are reportedly tied to the $400-million interest-free loan for the construction of the TAZARA railroad from Dar Es Salaam in Tanzania to the Zambian copper areas. A long-term collateral contract is reported in existence that provides for export of 50,000 tons of copper annually to China. Actual shipments from Zambia in recent years have been considerably smaller.

China may have stockpiled copper in recent years also in anticipation of high copper prices in 1974, which have since fallen cansiderably. If total imports during recent years added considerably to those stockpiles, future imports may decline. In any event, the various arrangements China has with the various copper supplier countries give it considerable flexibility in the copper market.

ALUMINUM PRODUCTION AND IMPORTS

China is known to possess huge deposits of aluminous ores, although their quality is very low. Total aluminum resources have been estimated to be in the order of 1 billion tons, but exploitable reserves are believed to be about 100 million tons. China's bauxite production is estimated to be about the same magnitude as that of either France or Hungary, but its primary aluminum production is only 320,000 tons (see Table 9.6).

Table 9.6 Production of primary aluminum and bauxite in selected countries in 1975 (All figures in metric tons)

Country	*Primary Aluminum Production*	*Bauxite Production*
United States	3,519,000	1,830,000
Soviet Union	2,450,000	6,600,000
Japan	1,016,000	—
West Germany	678,000	1,000
Norway	590,000	—
France	383,000	2,950,000
China	320,000	2,560,000[a]
United Kingdom	308,000	—
Australia	223,000	20,320,000
Spain	210,000	9,000
Romania	204,000	350,000
India	167,000	1,273,000
Brazil	117,000	1,277,000
Hungary	70,000	2,890,000

Source: United Nations Statistical Yearbook 1976; U.S. Government Handbook of Economic Statistics 1976.

[a] Data for 1974.

To exploit Chinese bauxite deposits a special technology had to be developed with the assistance of the Soviet Union. This technology requires very large quantities of electric power, causing Chinese aluminum to be produced at a relatively high cost. Nonferrous metallurgy was estimated to consume almost 11 percent of electric power generated in China in recent years, and most of this amount was probably used in production of aluminum (see Fig. 8.1).

About twenty-five bauxite mines, alumina plants, and aluminum refineries have been identified in China in several provinces. The largest aluminum plant is at Fushun in Liaoning Province, and its output is estimated to be 100,000 metric tons per year. Two 40,000 ton/year plants are believed to be under construction at Kweiyang in Kweichow Province and at Sian in Shensi Province. A large plant with 20,000 to 30,000-ton annual capacity was also reported in operation near Lanchou in Kansu Province, presumably taking advantage of the large electric power plant from the hydroelectric scheme of San Men Gorge on the Yellow River.

China is probably hindered in expanding its domestic aluminum production by a shortage of electric power. As a result, imports of aluminum have increased very rapidly from only 5000 tons in 1965 to about 100,000 tons per year by 1974. In 1975 imports skyrocketed to about 400,000 tons, valued at a total of $270 million. At this level China imported about 10 percent of world aluminum exports. However, such large imports are believed to be temporary, in order to take advantage of low prices and an abundance of aluminum on world markets. Long-term aluminum imports are expected to continue at about 100,000 tons/year levels, although in 1976 imports were estimated at 200,000 tons.

Although Norway exported the largest amount of aluminum to China during 1970 to 1975, France, the United States, Hungary, Japan, and Canada are also supplying aluminum at comparable levels. A total of about twenty countries, including Iceland, have been exporting aluminum to China in recent years. This policy is also expected to continue without a single country becoming a dominant supplier (see Fig. 9.5). In 1976 the United States, Canada, and West Germany were the leading suppliers.

Most recent sales were reported in 1977 by Mitsubishi Light Metal Company of Japan, which sold 50,000 tons of aluminum ingots at about $1,030 per ton of an FOB basis. Icelandic Aluminum Company also sold 3000 tons in 1977 for a total of $2.9 million. Mitsui Aluminum, Sumitomo Chemical, and Alcan Aluminum Company were significant suppliers in 1976.

With a continuing industrial growth aluminum consumption is bound to increase, probably to about 600,000 tons per year by 1985. Until significant electric power capacity is developed in China or the aluminum-refining process is made more efficient, it is most likely that imports will both continue and increase in the future.

CHROMIUM, COBALT, AND NICKEL DEFICIENCY

Although China is well endowed with most minerals, it is specifically deficient in three strategic metals, chromium, cobalt, and nickel. Deposits

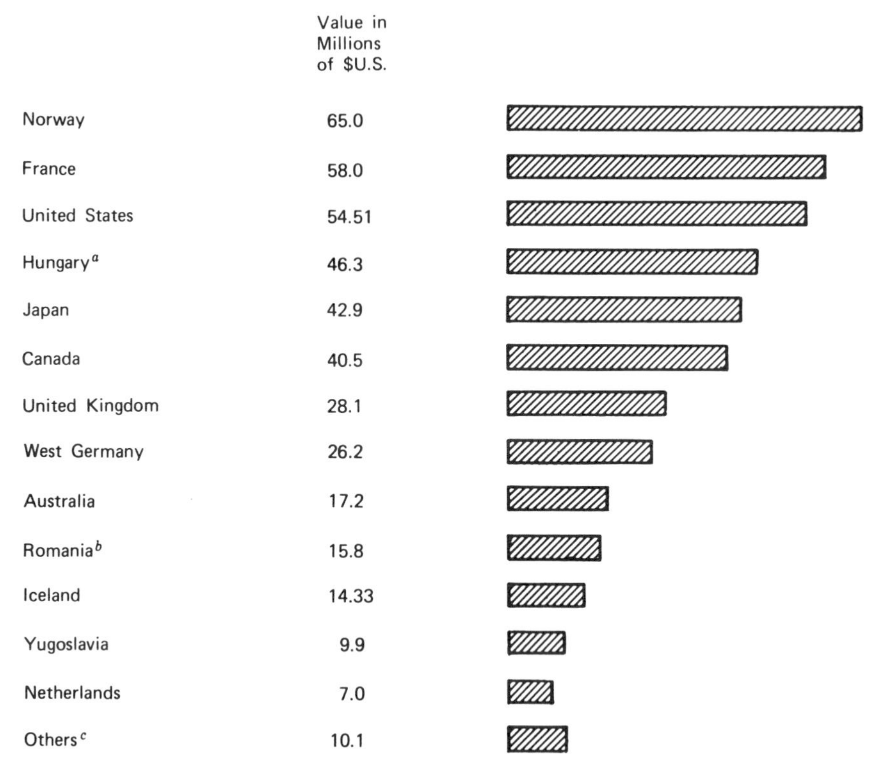

***Figure 9.5** Estimates of relative market shares of aluminum suppliers to China identified during 1970–1975.* (Source: *OECD Commodity Trade Statistics Series B for SITC code 684 and foreign-trade yearbooks of several countries.*) [a] *Data for total minerals and metals exports to China until 1974.* [b] *Data through 1973 derived from volume of aluminum shipments to China.* [c] *Includes Austria (2.3), Finland (0.2), Italy (0.7), Sweden (3.5), Switzerland (2.3), Soviet Union (1.1).*

of all those important ferroalloy metal ores are negligible, and Chinese geological research sources have not been expressing any hopes for significant discoveries of these metals in the future. Thus China is totally dependent on imports of chromite and cobalt concentrates, cobalt, and nickel metals. Unless significant domestic deposits of those metals are discovered, imports will continue to increase, probably in some proportion to the growth of Chinese steel production (see Table 9.7).

Viewing this situation in a different way, it would appear that exporters of equipment and products made of high-quality steels may hold an advantage over suppliers of other products. A selling argument may

Table 9.7 Growth in imports of strategic metals

Metal	*Unit of Measure*	*1957*	*1965*	*1970*	*1971*	*1972*	*1973*	*1974*	*1975*
Chromium	1000 Metric tons concentrate	20	90	140	160	240	250	220	NA
Cobalt	Metric tons of metal content	50	500	250	100	900	900	1400	NA
Nickel	1000 Metric tons	2	8	8	11	15	29	29	2

Source: *U.S.–China Business Review*, July–August 1976.

be developed around the fact that valuable strategic metals may be saved by such imports.

Chromium

Chromium is a component of all stainless steels that must contain at least 11 percent of this metal to prevent rusting. The growing Chinese production of stainless-steel products for expansion of chemical, petroleum, food processing, machine-building industries is bound to increase demand for imports of chromite the chromium ore. Lack of chromium can partially be offset by imports of high-quality steel products for final manufacture of stainless-steel equipment. This probably explains in part the high proportion of stainless-steel products that makes up China's steel imports every year.

Steel production is the largest consumer of chromium, but the metal is also used by the refractory and chemical industries. Typically for the United States in 1975, about 60 percent of the chromium demand was in metallurgy, 20 percent in refractories, and 16 percent in the chemical industry.

Chromium is mined and traded in the form of chromite, and the Soviet Union is the largest chromite producer in the world. South Africa, Albania, and Southern Rhodesia are the next three largest producers. The United States, western and eastern Europe, and to some degree Japan all must import chromium ores or chromium to maintain their high-quality steel production. Hence chromium is regarded as a highly strategic material.

China imported chromite and ferroalloys from the Soviet Union during the 1950s, but some industry observers believe that the Soviets stopped shipping chromium to China in the early 1960s. In 1962 only 2000 metric tons of ferroalloys moved from the Soviet Union to China, and shipments further dropped to 1000 tons in 1963 and 600 tons in 1964. Chromite became scarce in China, and magnesia–alumina brick was widely substituted for refractory chromite.

At that time Albania became the principal chromite supplier to China and shipped 66,000 tons of the ore in 1963. In recent years Albania is believed to account for two-thirds of all Chinese chromite imports, estimated at 220,000 tons of concentrate in 1974. Thus Albania is probably exporting about 30 percent of its chromite production to China every year. Iran and Turkey are other suppliers of chromite to China, and the total value of chromite imports in 1974 were estimated at $9.5 million.

There have been also reports that China purchased chromite from Southern Rhodesia when U.N. sanctions were invoked against that country in the 1960s. This is impossible to confirm because neither China nor Rhodesia have provided much foreign-trade data in recent years.

However, in 1962 China bought $1.5 million worth of South African products and increased its purchases to $6 million in the first 8 months of 1963. Whether these transactions included chromium is not known, but by dealing with South Africa China exposed itself to accusations by the Soviet Union of acting against the interests of black African states. It is difficult to believe that China would risk such denunciations unless some strategic materials such as chromium or uranium were involved.

Because of a changing situation in the Balkans, China can no more consider its Albanian source of chromium particularly secure in the future. Possible pro-Soviet changes in Yugoslavia after the death of Tito and the unpredictable effect this would have on the attitudes and policies of the Albanian government must now be taken into account, which can only mean that China should be receptive to alternative supply sources.

It is interesting to note here that during the 1960s China engaged in major foreign-aid project programs that led to the undertaking of the TAZARA railroad construction. This was the largest of Chinese foreign-aid programs and involved building of a railroad from Dar Es Salaam in Tanzania to the Zambian copper mining areas. By linkage with Zambian lines, this railroad also provided a connection to Rhodesia and South Africa, both of which are leading producers of chromium in the world.

Politically, China cannot engage in trade with South Africa because it is a champion of black African liberation movements throughout that continent. But in the long run China may have envisaged black majority rule in Rhodesia, and this could have decided it to invest in the TANZAM railway project. The railroad in fact provides access to alternative sources of chromium, cobalt, and copper through relatively friendly territories of Zambia and Tanzania, both of which received considerable Chinese aid in the past.

Because of its distaste for the necessity to rely on foreign sources of a strategic commodity such as chromium, China will most likely continue to develop additional sources of supply to lessen the dependence on any single country to a minimum. One possible new source is the Philippines, which can supply chromite as well as nickel and copper in return for Chinese oil. Madagascar, a minor chromite producer, is receiving some Chinese foreign aid. India, a substantial chromite producer who recently reestablished trade with China, is another potential partner for barter trade, particularly in return for petroleum products.

Cobalt

Cobalt is of importance in high-quality steel production and is used to produce magnetic alloys and hard alloys resistant to abrasion, corrosion, and extremely hot environments. Zaire accounts for about two-thirds of the world supply, and Zambia is the second largest producer of cobalt in the world. The Soviet Union, Cuba, Morocco, and Canada are the other significant producers.

By 1963 China was importing only 15 tons of cobalt from the Soviet Union, although its total demand is believed to have been several hundred tons. In recent years China has imported cobalt concentrate from Morocco and cobalt metal from Belgium, amounting to about 1400 tons of cobalt metal content.

In 1973, 200 tons of cobalt metal was imported from Zambia. As in the case of chromium, China will probably continue to develop all available sources of cobalt to ensure continuing and increasing supplies of this strategic metal. Total value of cobalt imports, however, is relatively small, and in 1974 imports reached $9.6 million.

Nickel

The most costly strategic metal imported by China is nickel. There is a small nickel mine at Panshih in Kirin Province, but its output is totally inadequate. In 1974, 29,000 tons of nickel were imported, and China spent $105.2 million on nickel imports mainly from Canada, which became the largest nickel supplier to China in 1972. In 1975, however, nickel imports fell drastically to only 2000 metric tons from an all time high of 29,000 in 1974 and 1973. This may be explained partially by a rise in nickel prices which peaked in 1976. China may have stockpiled nickel in previous years in sufficient quantities for the short term and backed off from imports as prices rose. There is also always the possibility that nickel imports from the Soviet Union may have been resumed because about 15% of Soviet exports to China amounting to about $35 million in 1976 remains unspecified in Soviet foreign trade statistics. Another possibility is new discovery and expansion of nickel production in China but there is no evidence that this has taken place.

Except for chromium, nickel is the most commonly used ingredient in the production of steel alloys and was in fact used as an alloying ingredient by the Chinese as long ago as the ninth century. As with chromium, the Soviet Union is also the largest producer of nickel in the world, followed by Canada, Japan, and New Caledonia. During the early 1960s China still imported over 1000 tons of nickel per year from the Soviet Union, but for more recent years Soviet trade statistics do not report nickel trade.

When a nickel shortage developed in China in the 1960s new sources of supply were developed. A $20-million contract was signed with the Societé Le Nickel of France in 1965 for a supply of 9300 tons of nickel from New Caledonian mines over a 4-year period. By 1970 Canada, Cuba, Albania, Japan, and the Netherlands sold nickel to China. In 1972 Norway also became a supplier, but the Canadian firms INCO and Falconbridge are now believed to be the principal suppliers to China.

As it does with chromium and cobalt, China must keep alternative sources of nickel available to feel secure. Thus expanding trade with the Philippines may be of significance because a new nickel mine and smelter there went on stream in 1975. Sino-Phillipine trade reached $90.6 million in 1976, and the 1977 trade agreement between the two countries indicates that the Phillipines will buy 900,000 tons of oil as well as Chinese machinery, equipment, and chemicals.

LEAD AND ZINC PRODUCTION AND IMPORTS

With the growing Chinese automotive industry and increasing pace of industrial construction, the demand for lead and zinc is likely to increase. Although China is a producer of lead and zinc, it also imports both metals in varying quantities from $10 to $30 million per year.

Almost thirty lead and zinc facilities, including mines, smelters, and refining plants, have been identified in China. The largest smelter is reported at Shaokuan, which is north of Canton in Kwangtung Province. Its capacity was originally designed to produce 18,000 tons of lead and 35,000 tons of zinc based on the Imperial smelting process. This capacity is believed to have been expanded significantly.

Chinese smelters are regarded as small in comparison to world standards. At Taoling in Hunan outputs of 10,000 tons of lead and 15,000 tons of zinc have been reported. At Shanghai only 7000 tons of refined lead have been produced annually. However, China may have opted for development of many regional smelters to satisfy local demand rather than invest in major central smelting facilities, which would be more visible. This policy appears to be effective in catching up with the heavy demand, which resulted in shortages and larger imports of those metals in earlier years.

China's refined lead production was estimated at 132,000 tons in 1975, making China the tenth largest lead producer in the world. Because China imported 20,000 to 35,000 tons of lead per year in the early 1970s, it is believed that its consumption demand is in the order of about 165,000 tons per year and growing (see Table 9.8).

Table 9.8 Production and consumption of lead and zinc in selected countries for 1975 (All data in thousands of metric tons)

Country	*Refined-lead Production*	*Refined-lead Consumption*	*Slab-zinc Production*	*Zinc Consumption*
Soviet Union	600	599	960	740
Japan	194	171	701	432
Canada	171	47	427	122
United States	691	661	364	694
North Korea	60	NA	345	NA
China	132	165	315	170
West Germany	236	158	270	227
Australia	171	97	182	65
Mexico	164	93	132	60
Brazil	60	55	24	80
India	5	44	25	70

Source: American Metals Market 1976; United Nations Statistical Yearbook, 1976.

Major lead suppliers to China are Peru and the United Kingdom. Peru by contract agreed to supply 5000 tons of lead per year, but in 1974 imports from Peru were as high as 13,000 tons, accounting for at least 10 percent of Peru's exports of lead. Other occasional suppliers of lead include Burma, North Korea, Japan, and West Germany.

Zinc production was estimated at 315,000 tons of slab zinc in 1975, making China the sixth largest zinc producer in the world. However, its consumption of their metal is estimated at only 170,000 tons per year. Zinc is the third most important nonferrous metal after aluminum and copper. Anticipating considerable future demand by its growing industries, China is probably stockpiling zinc for future use. Some zinc is still imported, probably as a result of contracts previously signed with Peru and aid repayments by North Korea. But Chinese zinc exports have been also slowly increasing from negligible amounts in 1965 to 2700 tons in 1975. It is believed that China could become a net exporter of zinc in the future if worldwide demand increases and prices are attractive enough.

PRODUCTION AND TRADE IN OTHER METALS

In comparison with iron and steel, copper, aluminum, and strategic metals, as well as lead and zinc, production of other metals represents relatively minor industries. However, China is a major producer of antimony, mercury, tin, and tungsten, accounting for more than 10 percent of world production in each of those metals (see Table 9.9).

Table 9.9 Production and trade in selected metals in China circa 1975

Metal	*Size of Deposits*	*Quality*	*Estimated Production in Metric Tons*	*Percent of World Output*	*Annual Trade in Millions of $U.S.*
Antimony	huge	excellent	12,000	17	15.2 exports
Bismuth	huge	good	275	6	negligible
Cadmium	unknown	unknown	96[a]	2	negligible
Gallium	large	unknown	NA	NA	0.5 exports
Gold	small	unknown	50,000[b]	0.1	some foreign sales
Magnesium	huge	good	1,000[c]	7	4.4 imports
Manganese	huge	good	300,000	3	1.9 exports
Mercury	huge	excellent	900	10	2.1 exports
Molybdenum	huge	excellent	3,300[d]	4	minor exports
Platinum	unknown	unknown	NA	NA	7.5 imports
Silver	unknown	unknown	25	0.2	negligible
Tantalum	medium	unknown	NA	NA	0.1 imports
Titanium	large	unknown	NA	NA	1.0 import/0.4 exports
Tin	large	fair	18,000	14	46.9 exports
Tungsten	huge	good	11,300	24	36.1 exports
Uranium	unknown	unknown	NA	NA	unknown

Source: United Nations Statistical Yearbook 1976; American Metals Market Yearbook 1976, as well as other publications listed in references to this chapter.

[a] Includes North Korean production. [b] In troy ounces. [c] In short tons. [d] Data for 1974.

Tin, tungsten, and antimony are the famous Chinese export metals and together account for most of the metal exports other than finished steel. During 1970–1974 China earned over $159 million from exports of tin, $140 million from tungsten, and over $40 million from antimony. In the final analysis, however, China is a net importer of metals by a large margin of about five to one. China also sells some quantities of mercury and manganese, but these do not contribute significant amounts to the total export income relative to the other metals.

Among other imported metals platinum is the most important. In 1973 about $27 million was spent on imports of platinum, and between 1970 and 1974 a total of $72.8 million of platinum was imported. The United Kingdom and other western European countries are believed to be China's major suppliers of platinum, which probably originates in the Soviet Union or Canada.

Magnesium is primarily imported from Norway and the United Kingdom in amounts not larger than a few million dollars per year. Titanium is both exported and imported but accounts for a small amount often less than $1 million per year. Some tantalum is also imported in quantities not exceeding one ton per year. Japan is the major supplier of all those metals to China.

Tin is the largest single Chinese nonferrous metal export, and in 1974 a total of 10,200 tons of tin, valued at $46.9 million were shipped. However, this amount is less than half the tin that China exported in 1957, when exports amounted to 24,000 tons and a large proportion went to the Soviet Union. In fact, in the 1950s and early 1960s China was the principal supplier of Soviet tin imports. Although the Soviet Union is the second largest tin producer in the world, its per capita consumption is still well below that in other industrialized countries, and 20 to 30 percent of tin consumed is imported every year. In 1975 this amounted to about 10,000 tons, valued at about $66 million.

China is now a very small supplier of tin to the Soviets and ships only about 500 tons every year, probably valued in the order of $3 million. In fact, total exports of metals from China to the Soviet Union were only $13.7 million in 1974, which was very significantly less than the $160 million exported in 1957. Soviet and Chinese metal-import needs, on the other hand, are complementary and present another economic opportunity area where exchanges of chromium, nickel, and cobalt for tin and oil from China may develop in the future. This type of exchange would save hard currency for both countries and is likely to develop if China is unsuccessful in developing satisfactory markets for its oil and traditional goods in the West and in the developing countries. Although Soviets will develop additional tin deposits, it is believed that they will continue to import large quantities of tin because their production costs are already higher than in other tin-producing countries. The United States, Netherlands, West Germany, France, and Japan also import some Chinese tin.

PRODUCTION AND TRADE IN NONMETALLIC MINERALS

Mining of nonmetallic minerals is dominated by the coal industry, which is by far the largest mining industry in China and one of the largest in the world. The estimated 427 million tons of coal produced in 1975 represent more than the combined volume of all other minerals mined in China. The Chinese coal industry is discussed in considerable detail under sections on the coal industry in Chapter 7.

In addition to coal, China is also a significant world producer of salt, magnesite, pyrites, graphite, fluorspar, and cement. At least fourteen other minerals are produced in commercial quantities, and China maintains regular export trade in a similar number of minerals (see Table 9.10).

Total trade in nonmetallic minerals including coal is in the order of $100 million annually, which is relatively small. China is a net exporter in this trade. In 1974 $72.0 million accounted for exports, and only $25.9 million represented imports of nonmetallic minerals. Both exports and imports are about two to three times larger than 10 years ago, but their growth is erratic.

Phosphates, sulfur, and diamonds account for the largest percent of import value. Coal, fluorspar, salt, talc, bauxites, barites, and precious stones are the largest exports. China would like to expand these exports, but global demand for many nonmetallic minerals is very low and it is difficult to sell large quantities of such minerals on a profitable basis.

Production of asbestos at 150,000 tons per year is believed to be low relative to China's industrial base. The largest asbestos deposit is located in Shihmien in Szechman Province, where over 50 percent of all asbestos is believed to be mined. Only very small quantities have been exported, primarily to Japan, but Poland and Mexico have also purchased Chinese asbestos in recent years.

At least four barite deposit areas have been identified in China, with a total production estimated at 200,000 tons per year. Barite resources are believed extensive, and production is believed to increase with expanding oil and gas-drilling activity. Exports of barites have increased during the 1970s, from 60,000 tons in 1970 to 125,000 tons in 1974. Japan is the largest customer, but West Germany, Poland, Romania, France, Italy, and the United Kingdom also import Chinese barites.

Production levels of borates are unknown, but most of it comes from a 40-km^2 (square kilometer) dry lake, Itsaidam, located in the Tsaidam Basin of Tsinghai Province. The lake is believed to hold 400,000,000 tons of resources, and total production of crude salts was estimated to be

Table 9.10 Chinese production and trade in nonmetallic minerals circa 1975

Mineral	*Size of Deposits*	*Quality*	*Estimated Annual Production in Tons*	*Percent of World Output*	*Trade in 1974 in Millions of $U.S.*	
					Imports	*Exports*
Anthracite	adequate	good	20,000,000	11	—	small
Asbestos	large	fair	150,000	3	—	0.5
Barites	huge	good	200,000[a]	5	—	5.1
Bauxites	huge	low	970,000	1	—	7.9
Borates	unknown	unknown	NA	1	—	—
Coal[b]	huge	good	427,000,000	19	—	17.0
Coke	adequate	moderate	24,300,000	7	—	small
Cement	large	unknown	46,900,000	4	—	some
Diamonds	small	unknown	NA	1	4.3	1.6
Flurospar	huge	good	300,000	7	—	16.7
Graphite	large	poor	30,000[a]	8	—	—
Gypsum	large	unknown	630,000[a]	1–2	—	—
Limestone	large	good	NA	5	—	—
Magnesite	huge	good	1,000,000	8	—	4.2
Mica	large	poor	NA	1–2	—	0.4
Phosphates	medium	low	3,400,000	3	15.0	—
Potash	unknown	unknown	400,000	2	—	—
Precious stones	large	good	NA		0.65	7.4
Pyrites	medium	unknown	2,000,000[a]	8	—	—
Quartz	unknown	unknown	NA		—	—
Salt	large	good	29,940,000	19	—	11.9
Silica	unknown	unknown	NA	NA	—	—
Sulfur	large	fair	130,000	1	5.5	0.3
Talc	huge	good	400,000	3	—	11.1
				Total	25.9	72.0

Source: U.S. Bureau of Mines; *United Nations Statistical Yearbook, 1976.*
[a] Data for 1974. [b] Includes anthracite and coking coals production.

62,000 tons in 1972. Japan imports several thousand tons of Chinese borates annually.

Chinese bauxites are of very low quality, but their deposits are huge. Total mine output in 1975 was 970,000 tons, destined primarily for domestic production of aluminum. In addition, some bauxite is exported as a refractory material or abrasives. Exports have increased in recent years, from 5000 tons in 1970 to 28,000 tons in 1974.

China is the fourth largest cement producer in the world, and its production is expected to reach 50,000,000 tons by the early 1980s. The cement industry is well developed in China, and though a large cement plant was recently purchased from France, China remains a net exporter of cement. For more details, see the section on building materials in Chapter 15.

Only one Chinese diamond mine has been identified in Changte in Yuangchiang Basin of Hunan Province, although deposits were reported in Shantung and Kweichow Provinces. In 1973 China announced the production of synthetic diamonds by the Academy of Sciences in collaboration with the Peking Grinding Wheel Factory. This may have been the reason for a decline in diamond imports, which dropped from an average of $18 million per year to $4.3 million in 1974. The United Kingdom supplied 90 percent of all diamonds to China. Belgium and some African countries are also believed to be minor suppliers. China is also an exporter of semiprecious stones, primarily consisting of jade, which is mostly sold in Hong Kong.

China accounts for about 7 percent of total fluorspar production in the world and, after coal, fluorspar accounts for the largest nonmetallic Chinese export. Fluorspar production is believed to be over 300,000 tons per year. Because Chinese consumption is estimated below 100,000 tons annually, large surpluses are exported. Traditionally China exported close to 200,000 tons per year since 1965 and as much as 360,000 tons valued at $16.7 million in 1974. Japan is the largest importer, accounting for almost 180,000 tons in 1974. The Soviet Union and Poland also buy Chinese fluorspar.

Magnesite production is estimated to have reached about 1,000,000 tons annually, and China accounts for 8 percent of world output. Traditionally China has been exporting 20,000 to 71,000 tons of magnesite annually to Japan.

Considerable deposits of phosphate rock exist in China, and annual production of phosphates is estimated at 3,400,000 tons. It is not sufficient to meet domestic demand for use in chemical fertilizers. During the early 1970s almost 1,000,000 tons of phosphate were imported every year, reaching a high of 1,800,000 tons in 1973. In 1974 imports fell to 400,000 tons, but prices of phosphate rock tripled and China spent $15.0 millions on those imports. Morocco supplies over 50 percent of phosphate rock, whereas Egypt, Tunisia, and the United States have been secondary suppliers in recent years. With higher phosphate prices it is believed that China has greater incentives to develop its domestic deposits or import larger quantities from countries like Vietnam, where it can barter Chinese manufactures for the mineral.

China is the third largest pyrites producer in the world after the United States and Spain. In 1974 pyrites production was estimated at 2,000,000 tons. Most of it is used in production of sulfuric acid and fertilizer. A certain proportion of pyrites produced in China is used for manufacture of elemental sulfur, which has been exported to the Soviet Union and other countries in previous years.

Between 1965 and 1970 China switched from being a net exporter to a net importer of sulfur. Imports of sulfur rose from 25,000 tons in 1970 to a high of 420,000 tons in 1973, valued at $6.2 million. At the same time exports of Chinese sulfur dropped from 45,000 tons to 1000 tons per year. Canada was a big recent supplier of sulfur to China, exporting 280,000 tons in recent years. Mexico, Iran, and Iraq are secondary sulfur suppliers. Increased sulfur recovery from China's oil-production and domestic sources as well as continuing imports are expected in the future.

Deposits of salt are widespread and abundant, and China is the world's second largest salt producer after the United States. In 1975 almost 30,000,000 tons of salt were produced, which is more than twice the output in the Soviet Union. Between 80 percent and 90 percent of Chinese salt is produced in shallow coastal ponds by evaporation. These are predominantly located in Liaoning, Hopeh, Shantung and Kiangsu Provinces. Exports of salt have been in the order of 1,000,000 tons per year during the first half of the 1970s and went primarily to Japan. If demand for salt increases, China appears to have considerable additional capacity that can be put into production.

Very high-grade talc produced in China has a ready market in Japan, which imported a total of 134,000 tons in 1974. Total talc production is estimated at 400,000 tons per year, including soapstone. Exports of talc have been increasing steadily from 83,000 tons in 1965 to 235,000 tons in 1974, valued at over $11.0 million.

In general trade in nonmetallic minerals occupies a minor position in China's foreign trade. Imports are less than 0.5 percent of the total value of China's imports, and exports are slightly over 1 percent of total exports of China. In spite of large production levels of some minerals compared with world supply, China's exports of all minerals are not a major factor in the trade of any particular country, and China's imports are insignificant. Alternative sources of all nonmetallic minerals produced by China are readily available in many countries.

REFERENCES

Business China. "Selling Steel Mills to China: Experience of DEMAG Consortium," *Business International Asia/Pacific,* January 21, 1977, Hong Kong.

China Letter, The. "Iron and Steel Special Report No. 58," *The Asia Letter Limited,* 1976, Hong Kong.

China Letter, The. "Lead and Zinc Special Report No. 64," *The Asia Letter Limited,* 1977, Hong Kong.

China Trade Report. "Ores and Metals," p. 3, March 1976, Hong Kong.

Clarke, William W. "China's Steel, the Key Link," *U.S.–China Business Review,* July–August 1975.

"Metal Statistics 1976," American Metal Market Fairchild Publications, New York, 1976.

Shelton, John E. "The Mineral Industry of the People's Republic of China," *Mineral Yearbooks,* Bureau of Mines, 1971, 1973, 1974.

Szuprowicz, Bohdan O. "People's Republic of China to Up Imports," *American Metal Market,* October 8, 1976.

Ta Kung Pao. "Metal Exhibition in the Trade Convention," May 1, 1972 (transl. by JPRS #56741), Hong Kong.

United Nations Department of Economic and Social Affairs. *"1976 Statistical Yearbook,"* 1977, New York.

U.S. Government Research Aid. "China's Minerals and Metals Position in the World Market," ER76-10150, March 1976.

U.S. Government Research Aid. "China: Role of Small Plants in Economic Development," A(ER)74-60, May 1974.

U.S. Government Research Paper, "China's Cement Industry," ER77-10704, November 1977.

Usack, Alfred H., Jr. and James D. Egan. "China's Iron and Steel Industry," in *China: A Reassessment of the Economy,* Joint Economic Committee, U.S. Congress, July 1975, p. 264.

Wang, K. P. "PRC, a New Industrial Power with a Strong Mineral Base," U.S. Bureau of Mines, Department of the Interior, 1976.

Wang, K. P. "The Mineral Resource Base of Communist China," in *An Economic Profile of Mainland China,* Joint Economic Committee, U.S. Congress, February 1967.

10

Chemicals, Petrochemicals, and Rubber

China's chemical industry appears to have developed more slowly than its other basic industries. This was probably due to the fact that significant petroleum production did not develop until the late 1960s. The coal-based chemical industry in existence involved relatively high production costs, although a considerable fertilizer production was developed using this raw material. The technological level of the Chinese chemical industry is comparably very low but is rapidly improving in quality, particularly with the purchases of the modern ammonia, urea, and petrochemical plants during the early 1970s.

Typically for a developing country, China imports three times as many chemicals as it exports, and about 95 percent of all chemicals imported come from capitalist countries like Japan, West Germany, and Italy. At least twenty-five different countries sold chemicals to China in recent years, but Japan is by far the dominant supplier, with a total market share twice as large as all the other supplier countries combined (see Fig. 10.1). Imports of chemicals in 1976 were $600 million, down 27% from 1975 levels.

Imports of chemicals and rubber account for about 15 percent of all Chinese imports and are second in volume only to iron and steel as the largest single import commodity. China is regarded as the world's largest importer of fertilizers, but chemical elements and compounds, synthetic fibers, and plastics are also imported in significant quantities (see Fig. 10.2).

The industry is supervised by the Ministry of Petroleum and Chemical Industries, which is headed by Minister Kang Shih-en. It is consid-

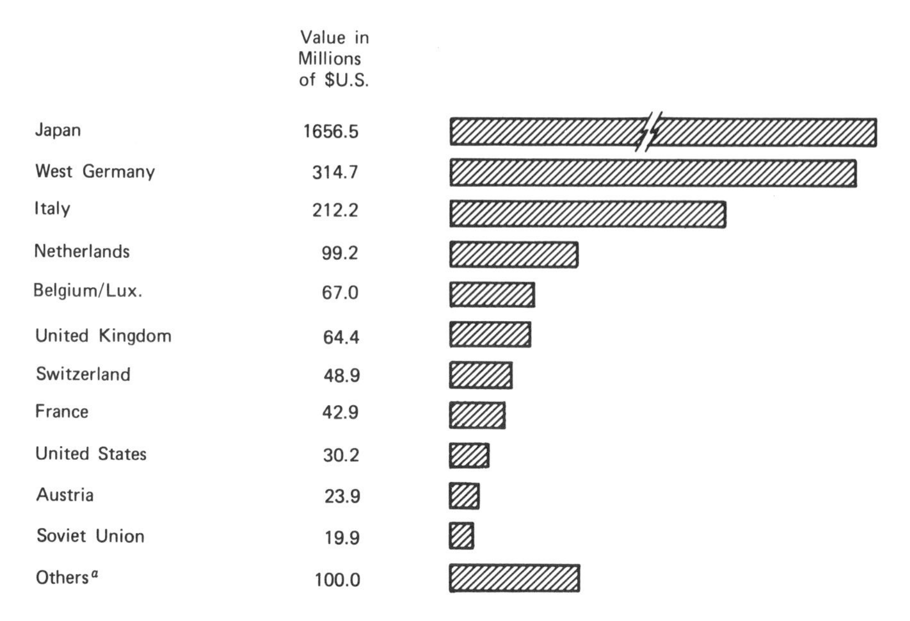

***Figure 10.1** Estimated market shares of major chemicals suppliers identified during 1970–1975. Data include value of chemical products under SITC commodity code 5 and synthetic textile fibers SITC code 266. (Source: OECD Commodity Trade Statistics Series B; United Nations statistics of exports of chemical products, 1975, and some individual country foreign-trade yearbooks.) [a] Includes Norway (12.3), Canada (10.0), Denmark (7.8), Sweden (8.1), Finland (3.6), Yugoslavia (2.5), Spain (3.6), Poland (1.3), Greece (3.0), Turkey (0.6), Australia (1.8), Czechoslovakia (0.555), Hong Kong (1.0), and estimate for Romania (10 to 50) derived from typical volume of chemicals exported.*

ered one of the key ministries in China because of its control over petroleum production as well as its responsibility to support agriculture by increasing the supply of fertilizers and other agricultural chemicals. Until 1970 the Ministry of Coal Industry was a part of the Ministry of Petroleum and Chemical Industries. This underscores the role coal has played as a basic raw material for the chemical industry prior to expansion of the petroleum industry.

Initially Chinese chemical industry began with caustic soda and a few basic fertilizer plants. Since 1949 new plants were built throughout China, but many were later consolidated into larger units. When petroleum production became significant the petrochemical industry came into being with some Romanian assistance.

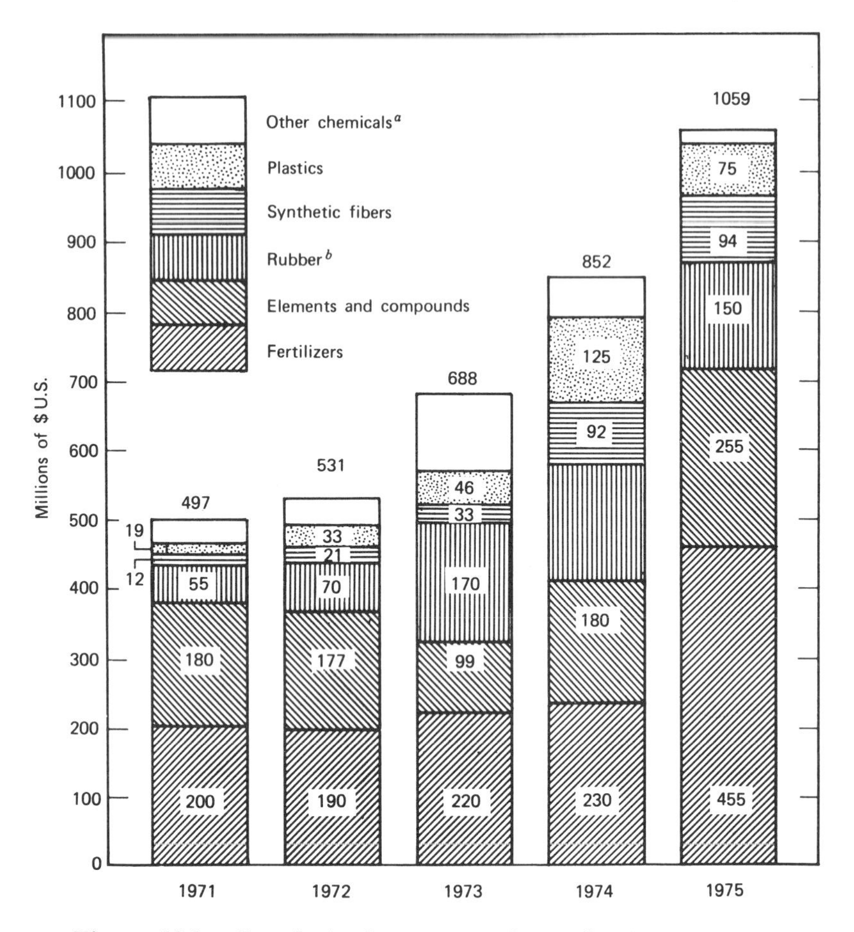

Figure 10.2** Growth in imports of chemicals.* (**Source:** ***Based on OECD Commodity Trade Statistics and the United Nations annual bulletins of trade in chemical products.) ([a] ***Includes dyeing materials, pharmaceuticals, essential oils, explosives, and pesticides.*** [b] ***Includes synthetic and natural rubber imports.***)

The synthetic fiber industry did not come into being until after 1960. By 1969 it was estimated that China operated nine synthetic fiber plants, producing dacron, nylon, vinylon, orlon, and saran. The petrochemical and fertilizer industries are now further developed by massive purchases of turnkey plants from Japan, France, the Netherlands, Germany, and the United States.

More than 50 percent of nitrogen fertilizer and 75 percent of phosphate fertilizer is manufactured in about 2000 small plants, producing 3000 to 5000 tons of synthetic ammonia per year. There are also suggestions that a program of development of small-scale synthetic fiber plants may come into prominence in an attempt to meet the expected demand of the future.

The Ministry of Petroleum and Chemical Industries is one of the few ministries that also operates large facilities to manufacture specialized equipment for petroleum and chemical factories. They are controlled by the Equipment Manufacturing Department of the ministry and produce high- and low-pressure vessels, large compressors, and instruments for the industry.

EMERGENCE OF CHINESE PETROCHEMICALS INDUSTRY

The Chinese petrochemicals industry developed during the 1960s as increasing domestic supplies of petroleum became available. There are also other important incentives for China to develop its own petrochemicals industry. In recent years China has been importing increasing amounts of chemical fertilizers, plastics, synthetic textile yarn, thread, and fabric, as well as natural and synthetic rubber. Annual imports of such products cost China almost $1 billion, yet practically all of them could be substituted by domestic petrochemical products using local petroleum and natural-gas resources if the industry were sufficiently developed. In the long run production of synthetic textiles and rubber will also free more land for the production of grains. This will create additional opportunities to save hard currencies by reducing imports of grains, cotton, and rubber.

During the 1960s, as China was developing its petroleum-refining capabilities, several plastics and synthetic fiber plants were purchased on a turnkey basis from Japan, West Germany, and the United Kingdom. These are believed to have served as prototypes for construction of additional Chinese-designed petrochemicals plants. In 1965 Lurgi of West Germany received a contract to construct an oil refinery that reportedly included the first cracking unit and olefin-extraction plant exported to China.

China is also believed to have received petrochemical technology assistance from Romania and Cuba. It is believed that Chinese engineers were given access to the former Esso and Shell refineries in Havana (now known as the Nico Lopez refinery), which contained catalytic cracking and platforming units. Similar units were subsequently constructed in China.

The present Chinese refining industry is comparable to American or Western refining industries of the late 1950s. This means that existing facilities produce a complete range of products and a variety of feedstocks for the petrochemical industry. However, a limitation to petrochemicals expansion may come from lack of production of some other industrial chemicals necessary for processing that are not derived from coal or petroleum.

Since 1972 China embarked on an ambitious expansion program of its petrochemicals industry. Fifty-seven petrochemical and synthetic fiber plants and another seventeen chemical fertilizer plants have been purchased, valued at over $1.6 billion. Most of these plants will use light olefins and aromatics as feedstocks derived from Chinese petroleum and natural gas. Industry observers estimate that these new plants will triple China's ethylene production capacity to about 3 million pounds per year. Ethylene is the most versatile petroleum product used in the petrochemical industry. It is also difficult to store and transport and is mostly consumed by the local petrochemical industry. In effect, ethylene-manufacturing capacity is a reasonable measure of the capabilities of a country's petrochemical industry.

When all the petrochemical plants recently imported come on stream, Chinese ethylene-production capacity could be in the order of 1,300,000 metric tons annually. This would make China the fifth largest ethylene manufacturing country in the world. Its petrochemical industry output capacity would be theoretically comparable to that of either Italy or United Kingdom (see Table 10.1).

On the other hand, petrochemical plants imported by China will manufacture only the most basic petrochemicals. Industry specialists

Table 10.1 Production of ethylene in selected countries of the world in 1975 (All values in metric tons)

Country	*Ethylene Production*	*Country*	*Ethylene Production*
United States	8,971,400	United Kingdom	958,900
Soviet Union[a]	4,900,000	Canada	429,500
Japan	3,399,000	Spain[c]	243,800
West Germany	2,140,100	Romania[d]	110,600
China[b]	1,300,000	India	47,600
Italy	1,128,000		

Source: *United Nations Statistical Yearbook, 1976.*

[a] Authors estimates based on average ratio of world ethylene production relative to world oil production, applied to oil consumption levels of the Soviet Union in 1975. [b] Estimate for Chinese capacity after installation of petrochemical plants purchased during 1972 to 1975 is complete. [c] Data for 1974. [d] Data for 1970.

pointed out that Chinese imports did not include plants for diabasic acids, fluorocarbons, or acrylics. Plastic plants were also regarded as being limited to basic resin production such as polyethelene, polypropelene, and polyvinyl chloride. Unsaturated polyesters, Teflon (tetrafluoroethylene), and plastics like ABS (acrylic-butadiene styrene or acrylonitrile-butadiene styrene), polyacetals, polycarbonates, and temperature-resistant varieties are not manufactured. It is also suspected that despite large synthetic fiber plant purchases to date, China will require still additional manufacturing capacity to support its huge textile industry (see Table 10.2).

It is now apparent that since 1975 China has concentrated on construction and installation of imported plants. During 1978 and 1979 most of those plants will become operational. It is expected that in future years China will once again resort to imports of advanced petrochemical turnkey plants to obtain additional processing technology and to further narrow the gap between its petrochemical industry and that of Western industrialized countries.

Major petrochemical centers exist in Peking, Shanghai, and Shenyang as well as at or near the major oilfields at Taching, Shengli, Takang, and Lanchou. Synthetic fiber plants have also been identified in Taiyuan, Tatung, Chungching, Fukien, Hantan, Hsinhsiang, and Kirin. Plastics are produced at plants in Peking, Tsingtao, Wuhan, Anyang, Canton, Hangchow, Huhehot, and Ishan. Other plants exist in Darien, Tzupo, Iyang, Tientsin, and Fencheng.

SYNTHETIC FIBERS PRODUCTION AND IMPORTS

This industry came into being after 1960 primarily as the cellulosic fiber industry, using large quantities of waste materials such as cotton waste and wood pulp. Development of the synthetic fiber industry began after the withdrawal of Soviet economic assistance. As a result, China obtained its basic synthetic fiber production technology from Japan, West Germany, and the United Kingdom by importing complete plants for the manufacture of vinylon, perlon, dacron, orlon, and saran.

By 1966 China was manufacturing an estimated 38,000 tons of cellulosic man-made fibers, and by 1975 this output increased to a total of 110,000 tons, slightly less than comparable production in India. By 1971 at least eleven plants in China had been identified as engaged in manufacture of cellulosic fibers (see Table 10.3).

In 1975 about 70 percent of all the synthetic fiber produced in China was still cellulosic, although production of the synthetic fiber

Table 10.2 Major contracts for basic petrochemical turnkey plants identified since 1963

Year	*Type of Plant Imported*	*Capacity in Tons/Year*	*Supplier and Country*	*Contract Value in Millions of $U.S.*
1963	Ethylene, hexanol, butanol	300,000	Melle & Speichim, France	8.5
	Oil refinery	200,000	ENI Group, Italy	5.0
	Polyester resin		Scott & Bader, U.K.	
1964	Acetylene		Japan	3.0
	Petroleum-cracking unit	50,000	Lurgi, West Germany	12.5
	Polyethelene HP	24,000	Simon Carves, U.K.	12.6
	Polypropylene		U.K.	7.3
1965	Acrylonitrile	10,000	Lurgi, West Germany	11.0
	Naphtha cracking		Norway	14.6
	Acrylic resin		Prinex, U.K.	8.4
	Oil refinery		Snam Projetti, Italy	9.0
1966	Aromatic chemicals	70,000	Snam Projetti, Italy	5.5
1972	Ethylene	300,000 }	Toyo/Mitsui	46.0
	Butadiene	45,000 }		
	Acetic acid		Kaisha, Japan	
1973	Ethylene, vinyl acetate	66,000	Kuraray/Bayer, Japan	26.0
	Polyester polimerization	25,000	Toray/Mitsui, Japan	49.0
	Aromatic extraction BTX	60,000	Sumitomo, Japan	6.0
	Vinyl acetate	90,000 }	Speichim/Lurgi/H&G/BASF	90.0
	Methanol	110,000 }		
	Polyethelene LP	60,000	Mitsubishi/Hitachi, Japan	22.0
	Acetaldehyde	30,000	F. Uhde/Hoechst, West Germany	4.0
	Polyethelene HP	180,000	Sumitomo, Japan	47.0
	Acrylonitrile monomer		Asahi Chemical, Japan	30.0
	Ethylene glycol	16,000 }	Nippon Catalytic	17.0
	Ethylene oxide	20,000 }	Hitachi, Japan	
	Vinyl chloride monomer		F. Uhde, West Germany	19.0
	Polypropylene		Mitsui, Japan	25.0
	Butyl rubber interest shown		Polisar, Canada	none
	Polypropylene catalyst	220	Toho Titanium, Japan	4.7
1974	Polyvinyl alcohol	45,000	Kuraray	19.0
	Reforming catalyst		Haldor Topsoe, Denmark	12.5
	Polyethelene HD LP		F. Uhde, West Germany	15.6
	Polypropylene	35,000	Snam Projetti, Italy	16.0
1975	Aromatics complex		Japan Gasoline Company	36.0
	Benzene	100,000	Linde AG, West Germany	20.0
	Dimethyltetraphthalate	90,000	Krupp-Koppers, West Germany	50.0
	Detergent		Mechanische Moderne, Italy	1.0
	Detergent alkalation		Eurotechnica, Italy	35.0
	Acetylene	30,000	Speichim, France	
	Ethylene	12,000	Mitsubishi, Japan	31.5
1976	Styrene butadiene rubber	240,000	Japan Synthetic Rubber	27.0
	Ethanol	100,000	F. Uhde, West Germany	20.0
	Diethylhexanol	50,000	BASF, West Germany	24.0
1977	Terephthalic acid TPA	36,000	Lurgi, West Germany	27.0

Table 10.3 Production of synthetic fibers in selected countries of the world in 1975 (All values in thousands of metric tons)

	Polymer-based Fibers			*Cellulosic Fibers*	
Country	*Total*	*Yarn*	*Staple*	*Yarn*	*Staple*
World	7460.0	3840.0	3620.0	1150.0	1860.0
China	45.7	13.1	32.6	49.0	61.0
United States	2670.4	1455.7	1214.7	166.1	173.7
Japan	1061.2	490.6	570.6	113.2	287.6
West Germany	618.2	309.1	309.1	52.7	63.0
Soviet Union	364.5	230.3	134.3	281.0	309.0
United Kingdom	361.1	184.0	177.1	62.5	138.9
Italy	321.3	123.2	198.1	41.5	48.3
South Korea	267.9	120.8	147.1	9.8	—
France	209.5	85.6	123.9	28.1	55.6
Mexico	147.9	97.8	50.1	15.5	17.2
Brazil	125.0	80.0	45.0	30.7	18.5
East Germany	113.0	49.0	64.0	32.9	135.6
Netherlands	112.7	67.8[a]	44.9[a]	27.8	—
Poland	110.6	51.9	58.8	28.3	67.4
Romania	84.0	17.0	67.0	14.0	50.0
India	33.2	19.7	13.5	52.6	66.8

Source: Based on statistics in *United Nations Statistical Yearbook, 1976.*
[a] Data for 1972.

pants based on polymers was increasing rapidly. Noncellulosic synthetic fiber production was estimated at 45,700 tons in 1975, almost 36 percent larger than that of India but still only slightly more than half of that of even Romania (see Fig. 10.3).

The noncellulosic synthetic fiber production based on polymers did not start growing until the early 1970s. It is believed that its growth may have been delayed by shortages of raw-material inputs until China developed its own sizable petroleum industry. The synthetic fiber industry also requires considerable electric power, which remains in short supply, and more sophisticated equipment with which China is only now familiarizing itself.

China is believed to have purchased the synthetic fiber plants during the 1960s in hopes of using them as plant prototypes. Failure to reproduce that technology in form of improved Chinese-built plants and a significant increase in output of petroleum products appear to have forced China into renewed massive imports of more advanced noncellulosic plants and technology during 1972–1975. As a result, however, China gained a 10-year advance in its petrochemical technology and is now put-

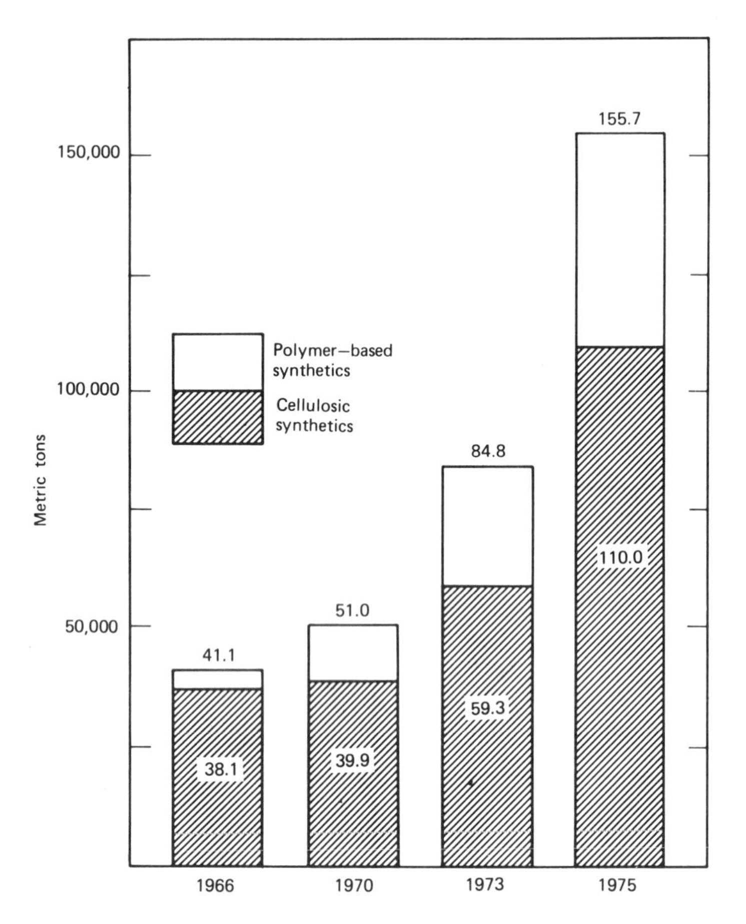

Figure 10.3 Growth in synthetic-fibers production. **(Source:** ***Based on data in the*** **United Nations Statistics Yearbook, 1976.)**

ting into operation some of the most modern and efficient textile fiber plants in the world.

It is interesting to note here that production of synthetic fibers in other communist countries such as the Soviet Union, East Germany, and Poland is also predominantly cellulosic-based industry. India is another country in that position, and one reason may be a limited supply of petroleum-based raw materials. Romania, which has its own petroleum industry, is the only exception among communist countries that has a relatively large polymer-based synthetic fiber industry in respect to total synthetic fiber production (see Table 10.3). Most Western countries, including even Brazil and Mexico, produce more polymer-based than cellulosic synthetic fiber products within their respective industries.

Since 1963 China imported almost twenty synthetic fiber plants of all types, valued at about $750 million. The largest of them all is the combined petrochemical and synthetic fiber project at Liaoyang near Shenyang where the French consortium headed by Technip and Speichim are completing a 2-million-ton capacity complex of twenty-one plants valued at $282 million (see Table 10.4).

Table 10.4 Major contracts for synthetic fiber plants identified since 1963

Year	*Type of Plant*	*Capacity in Imported Tons/Year*	*Supplier and Country*	*Contract Value in Millions of $U.S.*
1963	Vinylon fiber	18,000	Dai Nippon, Japan	30.0
	Vinylon fiber	11,000	Kurashiki Rayon, Japan	20.0
1964	Perlon		F. Uhde, West Germany	1.75
1972	Vinylon fiber		Kuraray, Japen	
	Chemical fiber plant		Kurashiki Rayon, Japan	18.9
	Polyester fiber	90,000	Toray, Japan	90.0
1973	Ethylene and poval	120,000	Mitsubishi, Japan	34.0
	Vinyl acetate and poval		Kuraray, Japan	26.0
1974	Synthetic fiber complex	56,000	Asahi, Japan	29.0
	Petrochemical complex	2,000,000	Technip/Speichim, France	282.0
	Polyester spinning	26,400	Toray, Japan	17.0
	Acrylic fiber		Japan Exlan	
	Nylon spinning		Rhone Poulenc, France	10.0
	Synthetic fiber		NISSO, Japan	14.0
	Poval	33,000	Kuraray/Bayer	26.0
1976	Polyester/polymer	80,000	Teijin, Japan	43.0
1977	Polyester fiber	40,000	Lurgi/Zimmer, West Germany	12.0

Source: Compiled by 21st Century Research from articles, reports, and other publications listed in references to this chapter.

Despite increasing production, China has been importing synthetic textile fibers for several years. During the late 1960s those imports were in the order of 50,000 tons per year. By 1975 imports were still about 40,000 tons, but their value has been going up sharply. In 1975 China spent $94 million out of a total of $263 million during 1970–1975. Japan is by far the leading supplier, with the United Kingdom and West Germany as secondary sources (see Fig. 10.4).

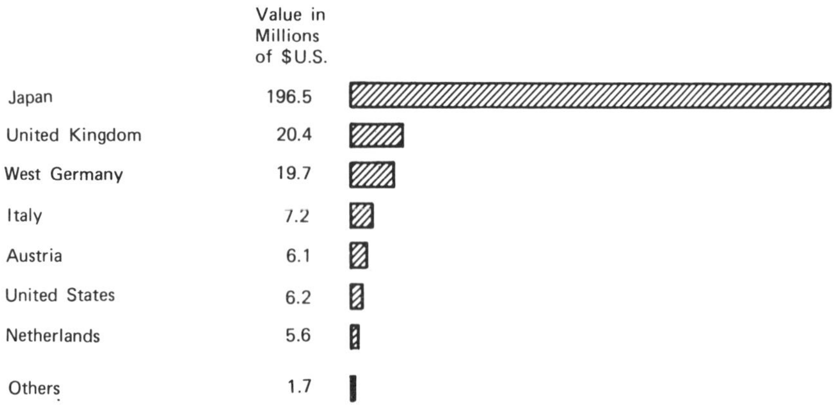

	Value in Millions of $U.S.
Japan	196.5
United Kingdom	20.4
West Germany	19.7
Italy	7.2
Austria	6.1
United States	6.2
Netherlands	5.6
Others	1.7

***Figure 10.4** Market shares of synthetic-fibers suppliers identified during 1970–1975.* (**Source:** *OECD Commodity Trade Statistics Series B for 1975 and previous years.*)

Several other countries, including Italy, Austria, the United States, the Netherlands, France, Yugoslavia, Spain, and Finland exported smaller amounts of synthetic fibers to China. These ranged in value from $7.2 million for Italy to only $100,000 for Finland during 1970 to 1975.

It is interesting to note that no single communist country has been supplying China with synthetic fiber products. Even Romania, traditionally the largest chemicals exporter to China from COMECON and having a sizable synthetic fiber industry, has not been exporting this type of product to China.

Because China invested such a considerable amount of hard currency in developing its own synthetic fiber industry it is logical to expect that as production gets under way in the new plants, imports of such products may be curtailed. China also maintains a significant trade in natural textile fibers and fabrics as well as products, which account for almost 20% of China's exports and were valued at $1,250 million in 1975. As the worldwide demand for synthetic textile fabrics continues to increase China will probably enter this market as an exporter of such products.

PRODUCTION AND IMPORTS OF FERTILIZERS

China is the third largest chemical fertilizer producer in the world after the Soviet Union and the United States. Despite the large production,

China is also the biggest importer of fertilizers in the world. During 1966 to 1975 at least 60 million tons of standard units of fertilizer were imported valued at $2.3 billion (see Table 10.5).

Table 10.5 Production of fertilizers in selected countries in 1975/1976 (in thousands of metric tons of nutrient content; i.e., nitrogen, phosphoric acid, potash)

Country	*Nitrogen Fertilizer*	*Phosphate Fertilizer*	*Potassium Fertilizer*
World	43,876	24,872	23,470
China	3,300	1,246	300
United States	9,262	6,655	2,099
Soviet Union	8,465	4,103	7,944
Japan	1,557	585	—
Poland	1,532	929	—
India	1,508	320	—
Romania	1,292	404	—
Netherlands	1,153	179	3
Canada	916	653	4,841
Brazil	160	507	—

Source: United Nations Food and Agricultural Organization (FAO).

The Chinese goal of the fourth Five Year Plan (1971–1975) was to achieve a total annual fertilizer supply of 35 million tons by 1975. Actual fertilizer production in 1975 was estimated to be 27.9 million tons, and the total supply including imports was about 33 million tons. This is actually equal to the total supplies already achieved in 1973, when production reached 24.8 million tons and imports were at an all-time high of about 7.5 million tons, or almost 30 percent of domestic production (see Fig. 10.5). In 1976 imports of fertilizers were 4.9 million tons.

Fertilizer Production

China has sufficient domestic raw materials for the production of all three types of fertilizers for the conceivable future. Coal and natural gas are the main sources of nitrogenous fertilizers and ammonia. Phosphate deposits provide most inputs for the phosphate fertilizers. Potassium salts are extensive in the Tarim and Tsaidam Basins of western China, and a large sea-salt manufacturing industry provides another source.

There are over fifty large fertilizer-manufacturing plants in China.

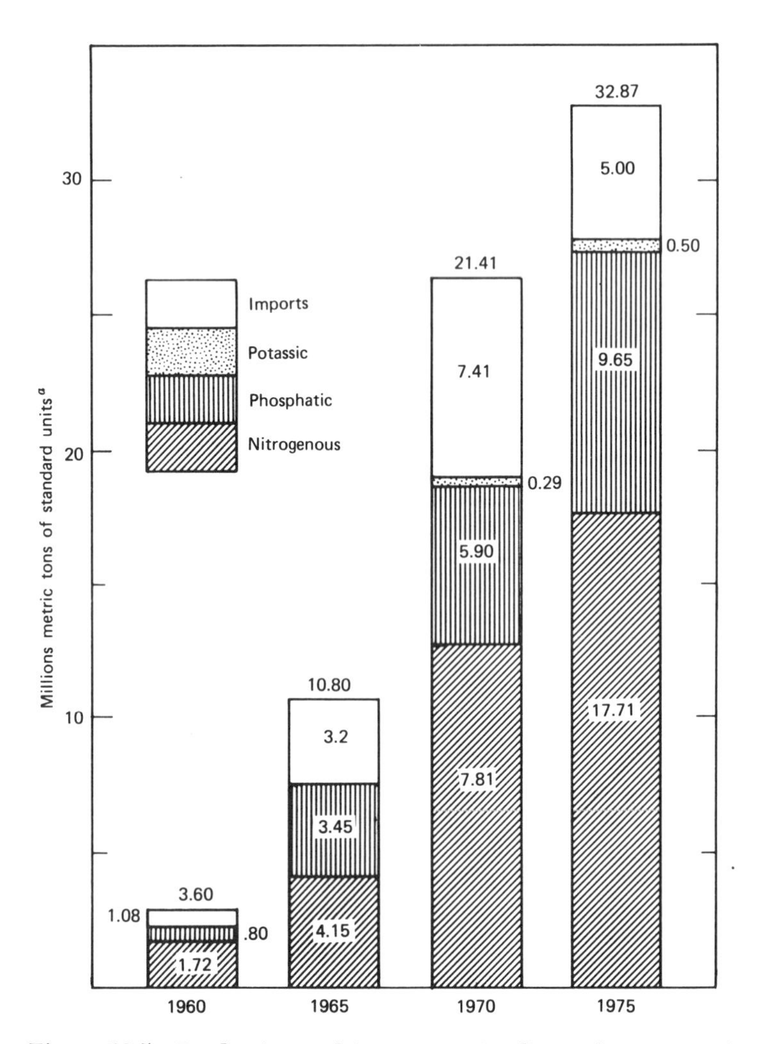

Figure 10.5 Production and imports of fertilizers. (Source: *U.S. Government Research Aid, "PRC: Chemical Fertilizer Supplies 1949–1974," August 1975.*) ([a] *Weight of chemicals containing equivalent nutrient content to standard fertilizer weight.*)

These now include thirteen of the world's largest ammonia–urea complexes ordered since 1972 from the Netherlands, France, Japan, and the United States. These plants should be coming on stream during 1978 and when fully operational will produce up to 3.5 million tons of actual nitrogen content annually, which is more than the 3.3 million tons of nitrogen-manufacturing capacity existing in China in 1975.

At least half a dozen large fertilizer plants with capacities over 300,000 tons per year have been identified in China for some time. A phosphate fertilizer plant of 400,000-ton annual capacity was completed in 1960 with Soviet assistance at Nanching in Kiangsu Province. An ammonium nitrate plant with 390,000-ton capacity was reported on stream in 1971 at Tsinan in Shantung Province. Another ammonium nitrate plant of 360,000-ton capacity was put into operation in Kirin as long ago as 1959.

An estimated 60 percent of all fertilizer made in China, however, comes from almost 2000 small plants that use local raw materials to produce aqueous ammonia and ammonium bicarbonate. Small plants account for over 80 percent of phosphate fertilizers production and close to 90 percent of all potassium fertilizers. Western observers often question the quality of the products of these plants, but the Chinese consistently claim that small plant output is suitable for its purpose. Nevertheless, there are shipment problems with ammonia, which is volatile and loses nitrogen quickly. Ammonium bicarbonate is unstable and decomposes quickly when it must be stored, as would have to be the case with fertilizer because it is a seasonal demand product.

Phosphate fertilizers are produced from domestic low-grade phosphate rock primarily by small plants throughout China. Many of these actually only crush phosphate rock to facilitate its distribution in the fields. Higher-quality phosphate rock imported from Morocco as well as from Egypt, Tunisia, and the United States is used to produce single or triple superphosphates by the large fertilizer plants.

At least two plants in Tsinghai Province reportedly manufacture potassium fertilizer from carnallite. In Fukien Province and other eastern coastal areas where salt is manufactured from sea-water potassium chloride and potassium–magnesium salts are extracted as fertilizer.

Ammonium nitrate is produced primarily for military and industrial explosives, but with a 30-percent nitrogen content it is also a useful fertilizer. Urea, which has 46-percent nitrogen content, was first produced in the late 1950s. It was not manufactured in commercial quantities until the mid-1960s, when a 175,000-ton/year urea plant was purchased from the Stork-Verkspoor firm in the Netherlands. Many standardized 40,000-ton/year urea plants have since been designed and built by the Chinese.

Role of the Small Fertilizer Plants

Despite criticism by industry observers, small fertilizer plants in China produce an estimated 50 percent of all nitrogen output. These plants are justified by savings in construction time and low investment. They require only one-third of the time required to build a large plant and use predominantly local materials. This relieves the transportation system from moving large quantities of construction equipment and from extensive distribution of fertilizer products.

The small plant may easily change processing technology to adapt to local or special input materials. This was demonstrated in 1973, when 44 percent of small nitrogen fertilizer plants switched to powdered coal and other local materials, making additional coke and coal available for industrial use where it was in short supply.

The relative importance of small ammonia plants will decline in the future as the large modern urea complexes come on stream. They probably will continue to be a major factor in supply of phosphates, whose production must increase to maintain the nutrient balance. To maintain that balance the production of phosphates must increase two and a half times its 1975 levels, that is, to about 21 million tons of standard units by 1980.

The small fertilizer plants are not a market for imported equipment. They are constructed and operated by county or commune authorities from locally available materials. Many standardized items are produced by numerous machine-building plants and workshops. Chinese press reports indicate that as many as 25,000 standard items have been manufactured to support small plant construction.

Imports of Fertilizers and Turnkey Plants

When the thirteen large imported urea plants become fully operational, it is likely that imports of nitrogenous fertilizer will not be necessary. China has imported increasing amounts of fertilizer since the early 1950s, when imports constituted over 65 percent of total supply in certain years. During the 1970s imports accounted for about 30 percent of total supply but rose to a high of 7.74 million tons of standard units in 1972. China then became the largest fertilizer-importing country in the world.

In 1974 sharp price increases in world markets compelled China to reduce fertilizer imports. Phosphate fertilizer production declined 11 percent because of quadrupling of phosphate prices in Morocco and a resulting reduction of phosphate rock imports. Overall fertilizer imports dropped by 33 percent in 1974. Although reduced in volume, the value of fertilizer imports doubled to a total $455 million in 1975. In 1976 prices for fertilizer fell by as much as 40 percent, but China's imports for that year amounted to only 4.9 million tons of fertilizer. In the early 1970s China sharply increased its imports of fertilizers from Romania from 62,000 tons in 1972 to 217,000 tons in 1973. Since Romania is a significant manufacturer of fertilizers it may be supplying larger quantities in the future in return for Chinese assistance in further development of its industries or other barter-trade arrangements.

Japan continues to be the foremost supplier; its shipments in 1975 were down to 1,480,000 tons of fertilizers from a total almost 2 million tons in 1973. Italy, the second largest supplier in 1973, had its sales cut down to only one third of that level in 1975. German and Finnish exports to China dropped by half, whereas those of Spain, Greece, and the United States stopped entirely (see Fig. 10.6).

China is clearly looking for alternative and cheaper sources of fertilizer while the large urea complexes are being readied. Recent interest in foreign aid projects for Morocco may also indicate the possibility of developing some form of barter in Chinese manufactures, foods, and textiles in return for phosphate rock.

Exports of fertilizers from Canada and the Netherlands actually in-

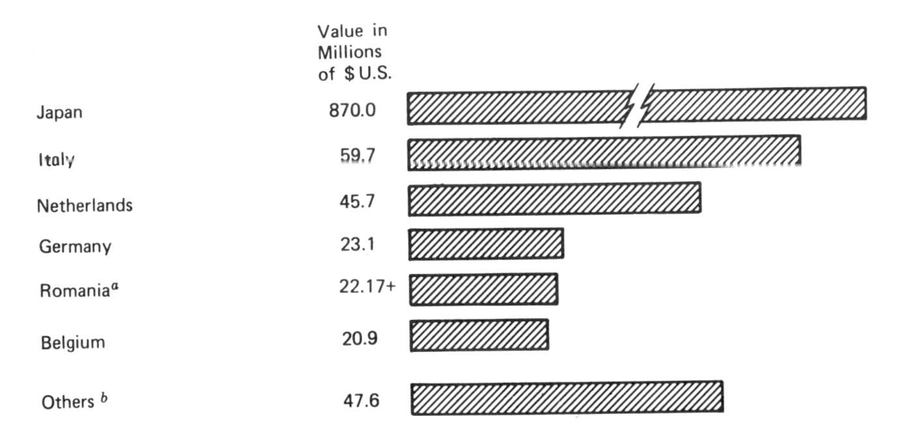

Figure 10.6 ***Estimated market shares of major fertilizer suppliers identified during 1970–1975. (Source: OECD Commodity Trade Statistics for SITC code 56 and yearly statistical data for Romania in 1973.)*** [a] ***Data based on volume of fertilizer shipped extended by average world market price of about $50 per ton. Data for 1974 and 1975 not available.*** [b] ***Others include Norway (12.1), Greece (10.2), Canada (9.2), France (5.7), United States (5.0), Finland (3.7), Spain (1.7).***

creased slightly. Canada sells China primarily potassium chloride, and the shipments increased from 114,000 tons in 1973 to 128,000 tons in 1975. Imports of potash are expected to continue to grow in the future until China expands its own potassium fertilizer manufacturing capacity to keep up with the proper balance ration with the other two nutrients. Although domestic potash deposits are considerable, their remote location and poor transportation facilities are not conducive to their immediate development. Imports of up to 2 million tons of potash may be required by 1980. Besides Canada, Belgium is also a supplier of smaller amounts of potassium fertilizers to China.

There has also been considerable emphasis in China on the use of organic fertilizers. This is believed to contraindicate any substantial increase in the supply of chemical fertilizers in the immediate future. In effect, as new urea plants start production their output will replace first of all the imports but may not be sufficient to provide for growth in demand resulting from further targets to improve the productivity in agriculture.

The Chinese chemical and machine-building industries will probably attempt to reproduce the latest imported plants technology. Judging by their performance during the 1960s, they may not have the resources and materials to do so before fertilizer demand once again increases to the point where large imports will have to be once again offset with additional plant purchases. This may lead to a new round of turnkey plant imports by the early 1980s at the latest (see Table 10.6).

The 1977 discussions have already begun with Nissan Chemical Company of Japan about phosphoric fertilizer technology sales, which may well turn out to be a turnkey plant purchase. China also showed recent interest in a very large ammonia–urea complex of the Nihon Ammonia Company of Japan, which may be another indication of future sales.

PLASTICS PRODUCTION AND IMPORTS

Polyvinyl chloride is the major plastic material produced in China. One of the original Chinese achievements is the synthesis of benzene, which is a basic material in the manufacture of plastics. Small plants for manufacture of synthetic benzene have been set up near plastic-manufacturing facilities in many areas of China. Major polyvinyl chloride manufacturing plants have been identified in Lanchow, Taiyuan, Peking, and at the Shanghai Liaoyuan Chemical Works. Industrial observers believe that

Table 10.6 Major sales of chemical fertilizer plants to China since 1963

Year	Type of Plant Imported	Capacity in Tons/Year	Supplier and Country	Contract Value in Millions of $U.S.
1963	Synthetic ammonia plant (Luchow)	105,000	Humphreys & Glasgow, U.K.	8.4
	Urea plant	175,000	Stork-Werkspoor, Netherlands	7.0
	Ammonia nitrate (for Albania)	110,000	Montecattini, Italy	14.2
	Synthetic ammonia plant parts		Italy	3.6
1965	Four ammonia plants	40,000 ea.	Humphreys & Glasgow, U.K.	23.52
	Chemical fertilizer plant	150,000	Montecattini, Italy	7.5
1972	Two fertilizer plants		Kagaku, Mitsui, Tojo, Japan	
	Two ammonia plants		Hitachi Shpbld. Nippon Shokubo	
1973	Three urea plants	480,000 ea.	Kellog Continental, Netherlands	34.0
	Three ammonia plants	330,000 ea.	MW Kellog, U.S.A.	70.0
	Ammonia plant	330,000 }	Toyo, Mitsui, Toatsu, Japan	42.0
	Urea plant	528,000 }		
	Five urea plants	480,000 ea.	Kellog Continental, Netherlands	56.0
	Three ammonia plants	330,000 ea.	MW Kellog, U.S.A.	130.0
1974	Three ammonia plants (Nanching)	330,000 ea. }	Heurtey Industries, France	118.0
	Three urea plants	574,000 ea. }		
	Ammonia catalyzer, H_2, methanol		Haldor Topsoe, Denmark	
1976	Synthetic ammonia converter	360,000	Nippon Kokan, Japan	
1977	Phosphatic plant technology		Nissan Chemical, Japan	
	Ammonia/urea complex interest	large	Nihon Ammonia Company, Japan	

Source: Compiled by 21st Century Research from various publications listed in references to this chapter.

China is not yet manufacturing commercial quantities of such plastics as Teflon or engineering plastics like ABS, polyacetals, polycarbonates, polyamides, and unsaturated polyesters.

Plastics like polystyrene and polyurethane have been reported in limited production. Some are being test-produced for use in certain consumer goods. There are also indications that China is interested in leather substitutes for manufacturing of shoes and other apparel.

Imports of plastics reached a high of $125 million in 1974 and consisted mainly of polysterene and polyvinyl derivatives, which account for about 80 percent of plastic imports by value. Japan once again is the largest plastics supplier, but by no means is it so dominating a supplier as in other chemicals. Italy, Belgium, and West Germany also relatively enjoyed a reasonable market share. Romania supplied 8600 tons of polyethelene granules during 1970 to 1973. It is probably continuing to export plastics to China, but data for more recent Romanian exports to China are not available. At least ten other countries exported some plastics to China, but the amounts have been small. Overall plastics account for less than 10 percent of total chemicals imported, and their market share appears to be declining (see Fig. 10.7).

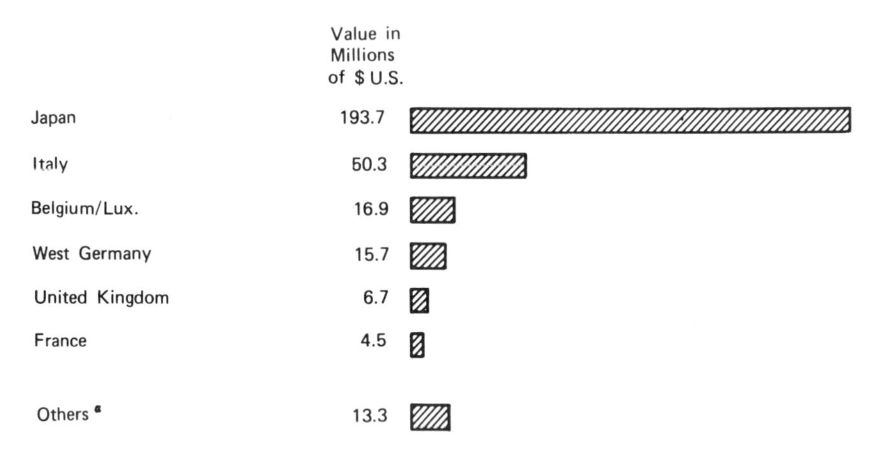

Figure 10.7 Market shares of major plastics suppliers during 1970–1975. (Source: *OECD Commodity Trade Statistics and* **United Nations Annual Bulletin of Trade in Chemical Products, 1973 and 1975.**) [a] *Others include Sweden (3.96), United States (2.7), Netherlands (2.6), Norway (0.5), Switzerland (1.5), Austria (1.1), Yugoslavia (0.84), Australia (0.1), Canada (0.004), Romania (8600 tons of polyethylene granules).*

INDUSTRIAL CHEMICALS PRODUCTION AND IMPORTS

For the manufacture of chemicals other than fertilizers, plastics, and synthetic fibers China appears to depend much more on local production in small plants located within user areas. Although quantities manufactured appear to be adequate for current needs, total output is not regarded as sufficient to provide for industrial growth in the future. Demand for sulfuric acid, for example, is bound to increase significantly and with it a need to develop larger more efficient plants, which may once again have to be imported from the West.

Over sixty chemical manufacturing plants have been identified in China, but the total number is unknown. Most produce relatively small quantities of caustic soda, carbide, detergents, pesticides, sulfuric acid, or industrial gases for local consumption. Their capacities, when reported, appear to vary from 2000 tons per year to 10,000 or even 20,000 tons per year of a particular chemical. In comparison with other industry sectors, only a very limited amount of information is available about production of these chemicals.

Elements and Compounds

One of the largest plants identified appears to be the Shanghai Electrochemical Works, which manufactured 75,000 tons of caustic soda annually as well as twenty other chemicals in 1972. It expanded from an original capacity of only 15,000 in 1959 and is now regarded as a medium-sized plant in the chemical industry, employing about 2300 workers.

Shanghai is the central point for several chemical plants. Its Wusung Chemical Works operates a large oxygen-generating plant and succeeded in extracting xenon and krypton from the air. The Liaoyuan Chemical Works claims to have 10,000-ton/year capacity in waste recovery from manufacture of caustic soda and chlorine products, among thirty other chemicals.

Near the Taching Oilfield the largest nitric acid plant in China was reported in operation in 1976. The Ningpo Sulfuric Acid Works in Chekiang Province was reported in 1969 to have 20,000-ton/year capacity. There are also sulfuric acid plants in Shanghai and Foochow, but details of their production are not known. The Yangshupu Coal Gas Plant makes sulfuric acid and caustic soda from waste gases. Hydrochloric acid was reported in production at the Canton Chemical Works, along with caustic soda, insecticides, and five other chemicals derived from residues in the electrolytic salt-production process.

The Hofei Chemical Works in Anhwei Province make carbide, oxygen, and solid emulsified DDT products. A few years ago the capacity of Chinese carbide furnaces was reported to be in the region of 205,000 tons per year. Carbide is produced in five standard models of furnaces, ranging in capacity from 3600 to 20,000 kVA (Kilovolt amperes). Nearly all carbide production systems include three-phase electric sealed furnaces. The quality of carbide manufactured was reported to be in the order of 80 to 83 percent of calcium carbide content.

Dyeing, Tanning, and Coloring Materials

Shanghai also appears to be the center of dyestuffs manufacture in China, and at least four dyestuffs plants have been identified there. The Shanghai Taihsin Dyestuffs Factory makes fast dyes for cottons, and other Shanghai plants are engaged in manufacture of high-grade dyestuffs. There is also a dyestuffs plant in Tientsin that apparently uses modern remote control machinery. The Kirin Chemical Industry Company manufactures dyestuffs as well as carbide and fertilizers. Some synthetic dyestuffs and pigments are made for exports, but China still remains a net importer of these chemicals.

Pesticides Production

During the early 1970s the insecticide industry claimed to operate 300 plants in China that were manufacturing 120 different products. The industry is supported by over twenty specialized institutes that employ a total of at least 2000 research workers. New developments included low-toxicity pesticides such as long-term phosphorous insecticide, against aphids and mites and other pesticides to prevent rice leaf wilt.

Salts of mercury, copper, and arsenic, or DDT provide the basis for most insecticides manufactured in China. A limited production of organic phosphate insecticides is believed to exist, but their use is not widespread because of their higher cost. On the other hand, the attention now given to improvement of agricultural productivity must eventually lead to more extensive use of modern pesticides, insecticides, fungicides, and herbicides that are of low toxicity to animals and are less damaging to the environment.

Production of insecticides has been reported at the Canton Chemical Works, the Kannan Insecticide Factory in Kiangsi Province, Nantung Insecticides Plant in Kiangsu Province, Pengpu Insecticides Factory in Anhwei Province, and Wuhu Farm Insecticide Plant, also in Anhwei Province. The Shihchi plant in Kwangtung Province was reported to produce herbicide ether, known as TOK, whereas the Tientsin Chemical Works, which makes forty-five different products, manufactured a reported 1400 tons of insecticides per year in the early 1970s.

Gum Rosin and Turpentine

China manufactures and exports chemicals such as tung oil, menthols, gum rosin, and turpentine. Tung oil is a by-product of pine sap and is used in the production of waterproof paints. The Hsinyi Resin Factory in Kwangtung is reported to be extracting oil of turpentine from resin by a new process developed at the plant. The Nantung Oils, Fats and Chemicals Plant in Kiangsu Province produces cottonseed oil, epoxy resins, and soaps.

Production of these chemicals is more than adequate because China traditionally exports tung oil and turpentine as well as gum rosin to several countries. In 1975 those exports amounted to $40 million and accounted for over 13 percent of all exports of Chinese chemicals that year. Japan was the main importer, but amounts varying in value from two to several million dollars have been exported to the United States, United Kingdom, West Germany, France, Italy, and the Netherlands.

Soaps and Detergents

The Suchow Synthetic Detergent Plant in Kiangsu Province was reported to have 10,000-ton/year capacity. It began operations in 1970 and is one of the largest identified in this product category. The Kalgan Chemical Plant in Hopeh Province, which manufactures 5000 tons of synthetic detergents per year, may be more typical in this industry. It was also completed in 1970 at a cost of $1.7 million.

In 1972 the Foochow Soap Factory in Fukien Province reported an output of 3000 tons of synthetic detergents. Another plant was also built in Chengtu in Szechwan Province in 1970. In the same year the Hsiku Ward Synthetic Detergents Plant went on stream in Lanchow in Kansu Province with an annual capacity of 2000 tons. Some plants that originally started as small soap factories may have changed in time to other products. The Wulanhaote factory in Kirin Province, which originally began as a soap and candle plant in 1964, was reported in 1975 to be engaged in manufacture of paints.

In 1975 China spent $36 million on purchase of a detergent plant from Italy that may serve as a model for a new generation of such plants to be built in China. Exports of soaps and detergents are negligible, but there is a small annual import of detergents from Japan amounting to about $6 million per year. This suggests that China did not so far solve the problem of an adequate supply of detergents, but this market is unlikely to increase and is of relatively minor importance, even in Sino-Japanese trade.

Medicinal and Pharmaceutical Products

China is believed to be basically self-sufficient in chemicals required for pharmaceutical production. Manufacture of herbal and traditional medicines as well as Western-type drugs is expanding. National output of sulfonamides, antibiotics, antipyretics, antitubercular drugs, hormones, vitamins, and agents against endemic diseases was claimed to have doubled between 1966 and 1971.

Many of the Western-type drugs produced in China were developed and evaluated in other countries. China, however, did develop some new products, notably synthetic insulin and a new antibiotic, Chuansanmycin, in 1972. Progress in this area will depend on Chinese access to latest discoveries in the form of foreign samples and acquisition of know-how for converting trial manufacture into mass production.

Much of the pharmaceutical production in China is carried out in relatively small plants, as is the case with many other chemicals. Tientsin has twelve pharmaceutical plants, with a total of 10,000 workers manufacturing 900 types of drugs. One plant specializing in herbal medicines and three other pharmaceutical plants have been identified in Shanghai.

Vaccines and biological sera are produced by specialized institutes. The Peking Institute for Biological Products has modern production facilities and employs 600 workers. It supplies vaccines for Hopei, Shansi, and Inner Mongolia. American visitors to that institute were informed that facilities similar in size and organization serve other parts of China. Some vaccines such as that for yellow fever are produced for export to foreign countries.

The importance of herbal and traditional medicines is underscored by formal research conducted under the guidance of the Academy of Chinese Traditional Medicine. The American scientists who visited China in 1974 felt that, despite the emphasis on herbal medicine research, effective progress in obtaining new principles or synthetic drugs from plants would require substantial investment in both advanced training and equipment.

China is a net exporter of medicinal and pharmaceutical products. In 1975 a total of $60 million of such products was exported. In 1976 exports dropped to only $40 million. Of this amount $20 million was imported by Hong Kong. The colony is supplied by China with a wide range of common drugs including antacids, streptomycin, paracetamol, and APC tablets, promethazine hydrochloride, vitamin C, calcium gluconate, and the pain killer sodium aminosalicylate. A total of $3.5 million of China's medicinal exports is in the category of opium alkaloids and other habit-forming drugs and their derivatives. The United States and West Germany were the largest importers of these products in 1975. In comparison, Chinese imports of medicinal and pharmaceutical products are relatively low. In 1975 China imported only about $5 million of pharmaceutical products from the West. Primary suppliers were Japan, France, West Germany, the Netherlands, the United Kingdom, Switzerland, and Italy.

Chinese traditional medicines are exported mainly to other Asian countries, Africa, and Chinese communities throughout the world. Less developed countries are the largest importers of Chinese medicinal products. Annual shipments to those markets are valued at about $25 million. Another $5 million is exported to communist countries, most likely to North Korea and Vietnam.

Imports of Industrial Chemicals

Total imports of chemical products other than fertilizers, synthetic fibers, and plastics amounted to $318 million in 1975 and represented 39 percent of all the chemicals imported that year. Of that amount $255 million represented imports of chemical elements and compounds (SITC code 51), which was the largest product group. Other significant product groups were pesticides, amounting to $25.6 million, and dyestuffs and coloring materials, amounting to another $23.6 million. Pharmaceuticals and detergents accounted for only $4.5 and $6.2 million, respectively. Imports of mineral tars, essential oils, cosmetics, and radioactive materials were negligible in comparison. This pattern of imports is also consistent with longer-term trends, as shown by cumulative imports in those product categories (see Fig. 10.8). Preliminary results for 1976 indicate that imports of elements and compounds were reduced to $210 million.

Organic chemicals represent the largest product group imported

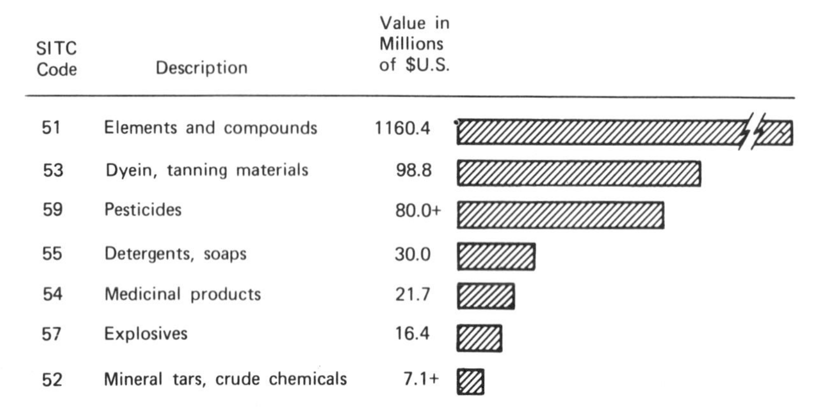

***Figure 10.8** Estimated proportions of chemical product import categories other than fertilizers, synthetic fibers, or plastics during 1969–1970.* (**Source:** *Based on U.S. Department of Commerce Market Share Reports and OECD Commodity Trade Statistics Series B for SITC codes as indicated.*)

within the chemical elements and compounds category. In 1975 imports of organic chemicals amounted to $163 million. These included hydrocarbons such as styrene, alcohols and phenols, organic acids and compounds, polyacids, nitrogen-function compounds, and organoinorganic heterocyclic compounds. Inorganic chemicals accounted for another $76 million, of which $50 million represented imports of ammonium chloride from Japan. Japan supplied 42 percent of all organic chemicals imports in 1975, and West Germany accounted for almost 30 percent of this product category. Italy, with over 11 percent of the total value, was the third largest supplier, which was consistent with its longer-term market share of this trade (see Fig. 10.9).

Imports of pesticides have been showing the fastest growth during mid-1970s, but their total value remains only 15 percent of the imports in the industrial chemicals category and only 3 percent of all the chemical imports excluding rubber. Japan supplies about 60 percent of all pesticides to China, and recently shipped products included Bassa (BPMC), Captan, Hopcide (CMOC), Phenazine, Mipcin (MICP), Padan (CARTAP), and Tsumacide (MTMC). Although pesticide imports are increasing, they will continue to represent a relatively small percentage of the total chemicals imports. Besides Japan, the Western countries of West Germany, France, Sweden, the United Kingdom, the United

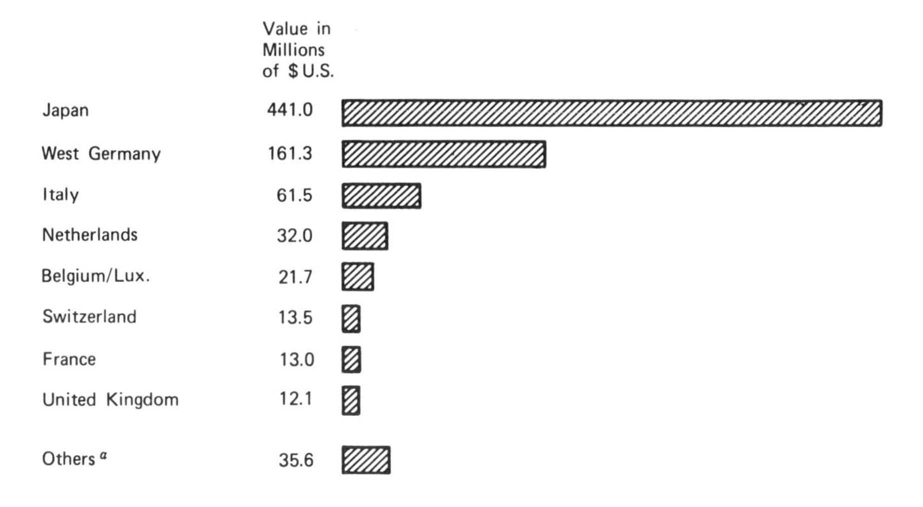

***Figure 10.9** Estimated market shares of organic chemicals suppliers during 1970–1975, SITC commodity code 512.* (Source: *Based on OECD Commodity Trade Statistics Series B and some foreign-trade yearbooks for individual countries.*) [a] *Others include Australia (0.1), Austria (3.8), Canada (0.4), Czechoslovakia (0.6), Denmark (5.2), Finland (0.1), Norway (0.2), Romania (3.4 for years 1970–1973 as derived from volume of chemicals), Spain (1.3), Sweden (2.9), Turkey (0.3), United States (8.1), Soviet Union (4.2 converted from rubles at varying exchange rates, data for Soviet code 302).*

States, and the Netherlands are also pesticide suppliers to China, but average shipments do not exceed $2.6 million for any country.

Imports of synthetic dyestuffs have also shown some growth again in recent years but they represent only about 3 percent of the total chemicals import market. In 1975 about $24 million of dyestuffs was imported, and in this product category West Germany was the leading supplier, with a 38 percent market share in 1975. France, Switzerland, the United Kingdom, and Japan are the other dyestuffs suppliers.

RUBBER PRODUCTION AND IMPORTS

China is both a producer and importer of natural as well as synthetic rubber. Estimates of total rubber consumption range up to 500,000 tons per year. This would suggest per capita rubber consumption in China in the order of 1.1 pounds per year, which is very far below 20 pounds for Japan and 30 pounds for the United States. About 60 percent of total consumption is believed to be natural rubber, mostly imported from Sri Lanka, Malaysia, Singapore, and Thailand. Rubber is considered a strategic commodity to China, and Chinese rubber trade is a significant factor in world rubber markets (see Fig. 10.10). During 1970–1976 China's imports of rubber averaged 231,000 metric tons/year. Volumes have ranged from a low of 195,000 in 1971 to a high of 290,000 tons valued at $150 million in 1976. About 10 percent of all natural rubber is imported from communist countries, probably Cambodia and Vietnam, and this percentage may increase in the future as rubber production in those countries recovers.

Synthetic rubber production was reported at 60,000 tons in 1963 in two rubber plants operating in Lanchow and Kirin. It is estimated that by the mid-1970s this production may have increased to about 150,000 tons per year. In 1976 China purchased a styrene butadiene rubber plant from Japan Synthetic Rubber Company with a capacity of 240,000 tons per year. It is the largest synthetic rubber plant by world standards, with comparable units known to operate only in Japan and in Texas. When this plant comes on stream China's synthetic rubber production capacity should be in the order of 400,000 tons per year. This will make China the fourth largest synthetic rubber producer in the world (see Table

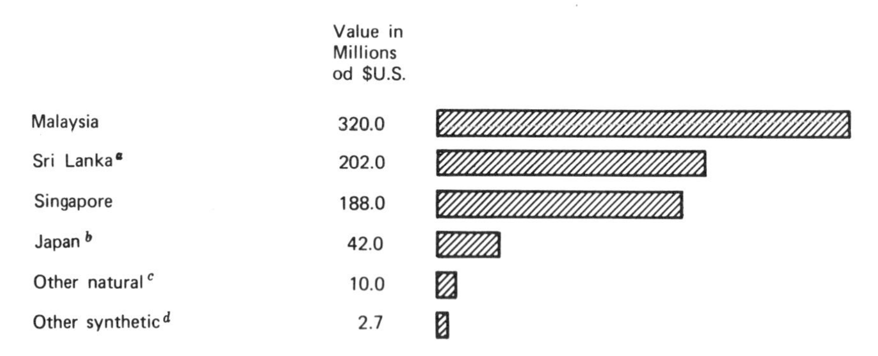

Figure 10.10 *Estimated market shares of natural and synthetic rubber suppliers to China during 1970–1975. (Source: Based on data in several publications of U.S. Government Joint Economic Committee, OECD Commodity Trade Statistics, and China Trade Report periodic exports statistics.)* [a] *Partially estimated by deriving value from volume of rubber exported to China under a barter agreement in exchange for rice.* [b] *Synthetic rubber only.* [c] *Communist areas in Asia, probably Vietnam and Cambodia.* [d] *Includes France (1.9), United Kingdom (0.4), Netherlands (0.2), Canada (0.1).*

Table 10.7 Production of natural and synthetic rubber in selected countries of the world for 1975 (All figures in metric tons)

Country	*Natural Rubber Production*	*Synthetic Rubber Production*[a]	*Estimated Consumption*
World	3,300,000	5,530,000	8,830,000
China	50,000	150,000+	350,000
Malaysia	1,477,600	—	
Indonesia	825,000	—	
Thailand	348,700	—	
Sri Lanka	148,800	—	
India	136,000	—	
United States	—	2,020,000	3,000,000
Soviet Union	—	1,450,000	
Japan	—	836,000	950,000
France	—	372,000	

Source: *United Nations Statistical Yearbook, 1976* and authors' estimates.
[a] Includes new synthetic production and all reclaimed rubber.

10.7). In 1973 China also showed interest in purchasing a butyl rubber plant from Polysars Company in Sarnia, Ontario, but that company was not interested in selling that technology. Exxon is the only other manufacturer of butyl rubber, and the technology for manufacture of this special highly resistant rubber is strictly controlled by those two manufacturers.

Natural rubber is grown in southern parts of China, notably on Hainan Island and in Yunnan, Kwangsi, and Fukien Provinces. Based on availability of cultivated land, up to 50,000 tons of rubber per year could be grown in Hainan alone. When all the other areas are considered it is not unreasonable to assume that China may have a capacity to grow up to 100,000 tons of natural rubber per year.

If Chinese rubber consumption increases at about 5 percent annually, the total rubber demand will exceed 500,000 tons in 1985. As in any industrializing economy, at least 15 percent of this total or 75,000 tons per year will have to be met with natural rubber. It is conceivable that China will expand its natural rubber production to at least that level; this would make it independent of any rubber imports once production of synthetic rubber reaches about 400,000 tons per year, probably sometime in late 1978 or early 1979. Based on natural rubber productivity of Malaysia, about 200,000 acres of land within appropriate climatic regions would have to be dedicated for that purpose, which is well within potential capabilities of China.

Whether imports of natural rubber will continue depends apparently on China's ability to expand its synthetic rubber production. This in turn depends on how quickly the new 240,000 tons per year rubber plant can be brought on stream. Since China has been spending at least $100 million annually for imports of rubber, the purchase of that plant for $27 million is a relative bargain. As a result of this new capacity, imports of natural rubber could be reduced or even eliminated altogether in future years. Theoretically, imports of natural rubber would not have to increase again until domestic supply of natural rubber becomes less than 15 percent of total rubber consumption in China. However, there are other important reasons why China may not want to eliminate imports of natural rubber, even though it will have the capacity to do so in the very near future. One reason is the possibility of using the land for growing food, but more importantly, there are special rubber-trade agreements with Southeast Asian countries who are the major rubber suppliers.

The ability to purchase rubber from Sri Lanka, Malaysia, Singapore, and Thailand without the necessity to do so would give China a unique political and economic leverage in Southeast Asia that it did not enjoy in previous years. In addition, those countries also purchase over $550 million of food and textile products from China every year, and trade with Malaysia and Singapore consistently shows a large surplus in favor of China. Continuing exports of rubber to China are the only means to offset some of the large deficits those countries are running in their trade with China. Imports of rubber from Sri Lanka have been conducted on a rice-for-rubber barter basis for the last 25 years. This arrangement came about as a result of the American embargo on exports of rubber to China at the time of the Korean War. Ceylon, whose commodity markets were seriously depressed at that time, resisted American pressure and continued selling rubber to China. It was rewarded soon after with a long-term rice-for-rubber barter-trade agreement. This trade now has some overtones of a foreign-aid program because China usually supplies Sri Lanka with 200,000 tons of rice at below world prices and buys the rubber at a premium over existing Singapore rubber quotations. In 1976, 67,000 tons of rubber was exported to China under those terms.

China's rubber trade is thus politically sensitive, as was also demonstrated during the 1960s when up to 70 percent of the natural rubber was being imported from Indonesia, after Sino-Malayan relations deteriorated in 1962. In its turn Sino-Indonesian rubber trade came to an abrupt halt in 1967 after Indonesia suspended relations with China in the aftermath of the attempted coup d'etat in 1965.

In 1975 China also began direct purchases of rubber from Thailand, and that country's embargo on rubber sales to China was lifted. In that year Thailand also imported 250,000 tons of Chinese oil. The prospects of oil-for-rubber barter may be another motivation for this new trade, although even before the embargo was lifted Thai rubber was purchased by China through Malaysia and Singapore. In the same year China also began purchasing rubber from communist countries in Asia, probably from Vietnam, which further diversified its sources of supply. It remains to be seen how expanded domestic rubber production and availability of Chinese oil for possible barter trade will affect China's relations with Southeast Asian countries. In any event, the ability to dictate terms in major world rubber markets provides China with a powerful new political and economic weapon not available to it in previous years.

REFERENCES

Chemical Purchasing. "International Buying," Interview with Julian Sobin, September 1974.

China Quarterly. "The Production and Application of Chemical Fertilizer in China," December 1975.

China Trade Report. "Far Eastern Economic Review," February 1975, p. 5.

China's Chemical Industry Surges Forward, Promotional brochure, Peking.

Crook, Frederick W. "The Commune System in the People's Republic of China, 1963–1974," in *China: A Reassessment of the Economy,* Part III, Rural and Agricultural Development, Joint Economic Committee, U.S. Congress, July 1975, p. 350.

Erisman, Alva L. "China: Agriculture in the 1970's," in *China: A Reassessment of the Economy,* Part III, Rural and Agricultural Development, Joint Economic Committee, U.S. Congress, July 1975, p. 324.

Hartley, William D. "Chinese Make Progress in Cutting Pollution," *Wall Street Journal,* November 21, 1972.

Hoose, Harned Pettus. "Petrochemicals," in William W. Whitson, ed., *Doing Business with China,* Praeger, New York, 1974.

Hydrocarbon Processing. "HPI Construction Boxscore: Far East," October 1974.

International Petroleum Encyclopedia 1977. "Ethylene and Aromatics from Crude," The Petroleum Publishing Co., 1977, p. 420, Tulsa, Oklahoma.

Joint Publications Research Service. "Carbide Production in China," (transl. from Chinese text by JPRS #65552), pp. 8–9. 27 August 1975.

K'o-Hsueh Shih-Yen. "Virtues of Small Ammonium Bicarbonate Fertilizer Plant Extolled," December 1971, (transl. by JPRS #57106), Peking.

Lasagna, Lauis. "Herbal Pharmacology and Medical Therapy in the People's Republic of China," *Annals of Internal Medicine,* **83**, 887–893, December 1975.

Liu, Jung-Chao. *China's Fertilizer Economy,* Edinburgh University Press, 1971.

Le MOCI, Moniteur de Commerce International. "Chimie-Petrochimie," No. 118, December 3, 1974, Paris.

Perkins, Dwight H. "Constraints Influencing China's Agricultural Performance," in *China: A Reassessment of the Economy,* Part III, Rural and Agricultural Development, Joint Economic Committee, U.S. Congress, July 1975, p. 350.

Sidel, Victor and Ruth Sidel. "Serve the People," Josiah Macy Jr. Foundation, New York, 1973, pp. 188–189.

Sigurdson, Jon. "Rural Industrialization in China," in *China: A Reassessment of the Economy,* Part III, Rural and Agricultural Development, Joint Economic Committee, U.S. Congress, July 1975, p. 411.

Sobin, Julian M. "Chemicals and Fertilizers," in William W. Whitson, ed., *Doing Business with China,* Praeger, New York, 1974.

Sobin, Julian M. *Bilateral Breakthrough,* Sobin Chemicals, Inc. Brochure, 1974.

United Nations. *Annual Bulletin of Exports of Chemical Products 1975,* U.N. Economic Commission for Europe, Geneva.

U.S.–China Business Review. The Magazine of the National Council for U.S.–China Trade, May–June 1975, p. 53.

U.S. Government. "People's Republic of China: Chemical Fertilizer Supplies 1949–1974," *Research Aid* A (ER) 75-70, August 1975.

U.S. Government. "China: International Trade, 1976–77" *A Research Paper* ER77-10674, November 1977.

Wang, K. P. *PRC a New Industrial Power with a Strong Mineral Base,* Bureau of Mines, U.S. Department of the Interior, 1977.

Wong, John. "Chinese Demand for South East Asian Rubber 1949–72," *The China Quarterly,* 1974, p. 490.

Yuan, S. Y. "China's Chemicals," *U.S.–China Business Review,* November–December 1975, p. 37.

11

Machine-building Industries

This section of China's economy is of crucial importance to the development of China as a modern economic and political power. Machine-building industries must provide the plant to operate all the modern industries in China as well as ordnance to equip one of the largest army, navy, and air forces in the world.

Machinery production has grown more rapidly than total producer goods output and almost twice as rapidly as the overall industrial production. It was singled out within the national industrialization program of the 1950s as an industry that was to make China self-sufficient in machinery and equipment within the span of three five-year development plans. This goal was never achieved because of the disruptions of the Great Leap Forward period and the Cultural Revolution. But it was not as unrealistic at the time as it may appear in retrospect. The Chinese machine building industry was developed with considerable assistance from the Soviet Union, and had this aid continued uninterrupted until about 1970, China's machine-building industry would have grown to be considerably larger than it is today.

Average annual growth rate for machinery production since 1957 until the early 1970s was 13 percent. It had declined by almost 60 percent in 1961 and again by 16 percent in 1967 but recovered rapidly once political disruptions ended. Between 1967 and 1973 machinery production increased at an average 18 percent per year. The rate of growth has slowed down to an average 10 percent during the 1970s, but the total machinery production of China is now estimated to be over ten times as large as 20 years ago (see Fig. 11.1).

Machinery and equipment represent about 30 percent of China's total imports and in certain industries account for a significant share of total consumption. Even so, 1975 imports of machinery and transportation

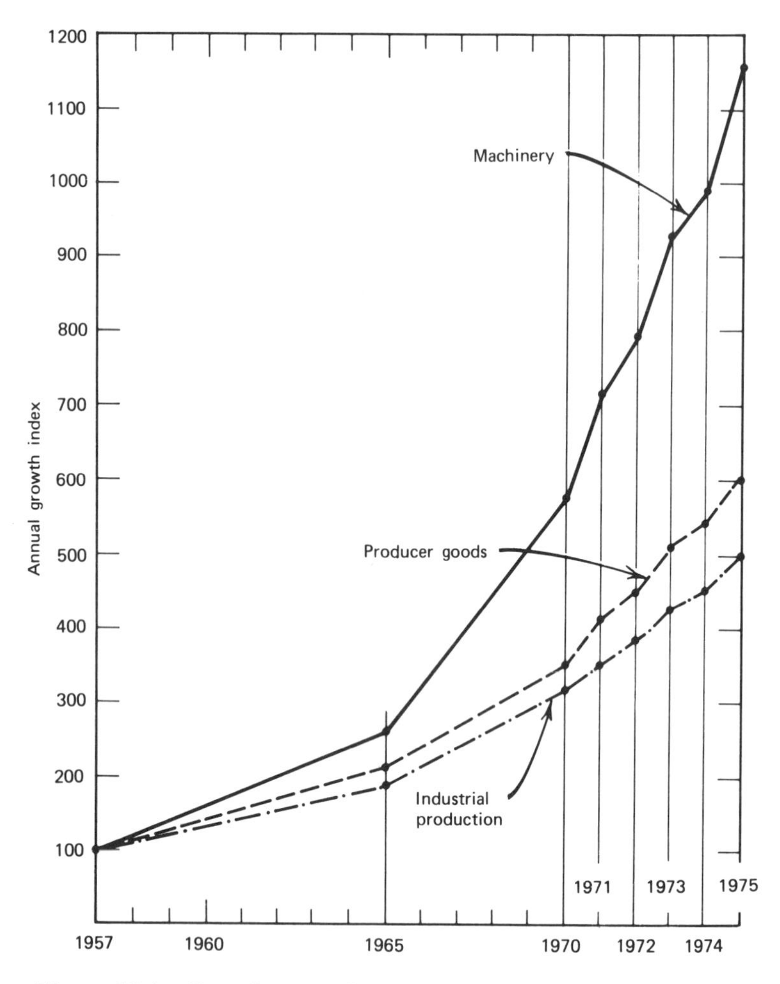

Figure 11.1** Growth of machinery, producer goods, and industrial production.* **(Source: "PRC: Handbook of Economic Indicators," U.S. Government Research Aid,** ***ER-76-10540, August 1976.**)**

equipment valued at $2,280 million were less than 1 percent of the GNP. Even if output of agriculture is ignored, imports of machinery account for only about 1 percent of the remaining GNP of China. If the machine-building industry is assumed to account for no more than 25 percent of total industrial output, which is comparable to the situation in the Soviet Union during the 1960s, the value of imports is still less than 5 percent of the total value of China's machine-building production.

Investment priorities in machine-building industries change according to the objectives of the five-year plan that happens to be in force. This leads to the sometimes confusing reports of door-handle factories producing computers and shipyards manufacturing hydraulic presses to produce equipment for chemical plants. What it actually means is that labor and equipment are reallocated to the construction of different products, probably because of immediate applicability of existing equipment or labor skills to the new priorities.

During the first Five Year Plan (1953–1957) development of machine building was concentrated on trucks and tractors, machine tools, mining and metallurgical equipment, and power equipment such as boilers, diesel engines, and turbines. The Great Leap Forward period (1958–1960) saw much greater emphasis on development of mining and steel-making equipment to support the massive move to open up small local coal mines and backyard iron and steel furnaces. The agricultural crisis that followed forced China to shift machine-building priorities once again to emphasize agricultural machinery, chemical fertilizer-production equipment, and petroleum-industry machinery.

In the late 1960s, even at the height of the Cultural Revolution, machine-building priority shifted to electronics. Telephone and instrument plants began producing computers and navigational equipment. After disturbances of the Cultural Revolution subsided a debate developed as to whether electronics or iron and steel should receive highest priority. During the early 1970s steel received top priority as the "key link," and the greatest expansion took place in plants manufacturing mining and metallurgical equipment once again. At present all machinery supporting the drive to increase agricultural productivity appears to have the highest priority.

Massive imports of turnkey plants and transportation equipment during the 1970s are perhaps the most telling symptoms of the shortcomings of China's machine-building industries. Momentarily these imports were estimated to be as much as 6 to 8 percent of China's total machine-building output. They also indicate that the Chinese machine-building industry failed to meet the full domestic demand for ships, trucks, aircraft, railway vehicles, metallurgical equipment, electric power-generating machinery, machine tools, and petroleum and petrochemical equipment. This is particularly significant because since 1971 military procurement, which relies heavily on output of the machine-building industry, is estimated to have been reduced by about 25 percent.

Growth in machine-building industries also depends heavily on innovation, a process that is not known to thrive in centrally planned economies. The Soviet Union and eastern Europe, despite significant achievements measured by volume of outputs in various industries, are constantly trying to import machines and equipment incorporating the latest technological advances that almost invariably are developed in the United States or western Europe. During its close cooperation with the Soviet Union China's machine-building industry depended for innovation on Soviet technology, not unlike the manner in which the communist countries depend on the West. Since the 1960s and its "prototype purchasing" programs from the West, China in fact is believed to have advanced to narrow the technological gap between several of its machine-building industries and their counterparts in the West.

MINISTRIES OF THE MACHINE-BUILDING INDUSTRY

At present there are seven different ministries that supervise machine-building production in China. Six of these are believed to be concerned with equipment designed for use by the armed forces, and their activities are shrouded in great secrecy. As a result, very little is known about some of the industries under the control of those ministries, although they represent a significant portion of all the machinery production in China (see Figure 11.2).

The First Ministry of Machine Building came into being in 1952 and supervises the production of machinery and equipment for general economic development. There are at least forty major product categories that are made by enterprises under the control of this ministry. According to one industry observer who studied the structure of this ministry during the 1950s, it consists of ten separate production control bureaus that supervise specific product groups.

The Second Ministry of Machine Building is believed to be responsible for all nuclear work in China since 1959. Originally this ministry came into being in 1952 to supervise the production of armaments and ammunition. In 1958 the Third Ministry took over some military production and was renamed the Second Ministry. Its vice minister at the time was Liu Chieh, who was China's chief representative at the Joint Nuclear Institute at Dubna near Moscow. His subsequent appointment in 1960 as minister led outside observers to believe that this ministry supervises China's nuclear research and weapons program.

The Third Ministry of Machine Building was recreated in 1960 and until 1963 was one of the only two ministries then engaged in defense production. It is believed to be engaged in the production of conventional weapons for the PLA and other paramilitary organizations in China. The nature of its production is officially unknown but is only

First Ministry of Machine Building	Light–industrial equipment Machine tools and tools Metals Metallurgical and mining equipment Power–generating machinery Locomotives and railway vehicles Automotive equipment Instruments and meters Electrical generating machinery Shipbuilding yards Radio and electronics
Second Ministry of Machine Building	Nuclear research Nuclear fuels production Nuclear weapons program
Third Ministry of Machine Building	Conventional weapons Armament production
Forth Ministry of Machine Building	Military electronics Telecommunications Air navigation
Fifth Ministry of Machine Building	Heavy weapons Tanks Artillery
Sixth Ministry of Machine Building	Naval ships Submarines
Seventh Ministry of Machine Building	Aircraft Missiles and rockets Jet engines Satellites

Figure 11.2 Areas of production responsibility of machine-building ministries in China.

presumed, based on the backgrounds of its top leaders. It was originally headed by Sung Jan-chiung, a former minister of the Second Ministry. Among his vice ministers were Liu Ting, former director of the Bureau of Arsenal; Cheng Han-tao, former deputy director of the Bureau of Arsenal; and Chao Chi-ming, a commander of the PLA Fleet. It is presently headed by Li Chi-tai, a former military commander.

The Fourth Ministry of Machine Building is believed to supervise military electronics production. This belief is also based on the fact that its minister, Wang Cheng, was formerly a deputy commander of the Signal Corps of the PLA and also served as vice minister of the Ministry of Posts and Telecommunications. He is also active in the China Electronics Society. Major electronics plants are probably under the control of this ministry. Other electronics plants that have proliferated all over China since the late 1960s are probably supervised by local county and municipal authorities. A few of the larger radio and television plants may be controlled by the First Ministry of Machine Building, particularly since the rapid expansion of radio and TV production during the 1970s.

The Fifth Ministry of Machine Building is believed to be involved in the production of heavy weapons and artillery. When it was established in 1963 it was headed by general Chiu Chuang-cheng, who was previously a deputy commander of artillery in the PLA. The plants of this ministry probably produce T-59 tanks based on Soviet 54A vehicles, the T-62 light tanks, the T-60 amphibious tanks, and a Chinese-designed armored personnel carrier (APC). The ministry may also be responsible for the manufacture of artillery units ranging from a 82-mm (millimeter) Chinese-designed recoilless gun to 152-mm units, as well as a range of antiaircraft guns, heavy mortars, and rocket launchers. Some magnitude of the manufacturing and maintenance effort required by this ministry may be gained by comparing pertinent Chinese and American armed forces. China is believed to operate up to 9000 tanks equal in number to those in the U.S. army. Its artillery inventory, estimated at up to 17,000 pieces, is almost three times as large as that of the United States and has 6000 heavy mortars, almost twice as many as the American army.

The Sixth Ministry, established in 1963, was originally headed by Vice Admiral Fang Chiang, who was formerly a deputy commander of the Chinese Navy. It is believed that this ministry now controls all naval and merchant-marine shipbuilding, although China's shipyards were once under the control of the First Ministry of Machine Building and at least seven large shipyards were supervised by the Ministry of Communications. A detailed discussion of China's shipbuilding production and imports is found in Chapter 13.

The Seventh Ministry of Machine Building was first reported in existence in 1964. Since its original minister was General Wang Ping-chang, who was previously the deputy commander of the Chinese Air Force, it is believed that the ministry controls the aerospace industry in China. This would include production of aircraft, rocket missiles, and earth satellites, placing the Seventh Ministry in the forefront of China's technological achievements. It also supervises a significant industry that must produce, maintain, and replace over 5000 aircraft of its air forces and at least 400 aircraft of the civil aviation fleet. A more detailed discussion of China's aircraft industry is found in Chapter 13, and additional data about Chinese aerospace achievements are presented elsewhere in the present chapter.

THE MILITARY–INDUSTRIAL COMPLEX OF CHINA

All the machine-building ministries except the First Ministry reflect PLA service arms such as Artillery, the Signal Corps, or the Navy. They are managed almost exclusively by former military commanders and in fact are conglomerates of industrial enterprises dedicated to the support of the military services. In addition, the PLA operates a Capital Construction Engineering Corps, a Railway Engineering Corps, and an Engineering Corps, which collaborate with other departments of the State Council. The primary mission of these services is to support the military capability, but they are involved in significant industrial activity that must have a great economic impact on many parts of the country (see Fig. 11.3).

In some areas of China it appears that PLA production and construction corps virtually run the economy. The Sinkiang Military Command, for example, was reported recently to operate 500 large, medium, and small factories and mines. These include power plants, iron and steel factories, coal and metal mines, and various plants producing chemicals, building materials, textiles, machinery, sugar, paper, leather, ceramics, cigarettes, food, and animal products. In all regions of China, PLA units try to raise their own food and operate small plants and mines to provide for their needs while any surplus is sold to the state. On the other hand, military equipment is often employed to support various economic activities. Air Force Transport Units provide cloud seeding, forest-fire fighting, insecticide distribution, supply, and mineral survey services. The PLA Navy units perform mosquito control. The antiaircraft artillery engages in hail cloud-dispersal programs, and in western China and Tibet PLA truck fleets make a significant contribution to civilian transportation.

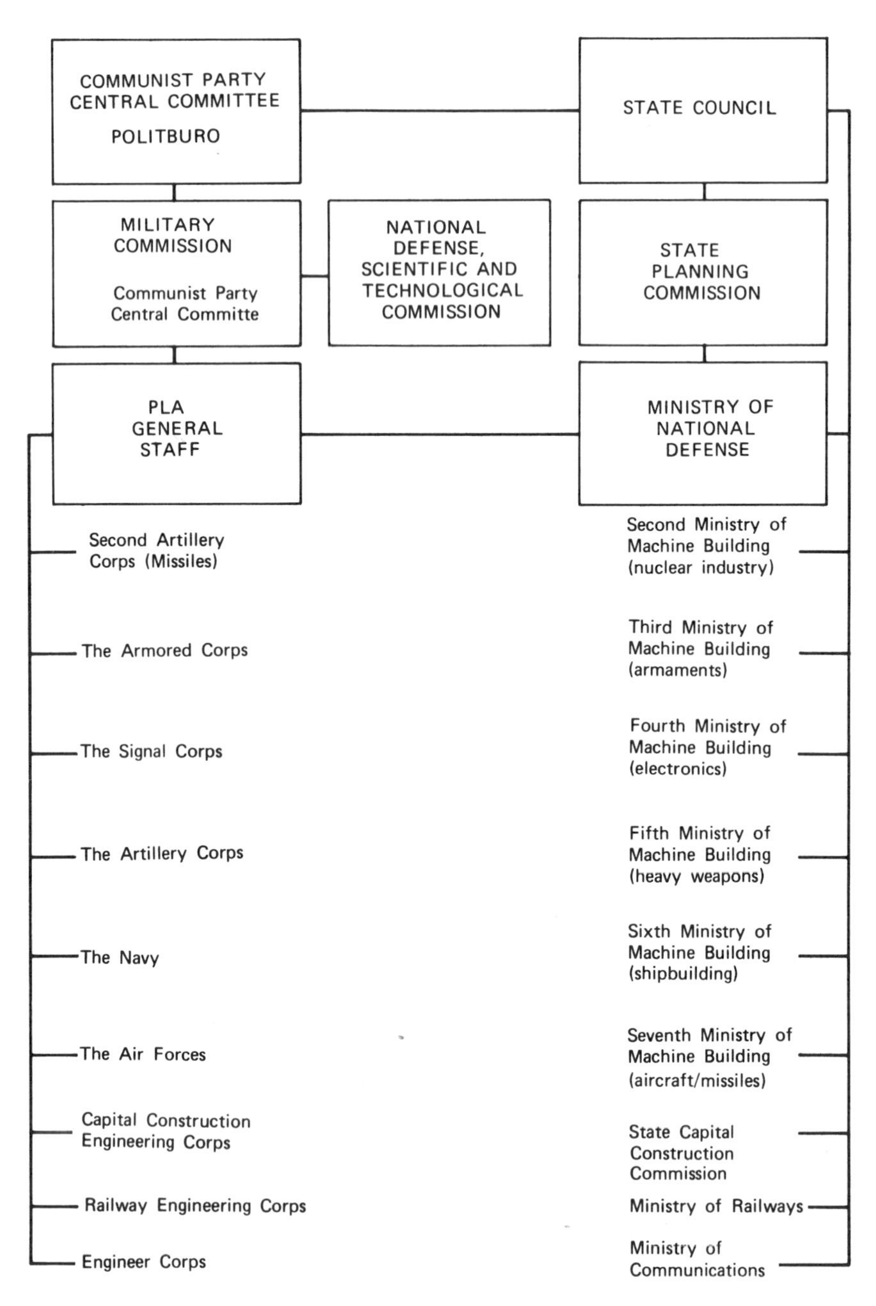

Figure 11.3 Military–industrial complex of the People's Republic of China.

This massive involvement of the PLA in economic activities has its basis in the Constitution of the People's Republic of China. Article 15, Chapter 1 of this law, adopted on January 17, 1975 at the Fourth National People's Congress, clearly states that "the Chinese People's Liberation Army is at all times a fighting force and simultaneously a working force and a production force. . . ." This mission of the PLA and also its relationship to the political apparatus of China make it quite unique among military institutions of the world.

The military–industrial complex of machine-building ministries also plays a very special role in China's industry. It is a major factor in technological innovation in China and was responsible for the development of advanced industries such as aerospace, electronics, nuclear materials, and shipbuilding. Military command and machine-building industries are also the only available executive training ground in China, combining exposure to the latest technologies with management experience. The present minister in charge of the all-important State Planning Commission, Yu Chiu-li, was originally a military leader and is a former minister of the Petroleum Industry. Similarly, the ministers of Posts and Telecommunications, Metallurgical Industries, Agriculture and Forestry, and Commerce all have military command backgrounds. It must be noted, however, that all these ministries control production and distribution of resources and supplies of crucial importance in military logistics.

For research and development, the machine-building industry relies on specialized institutes of the Chinese Academy of Sciences and other research institutes operated by individual ministries. Work of some of these institutes is discussed in various sections of this book in connection with specific industries they serve. However, there is also an Academy of Military Sciences, which is supervised by the Ministry of National Defense and is believed to influence the work of the defense-oriented ministries. This institution was established in 1958 for the specific purpose of accelerating the development of modern weapons in China.

The origins of China's military–industrial complex in its present form can be traced back to about 1960, when the Sino-Soviet relations deteriorated and the Third Ministry of Machine Building was reestablished to undertake the production of conventional weapons. Within only five years, by the end of 1965, four more machine-building ministries were organized, and the Chinese defence industry was already producing a variety of military equipment, including artillery, aircraft, missiles, submarines, military electronics, as well as nuclear bombs, the first of which was tested in 1964.

In the mid-1970s at least 50 percent of the total output of defense-oriented ministries was estimated to be equipment procured directly by the military services. When production capacity was available, defense factories have been known to produce equipment for the civilian sector of the economy on an irregular basis. A National Defense Industries Office has been identified in China, but its function is unclear. It is believed that through its regional offices it coordinates production for the civilian and military sectors of the economy.

In 1975 total defense expenditures of China were estimated at $21 billion, or about 7 percent of the GNP of $299 billion. About $16.0 billion or almost 22 percent of the value of all industrial production in China was believed to consist of military goods. This is comparable in size to total sales of General Electric or IBM, which are among the ten largest corporations in the world. If military procurement represents only 50 percent of production of all defense ministries, their combined output could be in the order of $30 billion annually, comparable in size to the Ford Motor Company, which is the fourth largest corporation in the world after Exxon, General Motors, and Royal Dutch/Shell. However, this total output represents six different ministries, each of which probably operates as a separate industrial conglomerate. Even then their average output would be about $5 billion per year, ranking each alongside such well-known industrial giants as Westinghouse, International Harvester, RCA, or Bethlehem Steel.

This is not surprising since China's total defense expenditure is the third largest in the world after the Soviet Union and the United States. But despite such large military expenditure and because of China's huge population, the per capita burden is relatively low; it is comparable to that of Brazil or Malaysia and several times lower than that of COMECON or Western countries (see Table 11.1).

MACHINE-TOOLS PRODUCTION AND IMPORTS

Metal-working equipment is basic to all civilian and military machine-building industries, and this was clearly understood by the Chinese leadership from the outset. Machine-tools production has shown impressive growth from sixteen plants manufacturing metal-working equipment in 1949 to an estimated 2000 by the mid 1970s, of which at least twenty are large modern enterprises. The output of machine tools is estimated to have increased from about 3300 units in 1950 to 90,000 in 1975. This is equivalent to about 150,000 tons of equipment manufactured in 1975, assuming that an average Chinese unit weighs at least 1.5 tons.

In 1975 Chinese machine tool production was comparable to that of Japan, where 139,000 tons of machine tools were produced, valued at

Table 11.1 Defense expenditures and military manpower of selected countries in 1976

Country	*Total Expenditure in Millions of $U.S.*	*Per Capita Expenditure in $U.S.*	*Percent of GNP in 1975*	*Numbers in Armed Forces*
China	21,000[a]	22	7.0	3,525,000
Soviet Union[b]	124,000	490	11 to 13	3,650,000
United States	102,691	477	5.9	2,086,700
West Germany	15,220	242	3.7	495,000
Japan	5,058	45	0.9	235,000
Israel	4,214	1,201	35.9	158,500
India	2,812	5	3.0	1,055,500
East Germany	2,729	158	7.8	157,000
Brazil	1,780	16	1.3	257,000
South Korea	1,500	42	5.1	595,000
Taiwan[c]	1,000	63	7.2	470,000

Source: Based on data presented in "The Military Balance 1976–1977," International Institute for Strategic Studies, London.

[a] U.S. Joint Economic Committee study estimate for 1975. [b] Data for 1975. [c] Data for 1974.

about $750 million. However, Japanese output was down from a previous high of 316,000 tons in 1970, whereas China's machine-tool production is increasing. China thus appears to be the fourth or fifth machine-tool manufacturing country in the world. If output grows at about 5 percent per year, China should be producing almost 250,000 tons of machine tools by 1985 (see Table 11.2).

By the early 1970s China's machine-tools industry was producing at least 1000 types of equipment, which satisfied an estimated 80 percent of domestic demand. The Second Production Control Bureau of the First Ministry of Machine Building is in charge of machine-tools production and distribution. Principal products manufactured are general-purpose machine tools such as: (1) lathes, (2) milling, grinding, and boring machines, (3) drill presses, and (4) gear-cutting machines. The output of those machines has been growing steadily, enabling China to reduce imports of general-purpose machines. However, it is believed that adequate capacity for manufacture of large-size and high-precision machines is still lacking, and production of special purpose tools is limited. Numerical-control machine tools are being developed but do not appear to be in mass production.

Many of the large plants have been built with Soviet, Czechoslovakian, Hungarian, and East German assistance. Shanghai Heavy

Table 11.2 Production of machine tools in selected countries in 1975

Country	*Unit of Measure*	*Machine-tool Production*
China	units	90,000
	tons of weight	150,000
Soviet Union	units	232,000[c]
West Germany[a]	tons	212,000[d]
Japan	tons	138,773
Italy[b]	units	112,000
United States	units	78,000
France[a]	tons	52,200
East Germany	units	19,500

Source: Japan Machine Tool Industry 1976; U.S. Government Handbook of Economic Statistics, 1976; GUS Statistical Yearbook of Poland, 1976.

[a] Data for 1974. [b] Data for 1971. [c] The Soviet Union also manufactures 50,500 units of metal-forming machine tools. [d] The United States also manufactures 25,600 units of metal-forming machine tools.

Machine Tool Works, which claims to be the largest machine-tools manufacturer in China, was built in 1958 with Soviet assistance. The well-publicized Shanghai No. 1 machine-tool plant, which employs 6000 workers, was expanded with Soviet aid. It is regarded as an industrial model because of its extensive training program; it also operates a Research Institute of Machine Grinding.

The Wuhan Heavy Machine Tool Plant is the foremost heavy machine-tools manufacturer. It specializes in large precision machine tools, including planers, lathes, drills, milling machines, and forge hammers. The plant itself uses at least 5000 machines in its workshops and makes 380 types of machine tools. It also manufactures machines for gear cutting and threading lathes. It was built with Soviet assistance.

Major machine-tool manufacturing centers appear to be Shanghai, Peking, Shenyang, Tsinan, Chungching, Huhehot, Kunming, Nanching, Sining, Tientsin, and Wuhan. Of those Shanghai and Tientsin appear to have the largest concentration, but numerous other municipal and small local machine tool plants exist throughout China (see Table 11.3).

Imports and Exports of Machine Tools

China relies to a large degree on imports for gear cutting, grinding, and milling machines but appears to have been steadily reducing its imports of general machine tools as domestic production increases. Total machine-tools imports in 1975 are estimated at about $70 to 80 million, although

Table 11.3 Major metal-working equipment manufacturing plants identified in China

Plant Name and Location	*Type of Equipment Manufactured*
Canton Metal Press Factory	60 to 160-ton metal-forming presses
Chengtu Measuring and Cutting Tools	49 kinds of cutting tools
Chengchou Abrasive Material	grinding tools and wheels up to 2-m diameter
Harbin Measuring and Cutting Tool	precision instruments and cutting tools
Kueiyang Machine Tools Plant	drilling, shaping, and planing machines
Kunming Machine Tools Plant	boring and universal milling machines
Lanchou Machine Tools Plant	produces about 1200 units per week
Loyang Machine Tools Plant	shaping and planing machines
Liuchou Machine Tool Plant	produces average 400 units per week
Nanching Machine Tool Plant	1500 precision lathes per year
Peking No. 1 Machine Tool Plant	large and numerical control machine tools
Peking No. 2 Machine Tool Plant	computer-controlled boring-milling machines
Peking No. 3 Machine Tool Plant	universal milling machines
Shanghai No. 1 Machine Tool Plant	largest producer of grinders in China
Shanghai No. 4 Machine Tool Plant	NC automatic vertical milling unit made
Shanghai Heavy Machine Tools Works	largest machine tools plant in China
Shenyang No. 1 Machine Tool Plant	medium-sized lathes for engine industry
Shenyang No. 2 Machine Tool Plant	boring and drilling machines
Shenyang No. 3 Machine Tool Plant	largest automatic lathes builder in China
Talien Machine Tool Plant	large transfer lathes
Tsinan No. 3 Machine Tool Works	high-precision vertical milling machine
Tsitsihar No. 1 Machine Tool Plant	shaping, milling, cutting machines
Wuhsi Machine Tools Plant	1800 grinding machines per year
Wuhan Heavy Machine Tool Plant	large precision machine tools
Yingkou Machine Tools Plant	gear-cutting machines

Source: Compiled by 21st Century Research from various articles, papers, and publications listed in reference section of this book.

imports in the overall metal-working machines category, which includes steel-rolling mills equipment, were in the order of $125 million. Preliminary results for 1976 indicate that China imported at least $67 million of machine tools from western countries. Japan was the leading supplier with $24 million, followed by West Germany and Switzerland which supplied $17 million each. Italy, France, the United States, and the United Kingdom were the other suppliers.

During 1965–1975 decade China imported at least $678 million of metal-working equipment from about twenty different countries. About two-thirds of this total was imported during the first half of the 1970s, with average annual shipments nearing $90 million. Almost 30 percent of that total were imports from COMECON countries, which continue to be among the largest machine-tools suppliers to China. Czechoslovakia, East Germany, Hungary, Poland, and the Soviet Union supplied more than 50 percent of machine-tools imports in 1972 and 1973. If imports of metal-working equipment such as rolling mills are excluded from market share comparisons, COMECON countries supplied about 25 to 30 percent of all machine-tools, imports in 1974 and 1975. If this trend continues, COMECON countries, particularly East Germany, Czechoslovakia, and Hungary, may once again become the leading suppliers (see Fig. 11.4).

Japan, which has been a major supplier since 1970, actually experienced a sharp reduction in shipments of machine tools to China in 1975. Exports fell by 67.3 percent when China purchased only 213 units in 1975, compared with 492 units in 1974. Grinding machines and boring machines made up almost 50 percent of all the machine tools imported from Japan in 1975. Japanese exports of metal-working equipment under SITC code 715 actually doubled in 1975 compared to 1974, but this is due to shipments of steel-making equipment for the steel-finishing plants in Wuhan, where the Nippon Steel Corporation is one of the major equipment suppliers.

West Germany was the leading machine tools supplier in 1975, exporting a total of $23.6 million of equipment to China. In addition, the West German consortium, under the leadership of DEMAG AG, is also constructing steel-finishing mills in Wuhan, and total German exports of

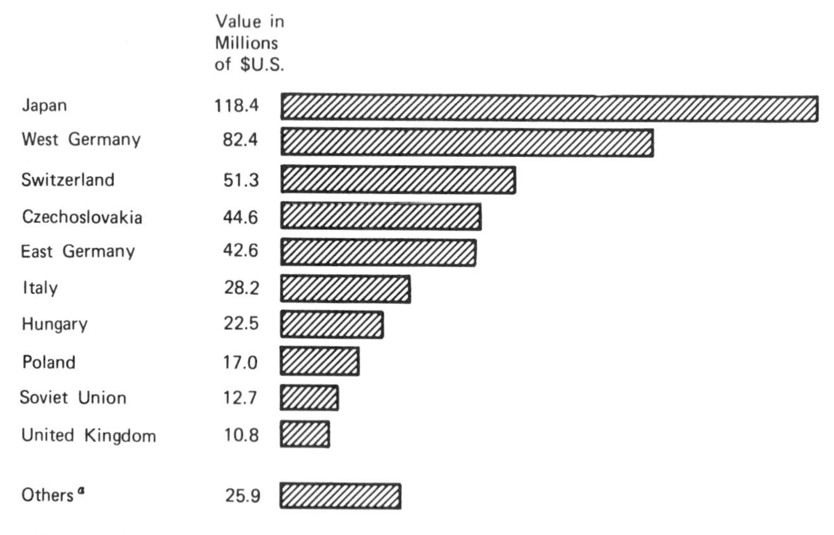

Figure 11.4** Estimated market shares of major suppliers of metal-working equipment to China during 1970–1975. All data for SITC export commodity code 715.* **(Source: OECD Commodity Trade Statistics Series B** *and foreign trade yearbooks of COMECON countries.)* [a] ***Others include France (8.4), United States (4.7), Romania (4.5), Sweden (2.4), Belgium/Luxembourg (2.3), Yugoslavia (0.2), Bulgaria (0.7), Denmark (0.6), Austria (0.2), Netherlands (0.1).

metal-working equipment rose to $39.1 million in 1975 as a result. Switzerland has recently also increased machine-tools exports to China, and with $13.5 million in 1975 was the third largest supplier after West Germany and Japan.

If China's machine-tools output can be assumed comparable to that of Japan in terms of value as well as volume, imports would represent about 10 to 15 percent of total annual consumption. This is considerably less than imports during the 1950s, when large numbers of machine tools were imported from the Soviet Union, Czechoslovakia, Hungary, and Poland, probably ranging between 30 to 40 percent of annual consumption.

While China's output expands the percentage of imports in total consumption will continue to drop, but the value of imports may remain at levels comparable to those in the mid 1970s. China will probably continue to import selected parts, accessories, tools, and dyes as well as special-purpose and high-precision machine tools because of its limited ability to produce high-grade and special steels and alloys. This situation in turn derives from China's deficiencies in domestic supplies of chromium, cobalt, and nickel, which are necessary to develop the high-grade and stainless-steel alloys required for manufacture of the more sophisticated equipment (see Chapter 9 for more details).

Industry analysts believe China will continue to rely on imports of high production and heavy-duty gear-cutting machinery, particularly with continuing expansion of its automotive and tractor production. Traditionally Japan, the United Kingdom, and East Germany supplied this equipment. In 1974 Gleason Works of Rochester in New York received an order worth $8.2 million for a complete set of gear-grinding and axle-manufacturing machines for six Chinese motor vehicle plants, which further confirms China's continuing import dependence for this type of equipment. Trade observers believe that much of future demand for machine tools will center on equipment such as: (1) very large and very small gear-hobbing machines, (2) spur, helical, and herringbone gear-hobbing machines, (3) bevel gear generators, and (4) gear cutters for the production of spur, helical, herringbone, internal and cluster gears, and gear-shaping machines.

Milling machines comprise another category of machine tools whose imports may continue. They are particularly important because of their versatility and adaptability to mass-production techniques. Shortages of milling machines were reported in applications requiring high-production and special purpose equipment, but domestic manufacturers produce several high-quality universal milling machines and will eventually satisfy most of China's demand as their output increases. Special steels technology is not such an important factor in this product group. Japan has been the major supplier of milling machines, but imports of Japanese milling and planing machines dropped from sixty-one units in 1974 to only twelve in 1975. This may signify that domestic production of this type of machine is catching up with demand. Hitachi and Enshu were the primary Japanese suppliers. China also purchased four very large milling machines from H. Ernault Somua of France in 1974 for over $2 million.

Grinding machines constitute the third important machine-tools category imported by China and represent the largest group of machine tools imported from Japan in 1975. The sixty-seven units shipped in 1975 still account for about 30 percent of all machine tools imported by China from Japan, but they represented less than half of the 144 units shipped in 1974. Imports of large-sized and high-precision grinding machines are expected to continue because domestic manufacturers are believed to have engineering problems with the production of such equipment. Besides Japan, Italy has also supplied grinding machines to China, primarily of the internal and surface grinding category. Major Japanese exporters identified in this trade are Tsugami, Okamoto, Yamada Koki, Toshiba, Toyo Kogyo, Nippei, Ichikawa Grinder, and Sanjo Machine Works.

China is also an exporter of machine tools, to North Korea, Vietnam, Albania, Hong Kong, and various Southeast Asian and African countries. Some minor sales have been also made to Australia and western Europe, and it is worth noting that Chinese economic and trade exhibitions in foreign countries usually devote a large part of their display area to exhibit Chinese machine tools of various types, including numerically controlled equipment. Chinese exports of machine tools remain relatively small because domestic demand continues to exceed production. Nevertheless, as the industry expands it should be expected that export of basic and less sophisticated machine tools from China will increase. Chinese machine tools may prove particularly useful in trade with Third World countries where they could be bartered for raw materials and minerals that China continues to import. Considerable political advantages may also accrue to China if it becomes another source of basic machining equipment that can be obtained without expenditure of hard currencies.

INTERNAL-COMBUSTION ENGINES

The Fourth Production Control Bureau of the First Ministry of Machine Building is responsible for manufacture of prime movers, including gasoline engines, diesel engines, and gas turbines. Aircraft engines are probably the responsibility of the Sixth Ministry of Machine Building. The majority of internal combustion engines produced in China are diesels, of which over forty different models, ranging from 3-HP units to 10,000-HP marine diesels, have been identified in serial production in China. About fifteen Chinese gasoline engine models are known to exist, rated from 2-HP units to at least two models in the 120-HP category. China also manufactures some coal-gas and donkey internal combustion engines.

Chinese internal combustion engine production is already a significant industry. Its output was estimated at about 4,000,000 HP per year in 1958. Between 1965 and 1972 output probably increased by about 400 percent to an estimated 20,000,000 to 25,000,000 HP per year. Judging by current Chinese output of equipment such as trucks, tractors, irrigation pumps, locomotives, and ships, production of internal combustion engines should be 30,000,000 to 40,000,000 HP per year. If output of such equipment continues to increase at similar rates, annual demand for internal combustion engines should exceed 60,000,000 HP by 1985.

At least thirty-five internal combustion engine plants have been identified in China. They manufacture mostly diesel engines, ranging from small 3-HP units for use in walking tractors to a 12,000-HP marine diesel engine that was first trial-produced in 1975 at the Canton diesel engine plant. Single-cylinder diesels are made up to 10 HP, whereas two-cylinder engines range from 10 to 40 HP. Five basic models of four-cylinder diesel engines have been identified, ranging in power up to 80 HP. Most diesels over 100 HP are six-cylinder engines, although at least four V12 engines are produced, the largest of which develops 400 HP. At least sixteen of the large low-speed units are marine diesel engines, which are also made in power ratings of 600 HP, 750 HP, 1000 HP, 1200 HP, and 10,000 HP.

Major gasoline engines production consists of 95-HP and 110-HP CA series six-cylinder units used in the "Liberation" trucks. The largest gasoline engines manufactured are two 120-HP models. Trucks over 7 tons are equipped with diesel engines. Several four-cylinder engines in the 70 to 90-HP range are made for use in jeeps and automobiles. Most internal engine production is concentrated in the Shanghai, Peking, Tientsin, Wuhan, Wuhsi, and Canton area, but there are small plants with limited diesel-engine production scattered throughout China (see Table 11.4).

Imports of internal combustion engines are about $26 million per year, and this amount is not expected to increase significantly. Actually, China obtains large numbers of internal-combustion engines in imported equipment such as trucks, locomotives, ships and other machinery. Large marine engines of the 12,000 HP class have been imported separately from Poland and Yugoslavia.

Many imports in the internal-combustion engines category are be-

Table 11.4 Major internal-combustion engine manufacturing plants identified in China

Plant Name and Location	*Type of Engine Manufactured*
Canton Diesel Engine Plant	12,000-HP marine diesel made in 1975
Changchou Diesel Engine Plant	12-HP diesel engines for farm use
Changchun No. 1 Motor Vehicle Plant	95-HP gasoline engines for trucks
Changcheng Motor Vehicles Plant	180-HP V8 diesel engines for 12 ton trucks
Hangchow Motor Vehicle Engine Plant	160-HP diesels for "Yellow River" trucks
Huafeng Machines Plant	12-HP internal-combustion engines
Ichang Diesel Engine Factory	3-HP engines made in 1966
Nanching Motor Vehicles Plant	NJ70 series of gasoline engines for trucks
Pangpu Diesel Engine Factory	20-HP diesel engines
Peking General IC[a] Engines Works	75-HP and garden gasoline engines
Shanghai Diesel Engine Plant	10,000-HP marine diesel engine made
Shanghai Penghu Engine Plant	7-HP walking tractor diesel engines
Shanghai IC Engine Works	gasoline engines for automobiles
Shihchiachuang Motive Power Plant	80-, 120-, 180-HP industrial diesel engines
Sian No. 3402 Plant	170-HP V12 diesel engines for trucks
Tientsin IC Engine Factory	70-HP gasoline engines for jeeps
Tsingtao Automobile Plant	high-speed engines
Weifang Diesel Engine Plant	diesel engines for factory use
Wuhan Diesel Engine Plant	7-HP hand tractor diesel engines
Wuhsi Diesel Engine Plants (five)	90-HP diesel first in 1958
Ningpo Motor Works	6- to 750-HP diesel engines

Source: Compiled by 21st Century Research from various articles, papers, and publications listed in the reference sections of this book.

[a] IC = internal combustion.

lieved to be spare parts for equipment already in operation. The West German diesel-manufacturing company MAN estimates that at least 100 large MAN marine-diesel engines are operating in China. They believe these are reconditioned marine diesels acquired within old ships bought by China for scrap value. This is confirmed by occasional Chinese orders for spare parts for the MAN diesel engines.

Imports averaged about $25 million annually during 1970 to 1975 period. East Germany and the Soviet Union are the leading suppliers. This is probably the result of previous imports of large quantities of internal combustion engines originating from those countries requiring spare parts. As China continues to import more trucks and equipment from France and other Western countries, the market shares will probably change in time as the imported equipment wears out and larger quantities of spare parts will be required from Japan and the West. This should be regarded primarily as an automotive aftermarket dependent on transportation and earth-moving equipment imports of previous years (see Fig. 11.5).

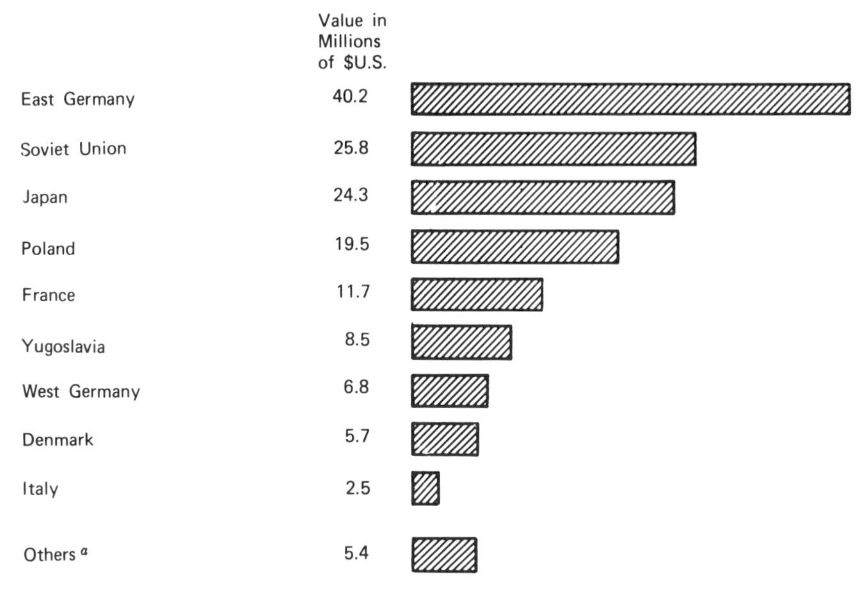

***Figure 11.5** Import market shares of internal combustion engine suppliers during 1970–1975, SITC code 711.5, excluding aircraft engines. (**Source:** **United Nations** bulletins of statistics on world trade in engineering products and OECD Commodity Trade Statisics Series B.)*
[a] Others include Sweden (1.8), United Kingdom (1.3), Switzerland (1.0), Morocco (0.7), Hungary (0.4), Finland (0.1), Norway (0.1).

HEAVY MACHINERY PRODUCTION AND IMPORTS

Machinery in this category consists of heavy equipment used by mining, metallurgical, and construction industries. Examples of products in this category include excavators, drilling rigs, hoists and winches, smelting furnaces, steel converters, construction cranes, bulldozers, loaders, graders, power shovels, and special vehicles. Production of such a variety of equipment is extremely difficult to assess and compare. Some discussion of China's manufacturing capabilities and imports in these categories is also found in Chapters 9, 15, and the section on the coal industry in Chapter 7.

During 1956–1960 it was reported that about 60 percent of all heavy-duty machinery was produced under the supervision of the Third Production Control Bureau of the First Ministry of Machine Building, another 23 percent was made by other Ministries of Machine Building, and 20 percent was made by special equipment-manufacturing departments in other ministries such as the Ministry of Metallurgical Industries or Ministry of Petroleum and Chemical Industries.

Very rough estimates of metallurgical equipment production have suggested that 125,000 and 300,000 tons of equipment were produced in 1965 and 1971, respectively. This estimate does not include mining or construction machinery, which together may equal the amount of metallurgical equipment produced in China.

It is possible, however, to develop an output estimate based on the assumption that production of heavy mining, metallurgical, and construction machinery consumes a reasonably stable percentage of crude steel output in a centrally planned economy. This ratio can be readily developed for Poland, where crude steel output in 1975 was 15,000,000 tons and production of mining, metallurgical, construction, and earthmoving equipment consumed 464,400 tons of steel, about 3 percent of total steel output. Poland is a particularly good example for comparison because its steel production is of the same order of magnitude as 26,000,000 tons in China and it also has a large coal industry with a corresponding heavy use of steel in the mining-industry sector.

Using a 3-percent factor, China's consumption of steel for the manufacture of mining, metallurgical, and construction equipment would amount to about 780,000 tons in 1975. If finished machinery of this type is priced at an average $2,000 per ton, the total value of this production would be in order of $1,560 million. China's imports of mining and construction equipment in 1975 were valued at $53.4 million and with metallurgical equipment from West Germany and Japan were about $90 million, which represents about 5 percent of the estimated demand. This is in line with Chinese claims in previous years that their metallurgical and mining machinery industries are 90 to 95 percent self-sufficient.

Looking at the 5-year record of imports of construction and mining equipment during 1970–1975, COMECON countries again appear as major suppliers. Strictly speaking, imports from Romania are mainly oil-drilling rigs, but these fall into mining-equipment SITC trade classification and thus are comparable with imports from other countries (see Fig. 11.6).

If on the basis of previous analysis it is assumed that production of this equipment will continue to consume 3 percent of China's steel output in the future, then mining, metallurgical, and construction equipment will amount to almost 1,000,000 tons in 1980 and about 1,350,000 tons in 1985. If the import-dependence ratio remains about the same, this market may increase to about $100 million in 1980 and $135 million in 1985 at

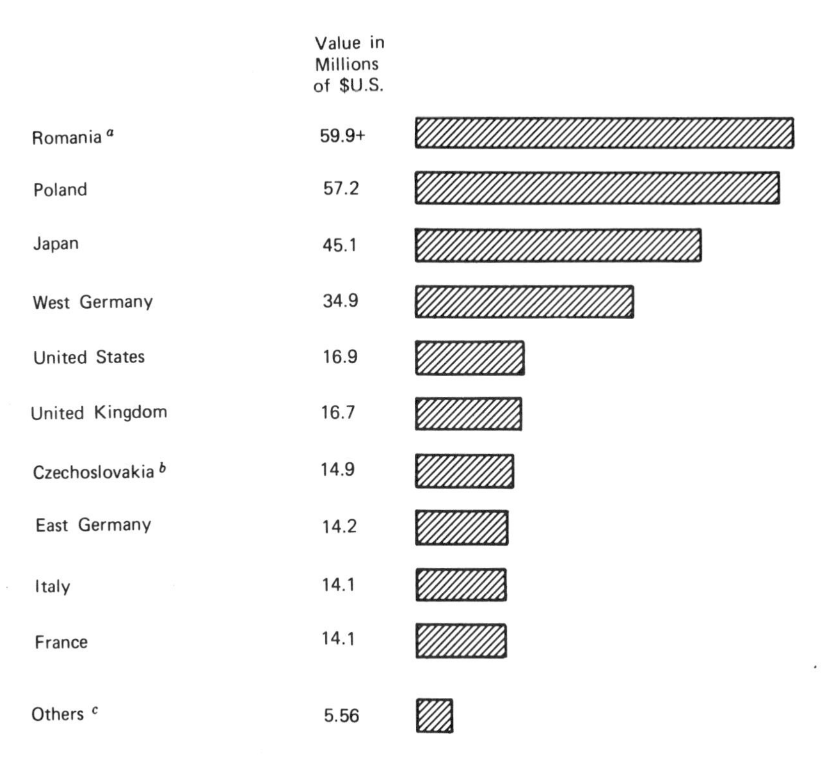

Figure 11.6 ***Import market shares of construction and mining equipment suppliers during 1970–1975, SITC code 718.4.*** **(Source: United Nations *bulletins of statistics for world trade in engineering products,* OECD Commodity Trade Statistics Series B, *and foreign-trade yearbooks of individual countries.*)** [a] ***Data only to 1973, which means that Romanian market share is probably considerably larger, possibly near $100.0 million.*** [b] ***Data available for SITC code 718 only.*** [c] ***Includes Finland (3.2), Sweden (1.2), Soviet Union (0.56), Austria (0.4), Belgium/Luxembourg (0.1), Netherlands (0.1).***

best. If imports become primarily prototypes for technological innovation in China's domestic production, shipments may remain at present levels or even decrease in the future. The preliminary results for 1976 indicate a total of only $18 million of mining and construction equipment imported from the west but none from the Soviet Union. Total level of imports in this category is not possible to determine until data from East European countries become available for 1976.

ELECTRONIC INDUSTRIES

Since about 1960 China developed an electronic industry that ranks among the top ten largest in the world. Chinese leaders realize the importance of electronics to the strategic mobility of their armed forces and industrial modernization. Consequently, during the late 1960s electronics received state investment priority and began growing faster than other industries. The emphasis shifted again to steel and agriculture in the 1970s, but electronics is now a well-developed industry, creating a sizable turnover of its own with an increasing output of consumer electronic products.

The total value of the output of China's electronic production is estimated to be roughly 1 percent of its GNP, or about $3,000 million annually. Chinese production is still several times smaller than the estimated electronics markets of the Soviet Union or Japan and only about 6 percent of the electronics industry of the United States (see Table 11.5).

The industry is controlled by the Fourth Ministry of Machine Building and between 50 percent and 66 percent of total electronics production is destined for military use. In contrast, about 25 percent of electronic output of the United States is procured by the armed forces, but this is equivalent to almost $12 billion per year or about six times the amount procured by the Chinese military. On a per-soldier basis, $5,769 is spent annually on electronics for every member of the armed forces in the United States, compared with only $560 in China and about $4,000 in the Soviet Union. China's current military modernization program should hence be expected to give new impetus to further expansion of electronics production.

Despite its present size, China's electronics industry is still regarded to be in the "takeoff" stages of its development with regard to the long-term potential of the Chinese market. Judging by the reported growth in output of some electronic products such as radios or television receivers, consumer electronics products increased at average annual growth rates of about 50 percent during 1970–1975. If the average annual growth rate for all electronic products is conservatively assumed to be only 10 percent as the industry becomes larger, China's electronics output may reach $7 billion by 1985, equivalent to about 1.3 percent of its projected GNP at that time. This would make China's electronics market comparable to that of West Germany today and probably place it among the five largest electronics markets in the world.

The Chinese electronics industry manufactures almost all types of standard components as well as special-purpose integrated circuits, transistors, and vacuum tubes. All of these are basic to an independent

Table 11.5 Major markets for electronic products and production of radio and television receivers in selected countries in 1976

Country	*Electronics Market in Millions of $U.S.*	*Market Size as Percent of GNP*	*Production of Radio Receivers in Millions*	*Production of Television Receivers in Millions*
China	3,000+	1.00+	18.00[a]	0.205[a]
United States	51,706	3.05	44.10	14.131
Soviet Union	27,000[b]	2.93	8.40	7.060
Japan	11,671+	2.12	16.70	17.037
West Germany	7,512	1.66	5.40	3.727
France	4,548	1.31	3.98	1.777
Canada	3,750[c]	2.00	0.75	0.510[a]
United Kingdom	2,533	1.15	0.679	2.122
East Germany	2,100[c]	3.00+	1.068	0.561
Italy	1,600[c]	1.00	1.80[d]	1.595[a]
Poland	1,500[c]	1.63	2.038	0.963
Brazil	1,100[c]	1.00	0.64[a]	1.450[a]
Netherlands	1,113	1.26	—	—
Sweden	1,032	1.39	0.19[d]	0.370[d]
Czechoslovakia	1,000[c]	1.72	0.16[a]	0.456
Spain	962	0.90	0.47[a]	0.530[a]
India	850	1.00	1.54[a]	—
Belgium	748	1.09	1.79[a]	0.579[a]
Switzerland	674	1.70	—	—
South Korea	580[c]	2.30[a]	4.28[a]	1.215[a]
Denmark	461	1.23	0.14[a]	0.064[a]
Norway	399	1.32	0.11[e]	—
Finland	259[a]	0.90[a]	0.17[d]	—
Austria	230[f]	0.60[a]	—	0.404[a]

Source: United Nations Statistical Yearbook 1976; Electronics, January 6, 1977; *U.S. Government Handbook of Economic Statistics 1976 and 1977.*

[a] Data for 1975. [b] Based on 1973 estimate of USSR electronics output as 60 percent of U.S. production. [c] Authors' estimate based on assumed percentage of GNP in each country. [d] Data for 1974. [e] Data for 1973. [f] Data for 1972.

electronics industry and are used to assemble final electronic equipment. End products include instruments, computers, communications equipment, and consumer electronics. Military equipment consists of special radio transmitters and receivers, radars, sonar, and avionics as well as missile and nuclear weapons control systems. Communications systems, and radio and television production and use are discussed in Chapter 14. Manufacture and trade of other electronic products are covered in later sections of this chapter.

Although electronics with its innumerable variety of products cannot be regarded as a truly basic industry such as agriculture, energy, or steel, the existence of a comprehensive electronics industry in a country is an indispensable factor today to becoming a modern industrial power. Electronics is an industry characterized by rapid technological change, very high labor content, and a relatively low demand on material inputs. Japan is the best example of exploiting those characteristics to its national advantage by developing a huge domestic electronics industry and extensive foreign trade in electronics products. Presumably the lesson of Japan has not been lost on China, which in addition has a domestic electronics market potential big enough in itself to support what may eventually become the largest electronics industry in the world.

Because of this vast captive and unsaturated market, China can become a profitable manufacturer of virtually any electronic product, thanks to this huge potential for economies of scale production. China has in fact a unique opportunity to expand its electronics industry as it becomes expedient to absorb any surplus labor released from other industry sectors when productivities increase. If output of consumer electronics such as color television receivers becomes sufficiently large, China will also be in a position to become a factor in the export market for such products at very competitive prices. It has the potential to become a leading exporter of consumer electronics and in the long run may pose a major threat to Japanese electronics exports.

Electronics as an industry grew rapidly from sixty major plants in 1960 to over 200 in 1971, with a total employment of 400,000. In addition, it was estimated that there were 500 small electronics plants employing another 50,000 workers mostly engaged in batch production of less sophisticated components and subassemblies. It is interesting to note that even during the height of the Cultural Revolution electronics suffered practically no disruption but continued its growth and many new products were introduced, including five different computer models, static electron accelerators, and direct digital controllers for industrial automation. Much of this growth is believed to have been the result of Soviet aid withdrawal in 1960, which in fact forced China to turn to Western countries for assistance. About $200 million of advanced electronic products were imported from Japan, West Germany, the United Kingdom, France, and Switzerland during 1960–1970, and this permitted China to avoid a lengthy and costly development period and organize its own electronics industry based on some of the most advanced technology in the world.

Although production of electronics in China is expected to continue, its rapid growth imports of electronics will continue to be relatively small, probably in the order of 1 percent of the total production. More likely than not, China will continue to follow a "prototype purchasing" policy in this industry because it permits a rapid expansion of the industry at relatively low research and development cost. Continuing purchases of such products in the most advanced electronic manufacturing countries will also keep China almost at par with electronics industries of the Soviet Bloc, which is always lagging a few years behind those of the West.

Imports, even if they were to continue at the 1 to 2 percent production value level, will not exceed about $150 million in 1985. On the other hand, as China's electronics industry expands the production of radios and television and other relatively simple consumer products, progressively less imported technology may be needed. In such a case imports may remain relatively stable, steadily decreasing as a percentage of total production. Only sales of large quantities of military equipment are likely to make electronics a significant part of the China trade.

ELECTRONIC COMPONENTS

Basic to the existence of a modern electronics industry is the production of semiconductor elements as components for use in the assembly of electronic equipment. This industry has been characterized by very rapid technological advances during the last 20 years. The state-of-the-art competition in this technology revolves around the degree of concentration of electronic circuits on microscopic chips made up of thin layers of semiconductor materials.

Reports of scientists who visited Chinese semiconductor laboratories in 1976 suggest that China is about 5 to 7 years behind the United States in development of large-scale integrated (LSI) circuits. Although China's integrated circuits industry is reported to suffer from lack of computer-aided design technology and poor masking, it has produced 1000 bit memory chips and shift registers and eight-bit microprocessors. Output yields are believed to be low, and the yield in one Shanghai integrated-circuit factory was estimated by a Western scientist to be 20 percent at best.

Japanese electronics industry executives who visited Chinese semiconductor plants in December 1975 reported that LSI circuits with capacity of 10,000 transistor elements were being manufactured in Peking. By comparison, leading LSI manufacturing countries such as the United States or Japan mass produce LSI circuits for practical applications with over 12,000 elements, and a 16,000-bit LSI memory was introduced by INTEL, Inc., the leading American LSI manufacturer in 1976. However, ability to produce a high-capacity LSI circuit does not necessarily mean capability to mass produce and test final chips in economic quantities. The process of mass production of such devices involves photochemical, thin-film, vacuum, and semiconductor materials technologies as well as precision computerized test equipment, most of which is under strict export controls in the West.

Countries lacking sophisticated semiconductor production technology may still produce and demonstrate high-density LSI circuits and receive scientific achievement acclaim. On the other hand, they are not necessarily considered capable of producing such elements at sufficiently high yield rates to obtain economic quantities or to present a competitive threat in the world semiconductor markets. Nevertheless, the various reports about China's semiconductor development suggest that China is fast approaching world highest performance levels in ability to develop high-capacity integrated circuits.

Components Production

Production of components was believed to account for about 30 percent of China's electronics output in the early 1970s but will probably decline to a lower share in the coming years as production of electronic equipment increases. Electronic-components markets constitute 26 percent of the total electronics market in Japan but less than 20 percent in western Europe and are below 10 percent in the United States.

At least 110 plants, of which seventy were identified as major factories, were engaged in component production at the end of 1971. At that time it was estimated that China was manufacturing several hundred million transistors and diodes, some integrated circuits, and an adequate number of tubes, passive components, and various parts required for assembly of electronic products.

Original electronic components production was established in the 1950s with the assistance of the Soviet Union and eastern European countries. Several major plants were built to manufacture radio tubes, resistors, capacitors, and other passive components. The Peking Electron Tube Plant, with a reported capacity of 15 million receiving tubes and 2 million transmitting tubes, dates from that period. Other plants built at that time include the Harbin Electron Tube Plant, Shanghai Electron Tube Plant, Peking North China Radio Equipment and Materials Plant, Peking Radio Equipment and materials Plant No. 2, and the Nanking Electron Tube Plant. Major concentrations of component production facilities also exist in Chengtu in Szechwan Province and in Sian, Canton, and Shihchiachuang in Hopeh Province. At least 150,000 workers were estimated to be directly involved in production of electronic components in 1971. Individual plants are believed to range in size from fifty to 5000 workers.

Components Imports

To support the production of solid-state components that developed after the Sino-Soviet split, China imported semiconductor manufacturing equipment from Japan and western Europe. By 1971 China was already displaying diffusion furnaces for semiconductor crystal manufacture at the Canton Trade Fair. Based on their specifications, Western observers assessed Chinese semiconductor production equipment to be comparable to that used in the West 5 years earlier. It is the opinion of electronics specialists that China will continue to depend to a large degree on samples imports of the most advanced semiconductor manufacturing equipment for production of integrated circuits and advanced types of transistors.

Actual imports of electronic components such as tubes, semiconductors, and photocells have been very small in volume but probably high in technological content. During 1960–1970 a total of only $3.4 million was imported averaging about $300,000 annually. During 1973–1975 imports in this category amounted to at least $5.8 million, but the increase is believed to be primarily due to temporary quantity imports of color television tubes for assembly into color television receivers. Since China is developing its own color television production the level of imports in this product category will probably continue to be low, designed primarily to provide samples of the latest semiconductor technology.

This becomes particularly obvious when China's imports of electronic components are compared with imports by countries with comparable electronic industries. Canada, Italy, East Germany, or the United Kingdom import anywhere from $43 million to almost $200 million of electronic components annually. Thus China must be regarded as practically self-sufficient in electronic components. Because of its large and growing market it is clearly one of the largest electronic components manufacturers in the world. As such, it should be regarded as a potential competitor to other component-exporting countries. To the electronic equipment manufacturers China may become the next large-scale components supplier. Sino-Japanese cooperation in this industry appears to be the most likely in the future.

INSTRUMENTS PRODUCTION AND IMPORTS

Products of this industry include a large number of different electronic, electrical, optical, and precision types of instruments, and it is almost impossible to determine production levels without actual statistics from the manufacturers. It is believed that in the early 1970s production of instruments was under way in at least fifty major plants, with a total employment of 90,000 workers. The largest concentration of instrument-manufacturing facilities is in Peking, Shanghai, Nanching, and Tientsin, where over 50 percent of all the plants are located. Their output is estimated to account for two-thirds of the total instruments production. At least 50 percent of all electronic instruments produced in China and imported from abroad are probably used in support of various military projects.

Western industry observers indicate a 5-year technology gap in Chinese instrument designs, particularly in more specialized equipment and in the high-frequency-range products. However, China is constantly introducing new products and during 1960–1970 completely modernized this industry by introducing transistorized instruments in most electronic equipment. Because many of the instruments now produced in China show a close resemblance to corresponding Western models, it is also believed that China follows the "prototype purchasing" policy in this industry more than in most other equipment-manufacturing areas. Much of the development work and instrument production is conducted by specialized institutes of the Chinese Academy of Sciences and the leading universities.

International trade in instruments is reported under two different categories. Electrical measuring and control instruments are identified by SITC commodity code 729.5. Professional and scientific instruments are normally classified under SITC code 861 and include microscopes, cameras, medical, surveying, measuring, and testing instruments and fluid-control instruments. Many instruments in both categories are transistorized and most are operated using some electrical or electronic circuitry.

During 1960–1970 imports of professional and scientific instruments amounted to $166 million, or about $10.5 million per year. During 1970–1975 imports increased to $172 million or about $34 million per year. Imports of electric measuring and controlling instruments totaled about $61 million for 1960–1970 and increased to $77 million for the 5 years up to 1975, or about $15 million per year.

During 1975 imports of all instruments amounted to about $80 million, which represents about 2.5 percent of the estimated value of China's electronics production. In 1976 imports of precision instruments were $60 million. In countries such as Japan, western Europe, and the United States instruments constitute approximately 5 percent of the total electronics market. If the same ratio is assumed valid for China, imports appear to supply 30 to 50 percent of China's electronic instruments demand. If import dependence continues at this rate, China's total market for electronic instruments could reach about $175 million per year in

1985. More likely, China will develop larger domestic production and continue to import samples of the latest available instruments from all countries until it masters its own research and development program. China's future policy in this industry is perhaps best illustrated by its achievements in the field of medical and surgical instrumentation and apparatus. As much as 80 percent of such equipment was imported in 1953, but within 10 years domestic production was developed to the point where it could satisfy 90 percent of China's demand.

Until 1960 the Soviet Union was the main supplier of all instruments to China, shipping an average $10 million every year and the Western countries had no market whatsoever. Soviet shipments gradually decreased, and by 1967 Japan and West Germany became major suppliers. During 1970–1975 Japan and Switzerland were the leading suppliers with West Germany and United Kingdom appearing to play a secondary role. Whereas the Soviet Union continued to be a minor supplier, East Germany and Czechoslovakia reentered the Chinese instrument market in 1975 (see Fig. 11.7).

Some observers also believe that China may soon become an important exporter of instruments to Third World countries in particular. There is an unmistakable stress on instruments at various Chinese exhibitions, both in China and abroad, which supports this belief (see Table 11.6). The several specialized instrument exhibitions organized by Western countries in China in the last few years must be regarded not only as marketing events promoting sales to China but also as competitive analysis orientation sessions for the Chinese instrument manufacturers.

COMPUTER PRODUCTION AND IMPORTS

China's total inventory of electronic computers is probably in the order of 1000 units, although many of these are relatively small and slow machines. About 20 percent of the total may still consist of first-generation machines already obsolete by world standards. In contrast, about 170,000 general-purpose computers, excluding minicomputers, are believed in operation in the United States and at least 39,000, in Japan. The Soviet Union has about 12,500 computers installed in the civilian sector of its economy, and there are 350 computers operating in India. It is believed that 50 percent of all computers made in China are used for military applications such as nuclear and missile programs and in weapons-systems design and shipbuilding.

Installations of general-purpose computers are not the best indication of how modernized a country's industry might be. The number of

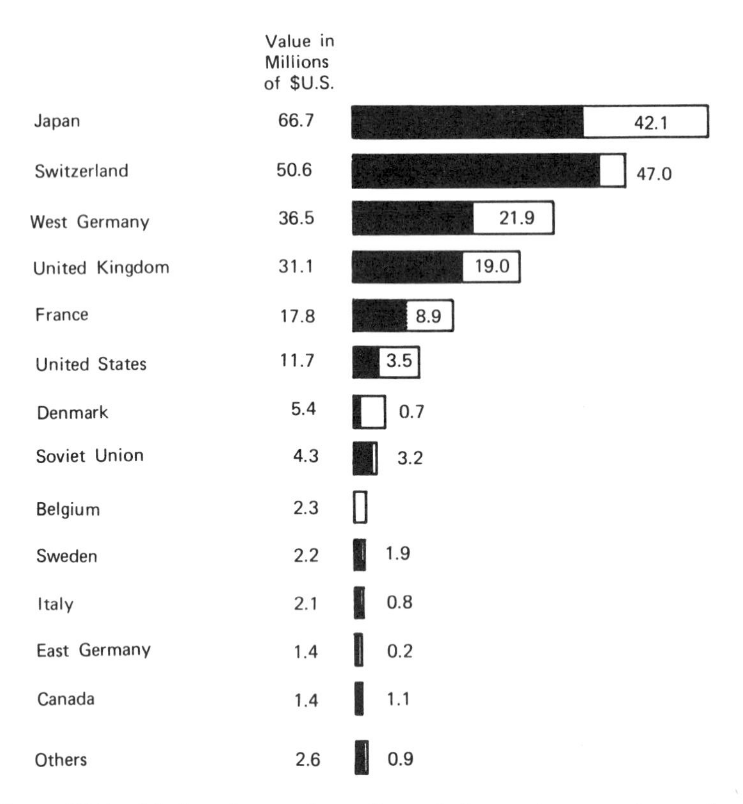

Figure 11.7 Market shares of suppliers of electronic measuring and control instruments (SITC code 729.5) and professional scientific instruments (SITC code 861) during 1970–1975. (Source: *OECD Commodity Trade Statistics Series B;* United Nations *bulletins of world trade in engineering products; and U.S. Department of Commerce Market Share Reports COM-74-10900-5076.*) *(Imports of scientific and professional instruments are shown by the dark portion of the diagram with value indicated at end of bars.)* [*Others include Czechoslovakia (0.5), Austria (0.6), Poland (0.2), Norway (0.2), Hungary (0.2).*]

computers available per million of nonagricultural labor force in a country is a more precise indication of computer usage. On that score China rates rather low, with only about seventeen computer systems per million, even if all machines involved in military production and readiness are considered. However, China's computer use is still considerably higher than that of India and Sri Lanka, where less than ten computers are available per million of nonagricultural labor. On the other hand, China's computer utilization appears to be comparable to that of Taiwan.

Table 11.6 Characteristics of selected Chinese instruments often shown at Canton Trade Fair and Chinese exhibitions in foreign countries

Model	*Description of Instrument and Basic Operating Frequency Range*
SB-10	Oscilloscope, bandwidth 10 Hz to 5 MHz, less than 3 dB
SB-14	Oscilloscope, bandwidth 0 to 100 kHz, 3 dB
SB-17	Transistorized oscilloscope, bandwidth DC to 5 MHz, less than 3 dB
SBE-20	Dual-trace oscilloscope, bandwidth DC to 500 kHz, less than 3 dB
SB-250	ULF Dual-trace oscilloscope, bandwidth 0 to 700 Hz, less than 5 percent
SBM-10	Transistorized multipurpose oscilloscope, DC to 30 MHz, 3 dB
SBE-7	Dual-trace oscilloscope, frequency response 0 to 15 MHz, 3 dB
SR-5	Double-beam oscilloscope, bandwidth 0 to 60 MHz, 3 dB
SR-6	ULF Double-beam oscilloscope, frequency response 0 to 1 MHz; 0 to 500 kHz
SR-8	Dual-trace oscilloscope, bandwidth DC 15 MHz, 3 dB; AC 50 Hz to 15 MHz
Q67	ULF Power transistor tester, 400-, 600-, and 1000-MHz frequencies
Q6S6	HF Power transistor tester, frequency ranges 10 to 80, 30 to 240, 100 to 800 MHz
QG6	UHF Transistor tester, operating frequency 400 MHz + 1 percent
JS-7A	Transistor tester, no details
—	Automatic integrated-circuit tester for TTLs, input threshold 2 to 3 V
GSY-2	Silicon single-crystal service life tester, 5 Hz to 2 MHz, less than 3 dB
XFD-6	LF Signal generator, frequency range 20 Hz to 200 kHz
XFG-7	HF Signal generator, carrier frequency 100 kHz to 30 MHz
XFS-8	AF Signal generator, frequency range 6 Hz to 100 kHz
XFS-9	AF Signal generator, frequency range 20 Hz to 20 kHz
212	Transistorized HF signal generator, frequency range 400 kHz to 130 MHz
XD-4	VLF Signal generator, frequency range 0.00001 Hz to 159.9 Hz
PPJ-10	Frequency counter (ECL circuits claimed), range ≦ 100 MHz
E311	Frequency counter, range 10 Hz to 1 MHz
FB 3	Frequency counter, transistorized, range 10 kHz to 300 kHz
PB 2	Decade frequency counter (to 9,999,999), range 10 Hz to 1.2 MHz
E-3002	Phase frequency–voltage counter, range 20 Hz to 200 kHz
E-312	Frequency counter, range 10 Hz to 10 MHz
E-325	Universal counter, range 10 Hz to 300 MHz
E-323	Universal counter, range 10 Hz to 30 MHz

Source: China Machinery Import and Export Corporation catalogs 1973; U.S.–China Trade Council Special Report No. 2; Canton Trade Fair Exhibit Reports; notes made by 21st Century Research at Chinese exhibitions.

Hz = Hertz; MHz = megahertz; kHz = kilohertz; dB = debye units.
ULF = Ultralow frequency; UHF = ultrahigh frequency; VLF = very low frequency; AF = alternating frequency; HF = high frequency; LF = low frequency; ECL = emitter coupler logic; TTL = transistor to transistor logic; AC = alternating current; DC = direct current.

However, there are at least 175 minicomputers in Taiwan, and if these are added to the total its computer availability per million of non-agricultural workers is over sixty, or three times the number in China. For direct comparisons in this discussion, however, minicomputers have not been included (see Table 11.7).

Table 11.7 Estimated inventories of general-purpose computers (excluding minicomputers) in selected countries of the world

Country	*Estimated Number of Computer Installations in 1976*	*Value in Millions of $U.S.*	*Nonagricultural Labor Force in Millions*	*Number of Computers per Million of Nonagricultural Workers*
China	1,000	0.25	60	17
United States	170,000	34.0	92	1847
Japan	39,000	8.1	47	829
West Germany	21,000	4.4	25	840
United Kingdom	15,500	3.5	25	620
Canada	14,000	2.4	10	1400
France	13,000	3.6	20	650
Soviet Union	12,500	4.5	103	121
Switzerland	2,300	1.3	2.5	920
Brazil	1,800	0.5	17	105
India	350	0.1	60	6
Taiwan	76	0.02	4	19
Sri Lanka	9	0.001	2	4

Source: Japan Electronic Computer Company (JECC); 01 Informatique; *Asian Computer Yearbook, 1977;* International Data Corporation; *U.S. Government National Basic Intelligence Factbook, July 1977.*

At least twenty different Chinese-built computer models have been identified in China since 1958, when the first Chinese computer "AUGUST 1st" was announced in Peking, only 6 years after the Soviet Union built its first BESM-1 machine. The Chinese computer was based on design specifications for the Soviet URAL-2 model that were provided as part of the Soviet technical-assistance program to China. Soviet scientists now admit that the Chinese built their first computer based on these specifications even before the Russians managed to construct their own prototype.

In April 1977 China announced trial production of the first Chinese microcomputer, the DJS-050. It is built with Chinese-made MOS LSI circuits by the Anhwei Radio Works in collaboration with the Electronic Engineering Department of Tsinghua University and the No. 6 Research Institute of the Fourth Ministry of Machine Building. This development suggests that the new computer-manufacturing thrust will be directed at the production of minicomputers and microprocessors for general use in China and possibly for export.

Regular production of Chinese computers did not begin until 1962, but by 1973 the output of the industry was estimated at about 100 units per year. Some Chinese computers scientists who accompanied a small DJS-17 computer exhibited at the China Economic and Trade Fair in Cologne in June 1975 indicated that it is produced in Peking in "large quantities." In other cases it is difficult to judge in what quantities computers are produced in China, but based on the numbers in use and variety in models it is doubtful if any computer model has been produced in numbers greater than 100 units to date.

By 1965 the Chinese computer designers incorporated some semiconductor components in an improved version of the Soviet M-20 computer and moved into the second-generation computer production. The first fully transistorized machine, the DJS-21, was announced in the same year. This time China was only about 3 years behind the Soviet Union, which introduced its URAL-4, MINSK-2, and RAZDAN-2 second-generation machines in 1962. In 1970 the Model 111 computer was announced. It was the first Chinese integrated circuit machine and marked China's entry into the third-generation computer production. This time China appears to have been only 2 years behind the Soviet Union, which developed the third-generation NAIRI-3 computer in 1968.

In 1974 China announced the DJS-11 integrated-circuit computer with an operating speed of 1 million operations per second. This machine is comparable to the fastest Soviet Union computer, the BESM-6, which is a second-generation machine but has been in operation since 1966. Neither China nor the Soviet Union has been able to match the operating speeds of the American Control Data CDC-7600 computer, which has operating speeds in the order of 12 million operations per second and has been in operation since 1969 (see Table 11.8).

Both China and the Soviet Union are now racing to develop a "supercomputer" to match the latest American machines. These include Control Data STAR and Burroughs ILLIAC computers with processing speeds of 100 million operations, the CRAY-1, rated at 150 million, and Burroughs BSP and Texas Instruments ASC computers rated at 50 million operations per second. Although rumors persist that China is developing a supercomputer, there have been no announcements about the status of this work. In the Soviet Union a BESM-10 computer with speeds in the order of 12.5 million operations per second was expected to make its debut in 1977. It has been delayed and is now planned for 1980, but Soviet scientists also suggest that a true Soviet supercomputer with speeds

Table 11.8 Characteristics of Chinese digital computers

Year	*Computer Model*	*Word Length (Bits)*	*Maximum Memory Size*	*Rated Speed Operations per Second*
1958	"AUGUST 1"	NA	NA	5,000[a]
1962	DJS-1	NA	2 K	1,800
1963	DJS-2		2 to 4 K[b]	10,000
1964	Unidentified	NA	4 to 16 K	50,000
1965	Model 109C (Peking)	48	32 K	115,000
1965	Unidentified		1 to 2 K	10,000
1965	First solid state	NA	NA	6,000
1965	DJS-7	21	4 K	3,000
1966	DJS-6	48	16 to 32 K	100,000
1966	DJS-21		4 K	60,000
1968	Model C2 (Shanghai)	32	8 K	25,000
1970	Model 111 (Peking)	48	32 K (IC)[c]	180,000
1971	Model 709 (Shanghai)	48	32 K	110,000
	TQ-3	24	8 K	80,000
	TQ-11	36	16 K	50,000
1973	DJS-17	24	8 K	100,000
1974	DJS-11	48	130 K	1,000,000
	Unidentified	NA	NA	920,000
1974	DJS-18	48	65 K	150,000
1976	DJS-130 (Tientsin)	16	NA	500,000
1977	DJS-050 (Micro)	8	NA	NA

Source: Compiled by 21st Century Research from various reports in the press and from visits by specialists to Chinese institutes.

[a] Speed of URAL computer on which this model was based. [b] K = 1000 words.
[c] IC = Integrated circuit.

in the range of 50 to 120 million operations per second may also become operational that year. Strict export controls of advanced computer sales to communist countries make it very difficult to undertake prototype copying, and the results of this particular Sino-Soviet competition will reveal much about their respective high-technology innovation capabilities.

The weakest aspect of Chinese computer equipment that is also true of the Soviet Union and eastern Europe is a relatively poor selection of the viable peripheral and input/output devices that allow a computer system to perform optimally. So far China has not announced the existence of its own magnetic disk, which is a basic external memory device used with computer systems throughout the world and already in production in all COMECON countries. Some magnetic head-per-track drums have been observed in China, but these are regarded as physically large for their capacities of 60,000 words of storage. Magnetic tape drives made in China appear to compare most favorably with Western or

COMECON equivalents of the second-generation era. Although obsolete by Western standards, Chinese peripherals are nevertheless quite adequate for the computers on which they are used.

The Chinese do not use punched cards, and most of the original input data is prepared on punched paper tape, which is also a legacy of the Soviet computer industry where it is widely used. Some Chinese computers are also operated via terminals built in China and could probably be used with teletypewriters manufactured for the telecommunications networks. Computer output line printers made in China operate at 600 lines per minute and use Roman characters. Some slow-speed 15 character-per-line numeric printers are also available for engineering and scientific applications.

Development of digital computers in China was a direct result of the priority targets announced in 1956 within the Twelve Year Plan for development of science and technology. That same year the Institute of Computation Techniques was established at the Academy of Sciences. Within 3 years additional computer design and development institutes were operating in Shanghai, Shenyang, Tsinan, and Chengtu.

Now known as the Institute of Computer Technology in Peking, this organization is regarded as the leading technological gatekeeper and hardware developer. It is also known to be developing computer-assisted instruction (CAI) systems for teaching purposes, information-retrieval systems, and use of APL language for interactive problem solving. All these programs suggest developments leading to time-sharing of computers, a technique in use in the West since the 1960s. The Institute of Computer Technology in Shanghai is also a major hardware developer, as is the Computer Technology Research Center of Tsinghua University in Peking. The latter, a technical university, appears to stress practical applications of data processing in automation of plants and developed small computers for numerical control of machine tools.

The institutes work closely with established radio and telecommunications factories, which manufacture computers in larger quantities. The best known of those is the Peking Wire Communications Plant No. 738, the only factory in the People's Republic of China that is known to use modern manufacturing techniques such as machine inspection of components, automatic back-panel wiring and core testing, and computerized circuit checkout. This plant manufactures DJS-6, DJS-7, and DJS-21 solid-state computers and also made the DJS-1 and DJS-2 vacuum-tube machines in the early 1960s. Originally the plant began operations in 1957 to produce automatic switchboards and was constructed with Soviet assistance. In 1973 it was estimated to have been producing thirty digital computers per year.

Other noted computer manufacturers in China include the Shanghai Radio Plant No. 13, which specializes in process-control machines and Peking Radio Plant No. 3, which makes small digital machines. Peking Radio Plant No. 1, the Shanghai Electric Relay Plant, and Tientsin Electronic Instruments Plant are also involved in manufacture of digital and analog computers.

During 1975 the Shanghai Radio Plant No. 13 announced that it is also building large general purpose computers at the rate of two units per year. These operate at 1 million operations per second and may be the largest computers in production in China. The plant also builds large numbers of small machines that operate at 110,000 operations per second and may be the DJS-17 models already displayed at Canton and Cologne.

Computer Market Potential and Imports

Potential commercial computer users in China are primarily the large state enterprises that form the backbone of the industry and various research institutes of the Academy of Sciences, the universities, and the ministries. China's national services such as banking, insurance, travel, and trading organizations present another market in the long run.

The original industrial development program called for the construction of 600 major plants that will eventually become the end users of large computers for management purposes and minicomputers for process control. These can probably absorb up to 1000 medium-sized computers and two to three times that many minicomputers during the next decade. The other major computer markets are the 1500 research institutes identified in China during the late 1960s, some of which are very small and may have been discontinued. These probably have an ultimate potential for another 1000 small and very small computers during the near term.

The present potential thus appears to be about 2000 computers, several hundred of which are probably already in place. Assuming that China can expand its computer production to only 250 units annually by 1980, it should be able to supply its basic military and commercial computer needs during the 1980s. It will probably continue buying single units of large computer systems for special unique complex applications that will offer immediate hardware and software solutions to specific problems.

The demand for computers could expand dramatically if China introduced a program for installation of minicomputer-based accounting and management systems at each of its 70,000 rural communes and other local administrative units. However, to be of much value to the central authorities such a program would have to wait for a more modern and efficient telecommunications system. What may happen first is gradual computerization of the estimated 30,000 to 40,000 branches of the People's Bank of China. Just installing a single terminal at each branch could be worth up to $200 million, but again availability of telecommunications is the limiting factor. As for the terminals, China appears to have the capability to produce a suitable unit within its electronics industry. Hence Chinese computer imports have been few and far between and exhibit all the classical earmarks of a prototype-purchasing policy. Only a large military-aid program designed to rapidly modernize Chinese armed forces would create a significant market for computer imports.

Actually, there is also a real opportunity for the sale of second-hand IBM 360 or 370 computers. China could immediately supply most of its data-processing needs for the next 10 years with equipment comparable or superior to the COMECON RIAD computers, which are based on IBM 360 and 370 designs and are only now entering serial production and client service in the Soviet Union and eastern Europe. Such a move would save China the investment in production capacity and allow it to concentrate on development of minicomputers and its own supercomputer, which it so desires. Time will tell whether the Chinese will take advantage of this unique opportunity.

During 1964–1975 about sixteen different computers were sold to China. Ten of these were basically parts of seismic or geophysical exploration systems. Actual value of all equipment under the SITC code 714 exported to China during those years amounted to only $14.1 million worth of computers and other office machines. It is interesting to note that no two systems were alike and were purchased from companies in France, United Kingdom, and the United States (see Table 11.9 and Fig. 11-8).

Burroughs Corporation also received an order for a multiprocessor B-7700 system valued at $13 million, which was the largest single computer order placed by China so far, but this sale was vetoed by the U.S. Export Administration on the grounds of national security. The sale of two Control Data CYBER 172 computers as part of a seismic exploration system sold by the French Compagnie Generale de Geophysique was approved by President Ford in 1976, overriding objections from the Pentagon and was viewed as a deliberate gesture of support for the Hua Kuo-feng leadership in China. Also in 1976 China placed an order for two large computers from Hitachi in Japan to equip the Central Meteorological Bureau in Peking. Valued at $8.6 million, this order includes HITAC 170 and HITAC M160–2 computers, which could become the largest computers operating in China in the near future.

One competition that has been warming up is for the supply of a

Table 11.9 Typical sales of computers to China since 1964

Year	*Type of Computer*	*Supplier, Country, and Order Value*
1964	Unidentified	SERCEL, France; seismic exploration system
1965	ARCH 1000	Marconi-Elliott, U.K.; process-control system
	NADAC 100 (five units)	SEA, France; plant automation systems
1966	ELLIOTT 803	Elliot Computers, U.K.; medical research use
1967	ICL 1903 ICL 1905	ICL, U.K.; unknown destination
1973	UNIVAC	SERCEL, France; seismic research system
	RAYTHEON 704	Geospace, U.S.A.; seismic control, $5.5 million
1974	IRIS 60 MITRA 15 (two)	SERCEL, France; oil-exploitation management
	CDC 172 (two) CDC 170	CGG, France; oil-exploration ship. equip. $5 million
1975	INTERDATA	U.S.A., oil-exploration system, $23.0 million
1976	MITRA 30	CII, France
	BURROUGHS B7700	Burrough, U.S.A., license denied
	RIAD EC-1040	East Germany; reported one installation in China
	HITAC M 160-2 HITAC M 170	Hitachi, Japan, Central Meteorological Office
	Unidentified	Hungary; petroleum-control system
	MITRA 125 (two)	SFIM-CII; France; civil aviation use in China
	PDP 11/45 (two)	Digital Resources; U.S.A., seismic exploration
1977	UNIDATA 7738	Siemens, West Germany; power-station use
	IBM 3032	U.S.A. for Bank of China in Hong Kong, $20 million
	IBM 360	U.S.A., process-control application

Source: Compiled by 21st Century Research from various reports, articles, and papers listed at the back of this chapter as references.

system for the Bank of China and twelve other banks in Hong Kong owned and operated by the People's Republic of China. Both IBM and NCR have been reportedly negotiating with the Chinese for this order. This contract is regarded as a key step to possible negotiations with the People's Bank of China to provide centralized data processing for its estimated 30,000 branches throughout the country. Top IBM executives were planning to visit China in late 1977, but other IBM representatives were reported in Peking earlier that year.

Western computer sales to China are destined to remain primarily a political and security issue. China clearly has the capacity and know-how to manufacture small and medium computers for its basic research and administration needs and is primarily interested in imports of very large systems for specific applications. Normally these are likely to be vetoed on the grounds that they could be used for military applications and could jeopardize the security of the West.

On the other hand, exports of jet engines, helicopters, and certain avionics systems having clearly military uses have already taken place.

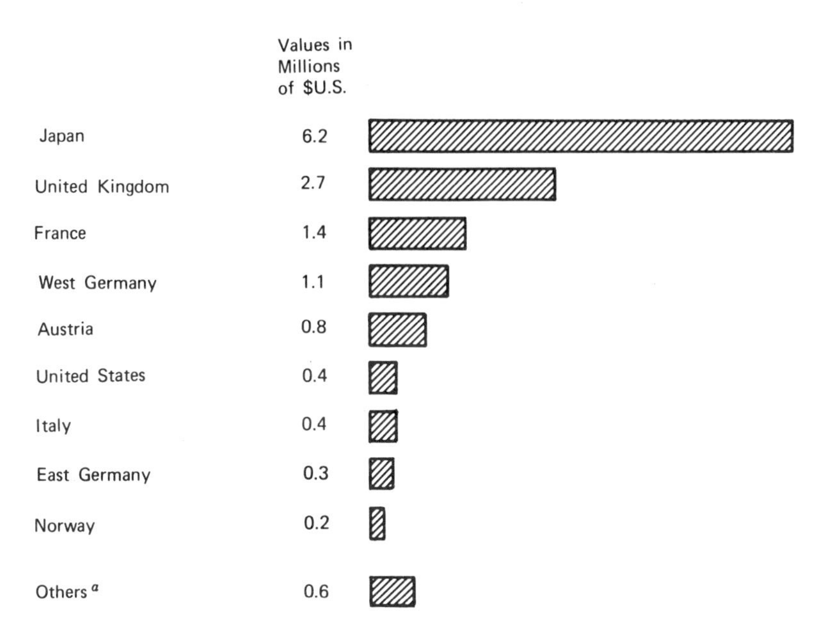

***Figure 11.8** Market shares of suppliers of office machines, computers, and calculators to China during 1964–1975 (SITC code 714).* (Source: United Nations *bulletins of statistics on trade in engineering products and* OECD Commodity Trade Statistics Series B.) [[a] *Others include Czechoslovakia (0.2), Denmark (0.1), Hungary (0.1), Switzerland (0.1), Netherlands (0.1).*]

As a result, China may be able to import large computers from Japan and western Europe, although such exports may be banned from the United States. There are also reports that an East German RIAD R-40 computer has been installed in China. Since China is also an importer of jet aircraft and helicopters from the Soviet Union it will be interesting to see what would happen if the Soviets finally build their supercomputer and decide to sell a copy to the Chinese.

AEROSPACE INDUSTRY

Aerospace industries normally encompass the design, production, and operation of aircraft, rockets, ballistic missiles, satellites, and the necessary supporting activities, which depend heavily on advanced electronics, computers, chemical fuels, materials, and telecommunications. Visible achievements in aerospace industries are good indicators of the level of advanced technology available or developed in a country. By the same token, imports of aerospace equipment or technology, particularly by centrally planned economies, are indicative of their shortcomings in high-technology development programs.

China's aircraft production, which is presently based on obsolete Soviet airframe and aero-engine designs and is in the process of modernization, is discussed in detail in Chapter 13. The critical first steps have been taken with the purchase of Rolls-Royce Spey turbofan jet engine licenses and manufacturing technology. China's aircraft industry is among the largest in the world and has to maintain and equip a fleet of over 6200 military and civil aircraft (see Table 11.10).

The aerospace industry provides China with strategic delivery systems for its nuclear weapons and defense systems against enemy attack.

Table 11.10 Composition of Chinese Air Force and Naval Air Force aircraft fleets in 1976

Type of Aircraft in Service	*Estimated Number of Aircraft*
Antonov An-2 military transport and liaison	200
Be-6 MADGE Maritime Reconnaissance aircraft	NA
Iliyushin Il-28 light bombers	300
Iliyushin Il-28 torpedo-carrying light bombers	100
Iliyushin Il-14 and Il-14 military transports	50
Lisunov Li-2 (Soviet DC-3 copy) transport	NA
MIG-15 fighters and trainers including F-2 aircraft	200
MIG-17 fighters including F-4 built in China	1,500
MIG-19 fighters including F-6 built in China	2,000
MIG-21 fighters including F-8 built in China	75
MIG-17, MIG-19/F-6 Naval Air Force fighters	500
F-9 Mach-2 supersonic Chinese designed twin jet fighter	100+
Mi-4 Hound helicopters (Navy and Air Forces)	350+
Super Frelons helicopters (French imports)	13
Messershmitt-Boelkow-Blohm BO-105 helicopters (West Germany)[a]	100+
Tupolev Tu-16 medium bombers	65
Tupolev Tu-2 light bombers (Navy and Air Forces)	100+
Tupolev Tu-4 medium bombers	few
Total combat aircraft estimate in 1976	4,950+

Source: International Institute for Strategic Studies, *The Military Balance 1976–1977*, London.

[a] Reported by Strategic Survey of IISS for 1976 as a contract to supply helicopters to China at an unspecified date in the future.

These systems include aircraft as well as land- and sea-based missile systems. The formal missile force in China, established under the name of Second Artillery Command, was announced in 1967.

Ballistic Missiles

Rocket testing is believed to have begun in 1959, when the Soviet Union supplied China with SS-3 and SS-4 medium-range ballistic missiles (MRBM) without nuclear warheads. By 1963 China was testing its own version of a single-stage liquid propellant MRBM based on the Soviet SS-4 vehicle. The Chinese MRBM is a transportable missile with a range of about 600 nautical miles and first became operational in 1966. There are also reports that subsequent versions of the MRBM were developed with a range of up to 1500 miles, capable of delivering 20-kton (kiloton) nuclear warheads. In October 1966 an MRBM vehicle was fired from the Shuang Cheng-tzu launching complex in Kansu Province carrying a nuclear warhead that was exploded at the Lop Nor testing range in the Sinkiang Uighur Autonomous Region in western China.

Up to 50 MRBMs are now believed to have been deployed at operational sites since 1970. Some missiles are believed to be in steel and concrete silos and some in caves from which they have to be rolled out before firing. Several MRBMs have been reported in Manchuria, presumably targeted against Siberian airfields, railway junctions, and the Soviet Far East naval base at Vladivostok. The new Soviet Baikal–Amur railway, now under construction to provide an alternative to the trans-Siberian line running along the Manchurian borders, is also within range of China's MRBMs. These missiles can also reach any target in Korea and at extreme range could probably reach Tokyo.

According to various reports China shifted its emphasis in about 1970 from MRBMs to the development of intermediate-range ballistic missiles (IRBM). The Chinese IRBM is also believed to be a single-stage liquid-propellant missile using storable liquid, but work on solid propellant is under way. The IRBM is apparently also used as the first-stage rocket booster for Chinese satellite launchings. Its range is reported to be 2000 to 3500 miles in a multistage configuration. Twenty to thirty IRBMs are believed to be deployed in permanent sites, consisting of concrete silos or special sites built into mountain sides. A new experimental missile-launching complex near the North Korean border in Kirin Province was reportedly test firing IRBMs into western Sinkiang.

At least some of the IRBMs are believed to be aimed at Moscow and other targets in European Russia, although they could only reach such targets at extreme ranges with limited deployment flexibility. On the other hand, if Chinese IRBMs are deployed in a country like Albania, where there are Chinese military bases, targets such as Moscow, London, and the Persian Gulf oilfields are well within the range of those missiles. This possibility is not as remote as it may appear at first glance. There are indications that the Chinese regard proliferation among less advanced nations to be beneficial since it inhibits major power dominance. On the other hand, China announced its readiness to join in action to destroy all nuclear weapons or to declare "nuclear-free" zones. If progress on this proposition continues to be slow or nonexistent, at least tacit approval by China of proliferation in the Third World cannot be dismissed out of hand.

Development of a Chinese intercontinental ballistic missile (ICBM) has been regarded as imminent for the past few years. At least one missile firing in China has been described as a "limited range" test of a long-range vehicle. Chinese satellite launchings since 1970 prompted repeated speculation about a Chinese ICBM. Its range is expected to be 6000 to 7500 miles with a 3-MT warhead; some observers believe several may become operational by 1980. The Chinese ICBMs are expected to be larger than the Soviet SS-9 or American Titan II vehicles, and silos for such missiles have already been reported under construction. In contrast, the Soviet Union now deploys 1477 ICBMs, whereas the United States has 1054, many of which carry multiple warheads.

It was believed recently that the Chinese ICBM development program was hindered by lack of adequate test range and sea-borne tracking and instrumentation systems. Therefore, much has been made of the launching of a 12,000-ton tracking ship, "Hsian Yang Hung," which was reported on a shakedown cruise as far as the Indian Ocean in 1971. China is also believed to operate ground tracking and telemetering facilities on the island of Zanzibar off the coast of Tanzania. Possibilities of other tracking stations exist in Pakistan, Sri Lanka, or Madagascar, all of which receive Chinese foreign aid.

China is also known to have built a ballistic missile submarine at the Dairen shipyards in 1964. It is similar to the Soviet "G" class submarines and has three missile tubes in its coning-tower superstructure. So far, however, there has been no indication that any submarine-launched ballistic missile (SLBM) has been built or tested. This situation may be indicative of the inability of the Chinese to develop solid-propellant rocket engines, which are mandatory for use in SLBMs. In contrast, the Soviet Union now deploys 909 SLBMs and the United States, 656 within their respective nuclear submarine fleets.

There has been no known effort in China to develop an antiballistic missile (ABM) system, and air-defense systems using Chinese-made radars are regarded as outmoded. A ballistic missile early-warning system providing 90 percent coverage against Soviet missiles has been reported. It is said to consist of advanced phased-array equipment and is deployed in western China. The system is said to provide warning of attack by Soviet IRBMs and ICBMs. Several hundred surface-to-air missiles (SAMs), believed to be modified versions of Soviet SAM-2 missiles, are deployed in forty to fifty different sites tied to the Chinese radar early-warning and control systems.

Missile-manufacturing facilities have been identified at Paotou, Sian, Harbin, Shenyang, and Chengtu. Solid and liquid fuels are probably produced at Liaoyuan, Taiyuan, Hsianhsing, and Dairen Missile Fuel Plant, and the Lanchow Missile Fuel Plant. Factories at Shanghai and Shenyang are also believed to have capabilities to produce air-to-air missiles. Missile-test ranges exist at Paotou, western Ninghsia, and Chang Hsiu Tien, where rocket engines are tested.

Satellites

On April 24, 1970 China became the fifth nation to successfully orbit an earth satellite using its own launch vehicle and facilities. The first Chinese satellite, CHICOM 1, weighed 381 pounds and was launched from the Shuang Cheng-tzu space-launching facilities in Kansu Province. The first Chinese satellite was not only the heaviest first satellite launched by any country so far but was also heavier than the combined payloads of the first satellites launched by the Soviet Union, United States, France, and Japan.

Since 1970 China placed six more satellites in orbit, three of which were launched during 1975. The CHICOM 3, launched in July 1975, was described by Chinese news media as a vehicle to support China's preparedness against war. The last known satellite as of this writing was launched on 7 December 1976 and was apparently returned to earth a few days later (see Table 11.11).

According to Western analysts, the primary objective of Chinese satellites is deployment of space-borne intelligence and communications systems. It is apparent from the altitudes and orbital angles relative to the equator that Chinese satellites are experimental forerunners of space photointelligence systems. Those arguments are supported by Chinese interest since 1975 in purchase of Western satellite cameras, advanced radar equipment, and large computers. Should the West further modify its perception of China's role in the stability of the Asian continent, it may

Table 11.11 Characteristics of Chinese Satellites

Date Launched	*Name*	*Weight in lbs.*	*Other Characteristics*
24 April 1970	CHINCOM 1	380	68.5° inclination to equator
3 March 1971	CHINCOM 2	487	
26 July 1975	CHINCOM 3	8,000	69.0° inclination to equator launched in support of China's preparedness for war
26 Nov 1975	CHINCOM 4		62.95° inclination to equator Reconnaissance/ capsule recovered
16 Dec 1975	CHINCOM 5		69.0° inclination to equator Stayed in orbit 6 weeks
30 Aug 1976	CHINCOM 6		
7 Dec 1976	CHINCOM 7		Reported recovered after 4 days

Source: Compiled by 21st Century Research from various references listed at the end of this chapter.

become a very significant market for advanced space-related weapons systems and electronic support equipment.

NUCLEAR INDUSTRY

The Chinese nuclear industry exists primarily to support a weapons program that was originally based on Soviet assistance. According to one American sinologist, the Soviet Union in a secret agreement during 1957 promised to provide China with a prototype nuclear bomb and details of its manufacture. This stimulated Chinese investment in nuclear industry, and in 1958 the first nuclear reactor was purchased from the Soviet Union. Although the secret agreement to transfer nuclear-bomb technology was repudiated by the Soviets in 1959, China proceeded to rapidly expand its nuclear-weapons program and its supporting nuclear industry. Some observers suggest that the Soviets changed their mind about transferring nuclear-weapons technology to China after Mao Tse-tung expressed his opinion that the use of nuclear weapons could be justified by a quick victory of world socialism. Apparently the majority of leaders attending the First International Meeting of the Communist Parties in the fall of 1957 in Moscow overwhelmingly rejected this idea.

Actually there appears to be more evidence suggesting that deterrence rather than aggression was China's primary objective in developing nuclear weapons. As long ago as 1957 Teng Hsiao-ping explained that the United States could station troops in Taiwan because China had no nuclear bombs or guided missiles to deter such action. As late as 1965 Andre

Malraux was told by Mao Tse-tung that China only needed six nuclear bombs because then neither side would dare to attack her. It is often declared that China will never be the first to use nuclear weapons, but it is also understood that China conceives of some form of nuclear proliferation among less developed countries because this inhibits their domination by major powers.

China entered the "nuclear club" on October 16, 1964, when it exploded a 200-kT nuclear device at the Lop Nor testing area. Some industry observers believe that this test would have been made at least 2 years earlier but for the industrial and agricultural crises that followed the Great Leap Forward.

Soviet sources stated in 1971 that China spent 12 billion yuan ($6 billion) on development of nuclear weapons. Other observers estimate that over $1.5 billion was invested in construction of the Lanchow Gaseous Diffusion Plant in Kansu Province, which has separation facilities for production of weapons-grade uranium (^{235}U). The Lanchow plant is also operating a large 300-MW nuclear reactor and draws electric power from the 1250-MW Liu Chia-hsia hydroelectric power station on the Yellow River. It was estimated to produce about 600 pounds of weapons-grade ^{235}U per year since 1963, but its ultimate capacity is believed to be at least 800 pounds of ^{235}U annually. There are also persistent reports that a second gaseous diffusion plant has been built and that work was proceeding to develop the gas-centrifuge uranium-enrichment process.

Plutonium is also produced in China at the Yumen nuclear reactor complex. This plant is equipped with a 600-MW reactor and is estimated to produce 400 pounds of plutonium annually. Sufficient stockpiles of plutonium are believed to have been accumulated at this site since it began operation in 1967 for the production of 200 20-kton nuclear weapons. A smaller plutonium plant has been identified in Paotou, based on a reactor of up to 200 MW. Its annual plutonium output is estimated to be in the order of only 20 pounds and is destined primarily for research purposes. Other reactors are believed in operation in Sian and Chungking.

China appears to have adequate uranium deposits. Three mines that began production in 1967 were estimated to have a combined daily uranium-ore output of 2500 metric tons. These are the Maoshan and Chushan mines in Chuannan County of Kiangsi Province and the Hsiachuang mine in Weiyuan County in Kwantung. Uranium-ore deposits were also reported at Chungkak and Sinkiang. Other minerals required for nuclear weapons, such as lithium, borax-wolfram, and beryllium concentrates, are also produced in sufficient quantities. Over forty chemical separation plants are reportedly involved in the extraction of uranium and thorium and in reprocessing of spent reactor fuels.

In the 1960s it was reported that Chinese uranium ore was only partly processed at the Chuchou Uranium Processing Plant in Hunan Province and was then sent to Czechoslovakia for further concentration. The Chinese paid for this service by giving the Czechs half of the uranium and as a result were able to produce ^{235}U cheaper and faster and relieved some of the heavy demand on their own Lanchow Gaseous Diffusion Plant, whose production only began in the spring of 1963. This apparently helped China to set off its first nuclear test in 1964.

Since then there have been at least twenty-one Chinese nuclear weapons tests with yields ranging from 20 ktons to 3 Mtons (megatons). The most recent nuclear test took place in September 1977. Some were tower-mounted explosions, but many were conducted by air drops from Chinese-built TU-16 medium jet bombers. At least one nuclear device was delivered and exploded as a warhead in a MRBM vehicle, and a few were underground explosions (see Table 11.12).

Table 11.12 Nuclear explosions conducted by nuclear powers since 1945

Country	*Explosions during 1945–1963*	*Underground Explosions during 1963–1976*	*Total Number of Explosions*
United States	293	321	614
Soviet Union	164	190	354
France	8	41[a] + 15	64
United Kingdom	23	4	27
China	—	18[a] + 3	21
India	—	1	1
Total	488	593	1,081

Source: International Peace Research Institute, Stockholm, 1977.
[a] Explosions conducted in the atmosphere.

Since China became a nuclear power it has made rapid progress in developing its nuclear capability. Industry observers point out that China took only 2.5 years to advance from its first nuclear test to the explosion of a 3-MT thermonuclear device on June 17, 1967. In contrast, it took the United States, the United Kingdom, the Soviet Union, and France 7.5, 4.5, 4, and 8.5 years, respectively, to make the same transition. China also reduced the weight of its original nuclear device from 20,000 to 2000 pounds in only 2 years. This was a necessary step in design of a warhead for its ballistic-missiles vehicles. China is believed to have already surpassed France in nuclear development and may also overtake the United Kingdom in the future.

Estimates of the number of nuclear reactors operating in China range from seven to forty, but the larger figure is probably based on the 1956 Soviet aid plan for China, which called for the development of thirty-nine "atomic centers." It is estimated that with those facilities China could now have a stockpile of several hundred nuclear weapons at its disposal.

China owes much of its success in nuclear weapons to the early access to some Soviet nuclear research at the Joint Institute for Atomic Research at Dubna near Moscow, where twenty-one Chinese students began participating in 1956. Soviet technicians also helped China to build the Lanchow Gaseous Diffusion Plant and the Lop Nor Testing Range. However, considerable credit must also go to the Chinese for the decisions to develop early a comprehensive nuclear research program of their own.

The first nuclear research reactor obtained from the Soviet Union began operations in 1958 at the Atomic Energy Institute of the Academy of Sciences in Peking. The reactor is a 10,000-kW enriched-uranium heavy-water (deuterium) type. The institute also produced a 2.3-MeV (megaelectronvolt) Van de Graff electrostatic accelerator and is heavily involved in new weapons design. It is regarded as the technology gatekeeper for China's nuclear industry. There is an atomic research center in the Tarim Basin area of Sinkiang that appears to be operated as a subsidiary of the institute in Peking.

A nuclear research center with a 7000-kW heavy-water nuclear reactor of Soviet design has also been reported in operation near Peking and is believed to produce 2 percent enriched uranium rods. Another experimental nuclear reactor is operated by the Institute of Physics of the Academy of Sciences in Peking. In 1974 this institute produced the first experimental thermonuclear device in a program of research to control thermonuclear reaction as a new energy source. The Physics Department of the Tsinghua University in Peking reportedly built its own nuclear reactor.

In Shanghai a Nuclear Research Institute of the Academy of Sciences develops isotope instruments and meters. In 1975 it was announced that a Nuclear Research Institute at an unidentified location put into operation a small TOKAMAK device to study thermonuclear fusion.

China planned to become self-sufficient in the nuclear industry and appears to have succeeded in doing so. There were some attempts on the part of Western suppliers to interest China in purchasing other nuclear research reactors, but no response was ever received.

REFERENCES

Brown, George S., General. "United States Military Posture for FY 1978," U.S. Joint Chiefs of Staff, 20 January 1977, GPO, Washington, D.C.

Cheng, Chu-yuan. *The Machine Building Industry in Communist China.* Aldine, Atherton, Chicago. 1972.

Cheng, Chu-yuan. "China's Machine Building Industry," *Current Scene,* July 1973, Hong Kong.

Chen, Chu-yuan. "The Machine Tool Industry," in William W. Whitson, ed., *Doing Business with China,* Praeger, New York, 1974, pp. 270–287.

China Export Commodities Fair. *Changchow's Growing Electronics Industry,* Canton, 1974.

China National Machinery Import and Export Corporation. *Export Catalogue,* Peking, 1973.

Chinese Academy of Sciences, the Fourth Company of the Electrical Engineering Institute. "Electron Beam Welding," *Scientific Experiments* October 1971 (translation by JPRS), pp. 16–17, Peking.

Chinese Diesel Engine. MACHIMPEX Export Brochure, Kwangchow, China, 1974.

Current Scene. "The PRC Economy in 1976," April–May 1977, Hong Kong.

Dernberger, Robert F. "The Economic Consequences of Defense Expenditure Choices in China," in *China: A Reassessment of the Economy,* Joint Economic Committee, U.S. Congress, July 1975, pp. 467–499.

Gelber, Harry. "Nuclear Weapons and Chinese Policy," *Adelphi Paper* No. 99, International Institute for Strategic Studies, London, 1973.

Heymann, Hans, Jr. "The Air Transport Industry," in William W. Whitson, ed., *Doing Business with China,* Praeger, New York, 1974, pp. 151–169.

Heymann, Hans, Jr. "Acquisition and Diffusion of Technology in China," in *China: A Reassessment of the Economy,* Joint Economic Committee, U.S. Congress, July 1975, pp. 678–729.

International Institute for Strategic Studies, The. *Strategic Survey 1976,* London.

International Institute for Strategic Studies, The. *The Military Balance 1976–1977,* London.

Jammes, Sydney H. "The Chinese Defense Burden, 1965–74," in *China: A Reassessment of the Economy,* Joint Economic Committee, U.S. Congress, July 1975, pp. 459–466.

MACHIMPEX. "Fork Lift Trucks," brochure by CMC, Kwangtung Branch, Canton.

MACHIMPEX. "Excavators," Brochure by CMC, Dairen Branch, Liaoning.

McGurk, D. L. "Electronic Products," in Wiliam W. Whitson, ed., *Doing Business with China,* Praeger, New York, 1974, pp. 289–296.

Minor, Michael S. "China's Nuclear Development Program," *Asian Survey,* June 1976.

Murphy, Charles H. "Mainland China's Evolving Nuclear Deterrent," *Bulletin of Atomic Scientists,* January 1972.

National Council for U.S.–China Trade. *Special Report No. 11.* "PRC, Foreign Trade in Machinery and Transportation Equipment."

Oberg, James, "China in Space," *Current Scene,* August–September 1977, Hong Kong.

Peking Review. "Shanghai Applies Electronic Techniques," August 25, 1972, Peking.

Reichers, D. Philip. "The Electronics Industry of China," in *PRC: An Economic Assessment,* Joint Economic Committee, U.S. Congress, May 1972.

Szuprowicz, Bohdan O. "Computers in Mao's China," *New Scientist,* March 15, 1973, London, p. 598.

Szuprowicz, Bohdan O. "Informatique, la Fievre Chinoise," *Usine Nouvelle,* October 1973, Paris, p. 168.

Szuprowicz, Bohdan O. "Automation in People's Republic of China," *National Investment and Finance,* India, December 1973, p. 201.

Szuprowicz, Bohdan O. "China's Computer Industry," *Datamation,* June 1975, p. 83.

Szuprowicz, Bohdan O. "Electronics in China," *U.S.–China Business Review,* May–June 1976, p. 21.

Szuprowicz, Bohdan O. "Canton Trade Fair Visit," *Purchasing,* September 1975.

Szuprowicz, Bohdan O. "CDC's China Sale Seen Focusing Western Attention," *Computerworld,* May 12, 1977.

Szuprowicz, Bohdan O. "Republique Populaire de Chine–Un Premier Pas en Micro-Informatique," *01 Hebdo Informatique,* May 23, 1977, No. 437, Paris.

Szuprowicz, Bohdan O. "The Chinese Micros Are Coming," *Pacific Computer Weekly* (Australia), May 27, 1977, p. 5.

Szuprowicz, Bohdan O. "How China Makes 'Foreign Things' Serve Her," *The Strait Times* (Singapore), July 25, 1977.

Szuprowicz, Bohdan O. "Etranger La Chase en Chine (avec IBM)," *01 Hebdo Informatique,* August 8, 1977, No. 448, Paris.

Szuprowicz, Bohdan O. "The Sino-Soviet Computer Race," *Pacific Computer Weekly* (Australia), September 30, 1977.

Tsu, Raphael. "High Technology in China," *Scientific American,* December 1972.

21st Century Research. *PRC, Industries, Markets, Imports, and Competition,* July 1976 and February 1977, North Bergen, N.J.

UNITAR. *Communications, Computers and Automation for Development* United Nations, New York 1971.

U.S. Department of Commerce. *Global Market Survey,* "Metalworking and Finishing Equipment," January 1975.

U.S. Department of Commerce. *Construction Equipment, a Market Assessment for the People's Republic of China,* March 1976.

U.S. Government. "Chiefs of State and Cabinet Members of Foreign Governments," *Research Aid,* CR CS 77-008, August 1977.

12

Agriculture

Agriculture is the pivotal sector of the Chinese economy. An estimated 80 percent of China's population, including about 700 million peasants, are engaged in agriculture. It must produce food and clothing for an estimated 950 million Chinese, as well as raw materials for industry and products for export. The most pressing and persistent problem in China is to maintain and increase the supply of food and agricultural products for a population growing at the rate of nearly 20 million a year. This is like adding the population of Canada, Argentina, or California to that of China every year.

About one-third of all economic output of China is represented by agricultural production. China is one of the largest food-producing countries in the world. It is by far the largest rice-growing area, producing 33 percent of all the rice in the world. It is the second largest producer of cotton after the Soviet Union and grows the second largest crop of soybeans after the United States. It is the third largest wheat producer after the Soviet Union and the United States and accounts for almost 12 percent of the total wheat supply in the world (see Table 12.1).

Production of grains that include rice, wheat, corn, barley, oats, and tuber crops was estimated at 260 million metric tons in 1975, or about 0.28 tons per capita. To merely maintain this consumption level in 1980 and 1985, grain production will have to increase to about 290 and 328 million metric tons, respectively. This is equivalent to a 26-percent increase in total grain output by 1985.

BASIC NEED TO INCREASE PRODUCTIVITY

It now seems unlikely that agricultural output can be further increased by opening new areas. About 11 percent of China's total land area is now

Table 12.1 Production of major agricultural commodities in selected countries in 1975/1976 (All figures in thousands of metric tons)

Country	*Wheat*	*Rice*	*Cotton*	*Soybeans*	*Raw Sugar*[a]
World	355,895	348,374	12,333	68,927	81,691
China	41,003	116,267	2,385	12,062	4,000
Soviet Union	66,224	2,009	2,648	780	8,200
United States	58,102	5,805	1,807	42,079	5,680
India	24,104	74,186	1,193	—	5,048
Canada	17,078	—	—	367	120
Australia	11,980	388	33	74	2,930
Argentina	8,570	351	160	485	1,367
Mexico	2,798	510	206	699	2,742
Egypt	2,033	2,423	382	—	537
Brazil	1,788	7,538	517	9,892	6,299
Japan	241	17,097	—	126	459
North Korea	150	3,700	3	290	—
South Korea	97	6,485	3	320	—
Nigeria	—	368	46	65	40

Source: United Nations Statistical Yearbook, 1976.

[a] Data for the year 1975 represent production of centrifuge sugar from beat and cane in terms of raw sugar.

under cultivation, mostly in the eastern half of the country. The potential for opening new lands in western or northern China appears limited because of their remote location, short growing season, and unsuitable physical features. The Chinese leaders are well aware of these limitations and have adopted a policy of increasing agricultural productivity by mechanization, increased use of fertilizers, and improved management of lands under cultivation.

An ambitious goal for the 1976–1980 Five Year Plan is to achieve production of grains and soybeans equal to 400 million metric tons by 1980. This could mean about 370 million metric tons of grains alone, which would increase per capita consumption by 25 percent to 0.35 tons per year. Western analysts are more inclined to accept a goal of 300 million metric tons by 1980 as more realistic. They point out that returns from additional fertilizer inputs may not be realized until considerable additional water-control and irrigation systems are also put into place. Even so, output would have to grow at an average 3 percent per year. This is somewhat higher than output growth during 1970–1975, which averaged only 2.3 percent per year.

However, there appear to be opportunities to improve agricultural productivity in all major crops grown by China. Wheat and soybean crop yields are still below world averages. Chinese rice and cotton are produced at above world average yield levels but are well under the yields achieved by the leading rice-producing countries. Japanese rice crops yield almost twice the amount per hectare achieved in China. The Soviet Union seems to grow twice as much cotton as the Chinese on a comparable amount of land (see Table 12.2).

Table 12.2 Comparison of average crop productivities in selected countries in metric tons per hectare in 1975

Country	*Rice*	*Wheat*	*Cotton*	*Soybeans*
World average	2.43	1.55	0.39	1.42
China	3.24	1.32	0.45	0.82
United States	5.11	2.06	0.49	1.81
Soviet Union	4.08	1.07	0.88	0.50
Japan	6.19	—	—	1.42
India	1.83	1.34	0.17	—
Canada	—	1.80	—	—
Australia	—	1.33	—	—
Brazil	1.46	0.60	0.23	1.83
Egypt	4.71	2.50	0.59	—
Mexico	2.37	3.75	0.76	1.68
Netherlands	—	4.94	—	—

Source: GUS Statistical Yearbook, 1976, Warsaw, Poland (GUS = Glowny Urzad Statystyczny).

China should be in a position to increase its major crop yields to the levels already achieved by Egypt or Mexico. This would increase its rice output by 45 percent, its wheat production by over 80 percent, and cotton by 70 percent and would probably double the soybean crop. This means that there is an overall short-term possibility of increasing China's grain output by 50 percent to the 400 metric tons targeted by the country's planners. It is only a question of how soon it can be accomplished.

The Chinese agricultural situation presents three separate markets to the foreign exporter. The largest is the market for grains, which totalled in value over \$5 billion since 1966 and averaged almost 5 million tons per year. Next in importance is the market for chemical fertilizers, which totaled almost \$2.3 billion during the same period. This trade, and expansion of domestic fertilizer production, is discussed in detail in Chapter 10. A third and relatively modest market is for imported agricultural machinery. It has been estimated only at about \$20 million per year during the first half of the 1970s.

PRODUCTION AND TRADE IN MAJOR AGRICULTURAL COMMODITIES

Agricultural commodities of significance in Chinese foreign trade include rice, wheat, corn, cotton, soybeans, and sugar. Tea, tobacco, fruits, and vegetables are also major Chinese export crops. Other grains produced in China include barley, oats, rye, sorghum, and millet. Tubers production includes white and sweet potatoes, manioc, and taro (see Table 12.3).

Rice, wheat, corn, soybeans, barley, and cotton are by far the most important crops in China's agriculture, using the bulk of cultivated land. Of the estimated 152 million hectares under cultivation, about 135 million are under grains. Agricultural production increased at an average annual rate of 4 percent since 1949 but has been growing only at about 2 percent since 1957.

Grains Production and Imports

The most significant agricultural division of China occurs along the Huai Ho River following an east–west line running from above Shanghai across

Table 12.3 Production of major agricultural commodities in China in millions of metric tons in 1975

Commodity	*Area under Cultivation in Hectares*	*Average Production, 1971–1975*	*Production, 1975*
Rice	34.50	115.40	122.00
Wheat	27.70	35.30	38.70
Soybeans	14.45	11.60	12.00
Tubers	14.00	27.50	28.90
Barley	13.61	20.06	21.06
Corn	11.03	30.64	33.08
Cotton	4.85	2.30	2.40
Potatoes	3.85	36.43	40.02
Oats	2.81	2.74	3.00
Turnips	2.50	1.18	1.25
Rapeseed	2.25	1.21	1.39
Peanuts	2.15	2.61	2.80
Tobacco	0.698	0.913	0.969
Sugarcane	0.47[a]	39.29[b]	40.03[c]
Sugar beets	0.27[a]	6.14	6.70
Tea	—	0.304	0.336

Source: U.S. Department of Agriculture 1976, Washington, D.C. and *GUS Statistical Yearbook, 1976,* Warsaw, Poland.

[a] Data for 1970. [b] Data for 1971–1974. [c] Data for 1974.

China below Sian into Tibet. South of that line rice is the dominant grain crop. North of it, in the drier northeastern provinces, wheat and coarse grains dominate. Sweet or white potatoes are grown in all regions.

Two crops of rice are harvested annually in Kwangtung Province, Kwangsi Chuang Autonomous Region, and southern Fukien. In parts of Hainan Island three crops of rice occur. North of those areas, up to the Huai Ho line, single crops are traditional. Szechwan Province in central China has the highest grain production in all of China and harvested a total of 30,896,000 metric tons in 1975, or about 12 percent of the national output.

Winter wheat is grown in the areas south and north of the Huai Ho line. These include Kiangsu, Anhwei, Hupei, and Szechwan Provinces and the cities of Shanghai, Wuhan, and Chengtu. Winter wheat, corn, millet, and kaoliang are also predominant in areas north of Huai Ho line as far as Peking. In Kansu, Inner Mongolia and most of Manchuria spring wheat, kaoliang, and soybeans predominate. Most grains are also grown in lesser quantities in numerous oasis areas in the Sinkiang arid regions of western China.

Except for disastrous harvests during 1959–1961, total grain production increased steadily from 108 million tons in 1949 to the estimated 260 million tons in 1975 or about 141 percent. At the same time, total land under grains increased from 101.6 to 135.0 million hectares, or only 34 percent. The Chinese, in fact, appear to have doubled their productivity from an average 1.06 to the current 2.0 tons of grains per hectare.

Since 1961 China has been importing an average 5 million tons of wheat and corn annually at an average annual cost of $500 million. Because of the magnitude of those imports they have been often singled out as proof of poor agricultural performance in China. But taken in proper perspective those imports represent less than 2 percent of China's total grain production, and only about 3 percent of wheat, corn, and miscellaneous grain output, excluding rice. In contrast, Mexico and Central American countries with similar per capita grain consumption levels must import up to 10 percent of their domestic output.

In 1976 China imported only half the usual annual grain volume, which is probably the result of relatively better harvest that year. Initial estimates at the time of writing put it in the range of 275 to 290 million metric tons. Thus China's grain-import dependence dropped to less than 1 percent in 1976. Grain contracts for 1977 suggest the possibility of imports again increasing to over 7 million tons, which will nevertheless represent a diminishing percentage of the total grain supplies as grain production may reach output levels of 290 to 300 million tons.

Chinese grain imports through 1985 are expected to continue at levels comparable to previous years and may become more erratic as they progressively account for a smaller percentage of total grains production. Grain imports are primarily from Canada and Australia, both of which have supplied wheat and corn to China since 1961 and account for 80 percent of all grains imports. Canada is the largest supplier, accounting for 50 percent of China's imports to date. Argentina, the United States, and France provided supplementary grains to China in certain years, but their cumulative market share so far is only 20 percent of the total (see Fig. 12.1).

The importance of grain imports so far is also underscored by the fact that China has been consistently accumulating huge hard-currency deficits in its trade with Canada and Australia since 1961. The general myth that communist countries insist on balanced trade with foreign partners is completely dispelled by the existing trade between China and Canada, Australia, or even West Germany. What remains to be quite true, however, is that trade between China and other communist countries is always balanced and prearranged by annual agreements.

There is little speculation about who will supply grains to China in the future because only Canada, Australia, Argentina, and the United States produce sufficient grain surpluses to satisfy Chinese demand on an ad hoc basis. Rather, the significance of China's grain import lies in their impact on China's foreign-trade balance. The larger the demand on hard currencies for import of grains, the smaller will be China's capacity to

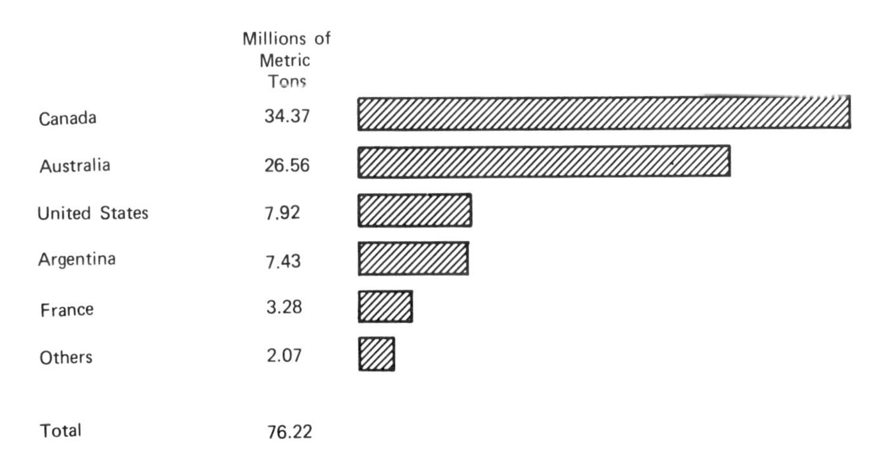

***Figure 12.1** Market shares of grain suppliers during 1961–1975 in millions of metric tons.* **(Source: U.S. Congress Joint Economic Committee.)**

import industrial equipment and other products from the West because although it permits itself even large trade imbalances with individual noncommunist countries, it strives to keep its overall trade in balance every year.

In 1975 China in fact became a net grains exporter, earning $745 million from exports of rice and other grains while paying only $680 million for imports of wheat and corn from Australia and Canada. Besides rice, China was also a small net exporter of corn in 1975, shipping an estimated 150,000 tons, of which 85,000 tons went to Japan. China also exports lesser amounts of sorghum and buckwheat, mainly to Japan and Hong Kong (see Fig. 12.2).

Rice is the largest single grain export, and China consistently ranked as the world's leading exporter. It was the world's largest rice exporter in 1973 and 1974, when exports reached about 2 million tons each year. Indonesia, Sri Lanka, Malaysia, and Cuba are the largest importers

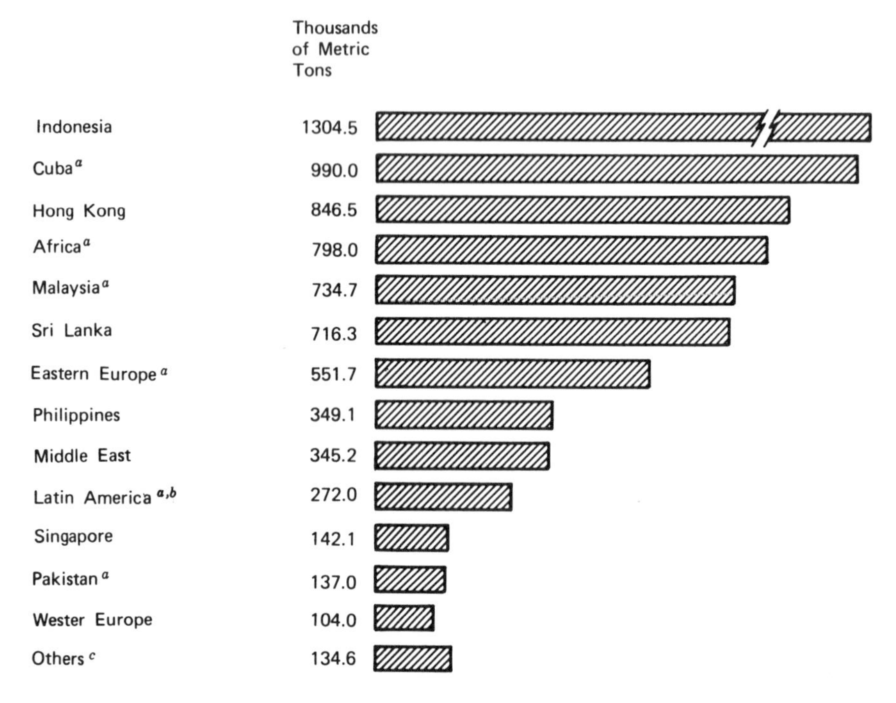

***Figure 12.2** Major importers of rice from China during 1971–1975.* (**Source: U.S. Department of Agriculture, 1976.**) (*[a] Data partly estimated or unofficial. [b] Latin America, excluding Cuba. [c] Other Far East countries.*)

of Chinese rice. Even at that level exports of rice account for less than 2 percent of China's total output, which gives it a considerable advantage in this trade. With increasing productivity and willingness to enter into barter-trade agreements with developing countries, China is a major factor in global rice trade, growing in importance. Its ability to barter rice for badly needed natural rubber from Sri Lanka during the American embargo proves its readiness to use this commodity as a political weapon when required.

Cotton Production and Trade

China is the second largest producer of cotton in the world after the Soviet Union and produces 30 percent more cotton than the United States. The 1975 cotton output was estimated at 2,400,000 tons, down by about 150,000 tons from a previous high in 1973. China has been supplementing its cotton demand by importing an average 240,000 tons per year, with peak imports during 1972–1973, 425,200 tons of cotton were imported from at least ten different countries. This was equal to almost 18 percent of all cotton output in China and nearly twice the average annual import rate of about 10% of domestic production (see Fig. 12.3).

China's demand for cotton varies with world market for cotton textiles and Chinese textile exports. In 1975 China even exported 50,000 tons of raw cotton to Hong Kong, while reducing its cotton imports to 150,000

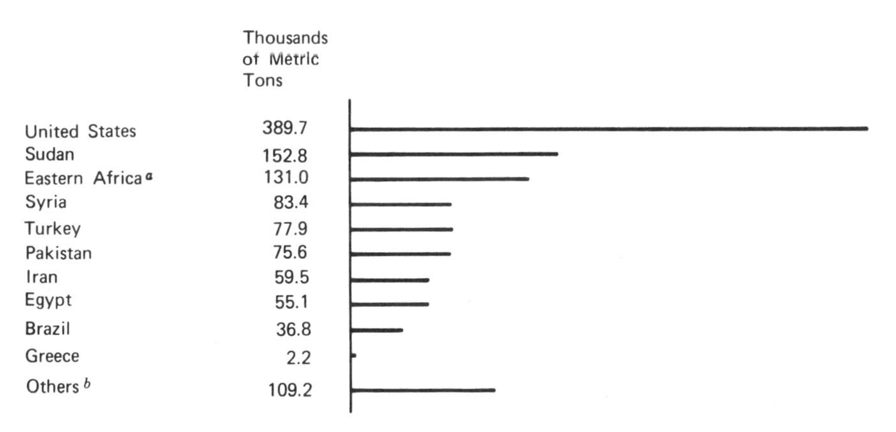

***Figure 12.3** Market shares of cotton supplier countries during 1970–1975 in thousands of metric tons.* (**Source: U.S. Department of Agriculture, 1976.**) (*[a] Kenya, Tanzania, Uganda. [b] Afghanistan, Colombia, India, Mexico, Morocco.*)

tons. China has been an importer of cotton in quantities ranging between 70,000 to 170,000 tons for at least 15 years while at the same time it exported about 10,000 to 20,000 tons annually.

China's capacity to increase its cotton production depends on its willingness to allocate more land to growing this crop. In 1955 China had 30 percent more land under cotton at about half the present productivity. Although China's average cotton yields are above the world's average and even comparable to those in the United States, they are still below such cotton-producing countries as the Soviet Union, Egypt, or Mexico.

Most likely China will continue imports of cotton in the future when demand for Chinese textiles in foreign markets justifies such purchases. China may also increase its cotton yields with additional agricultural inputs. It is questionable whether additional land will be returned to growing cotton, however, because: (1) other, more valuable crops can be raised for exports that will justify imports of cotton when required; and (2) China's textile exports, which are close to $1 billion annually, may also contain an increasing percentage of synthetic textiles in the future. This in fact may reduce the overall demand for cotton unless world textile markets expand.

By being able to absorb a certain amount of imported cotton from the Third World countries in Africa and Latin America, China also maintains another valuable political weapon. In times of weak world cotton markets this posture allows China to offer better terms for cotton to some hard-pressed countries in return for oil as one example. Such action would also bring China additional political influence and international prestige.

Soybeans Production and Imports

China is the second largest soybeans producer in the world after the United States and is a traditional exporter of soybeans to Japan. The 1975 soybeans production was estimated at 12 million tons. Actually, this production level was already reached when a comparable amount of land was under soybeans cultivation. Soybeans output was reduced during the 1960s, but new emphasis on this crop began after 1971 and output expanded to current levels.

As a result of a poor crop year in 1972, China purchased 570,000 tons of soybeans from the United States in 1974, becoming a net importer. Traditionally, however, China is an exporter of this crop and has been shipping about 300,000 to 400,000 tons to Japan every year in the 1970s. China is expected to increase its soybeans output to 20 million tons per year, but this also depends on weather. Imports of soybeans ceased in

1975, but in 1977 China again purchased about 380,000 tons from the United States for $122 million.

Sugar Production and Imports

China is the sixth largest sugar producer in the world after the Soviet Union, Brazil, Cuba, the United States, and India. Raw-sugar production estimates range from 2.5 to 4.0 million tons in 1975, which could be nearly half the Soviet output and 70 percent of that in the United States. Sugarcane as well as sugar beets are grown in China and it is believed that growing areas for both are increasing.

Production and import levels suggest that per capita consumption of sugar in China is in the order of only 12 pounds per year. This is about eight to ten times less than per capita consumption in the West or in other sugar-producing countries. Since 1960 China imported an average 680,000 tons of sugar annually, which was close to 25 percent of domestic production during the 1960s. Until 1972 virtually all sugar imports came from Cuba.

Throughout the 1960s China was also an exporter of sugar. In 1964 and 1965 Chinese sugar exports were in the order of 500,000 tons, which was in fact larger than Cuban sugar imports for those years, and coincided with significant increases in world sugar prices. China appears to have been a large-scale trader, receiving Cuban sugar in barter for rice and selling it on the open market, probably for hard currencies. Because Cuba was under American trade embargoes at that time, China in fact acted as an intermediary absorbing at least 10 percent or more of Cuban sugar production. In 1961 China imported 1.5 million tons of Cuban sugar, probably equal to 50 percent of its domestic production at that time.

Toward the late 1960s China became a net importer of sugar, and exports diminished to the 150,000-ton levels. Since 1972 China began importing sugar from Brazil and Australia and entered into negotiations with Guyana and Jamaica. Imports from Cuba steadily diminished to 530,000 tons in 1970, 200,000 tons in 1975, and only 150,000 tons in 1976. In that year China imported 150,000 tons from Australia and contracted for a reported 700,000 tons with the Philippines for deliveries in future years. In 1976 520,000 tons of sugar were imported for a total of $200 million. Australia was a major supplier providing a total of $60 million. It is expected that China may import as much as 1,000,000 tons of sugar in 1977.

Sugar prices skyrocketed in 1974, only to fall back to the 1973 levels by 1976. Whether this affected the Sino-Cuban rice-for-sugar barter trade is not known, but relations cooled between the two countries. Some observers suggest that China's reduction of Cuban sugar imports was also motivated by Cuba's intervention in Angola in support of the Soviet-backed revolutionary movement that eventually came to power.

Relatively lower sugar crops in Cuba in 1970 and 1971 may have also been a factor, but China still appears to have imported about 750,000 tons of Cuban sugar in 1972. In recent years sugar prices have slumped, and much sugar is traded below its cost of production. It is certainly conceivable that under the circumstances China may have used its considerable market power to deprive Cuba of a significant export market equal in size to 10 or 20 percent of Cuban sugar output. The fact that it began doing so since 1972 when Nixon visited Peking may be more than a coincidence.

AGRICULTURAL MACHINERY PRODUCTION AND IMPORTS

In 1976 there were an estimated 1600 agricultural machinery plants manufacturing over 1300 different products. Conventional and garden (walking) tractors, combine harvesters, and powered irrigation equipment are the most important of those. In addition, 95 percent of the 2126 counties in China have been reported to operate their own plants and workshops for manufacture and repair of farm implements. Up to 80 percent of all farm equipment manufactured has been estimated to come from small local factories.

Large tractor and machinery plants are under the supervision of a production control bureau of the First Ministry of Machine Building. Originally this was under the control of the Ministry of Agricultural Machinery established in 1959. It was renamed the Eighth Ministry of Machine Building in 1964 and around 1970 was merged into the First Ministry.

Production of Tractors

In 1975 the tractor inventory of China probably consisted of about 280,000 tractors, predominantly of the 54-HP category, each developing 36-drawbar horsepower. In addition, there were an estimated 580,000 two-wheel garden tractors ranging from 2 HP to 12 HP each. This total inventory is equivalent to about 350,000 conventional tractors in terms of tractive horsepower used in cultivation.

Measured in terms of hectares of arable land per tractor, China would need almost four times as many tractors to equal the mechanization rate in the Soviet Union. However, it is almost two and a half times better off than India although considerably behind such countries as Brazil, Argentina, and Mexico (see Table 12.4).

Table 12.4 Comparative use of tractors in selected countries of the world in 1975

Country	*Arable[a] Land in Million Hectares*	*Total Tractors in Use*	*Tractor Usage in Hectares per Tractor*
China	105.6	350,000	301.7
Soviet Union	207.1	2,400,000	86.3
United States	177.9	4,109,000	43.3
India	156.8	215,000	729.3
Australia	46.1	335,000	137.6
Canada	39.9	645,000	61.8
Brazil	34.1	254,000	134.2
Pakistan	32.1	38,000	844.7
Argentina	30.4	188,000	161.7
Mexico	23.7	140,000	169.3
Nigeria	22.2	7,500	2,960.0
France	19.3	1,345,000	14.3
Thailand	12.3	19,173	641.5
Bangladesh	9.4	2,300	4,087.9
West Germany	8.2	1,441,778	5.7
Japan	5.9	350,000	16.8
Vietnam	4.6	4,500	1,022.2
Netherlands	2.3	180,000	12.7
South Korea	2.3	450	4,888.9
North Korea	2.1	10,400	301.9
Mongolia	1.5	8,100	185.2

Source: United Nations Statistical Yearbook, 1976; U.S. Government National Basic Intelligence Factbook, July 1977; *GUS Statistical Yearbook, 1976,* Warsaw, Poland.

[a] Data are not always available for arable land and may represent cultivated land, cropland, or farmland. Resulting tractor usage in this table is only roughly comparable in magnitude of units available in the country per hectare and may differ widely within each country from one region to the next.

Most production and inventory data about tractors in China is available for standard 15-HP units. A typical conventional tractor is about 2.4 standard units, and a typical garden tractor is only 0.25 standard units. Production of conventional tractors has been increasing at about 10 percent per year. Production of garden (walking) tractors, which only began in 1964, was estimated to have grown by 30 to 60 percent between 1974 and 1975. Measured in terms of standard 15-HP units, total tractor horsepower inventory probably increased from 292,600 to 835,000 standard units during 1970–1975. This estimate takes into account up to 5 percent retirement rate for conventional tractors and an average 2 percent retirement rate for the cheaper but much newer garden tractor inventory (see Fig. 12.4).

If similar production growth rates are maintained, China will have a total of almost 2 million standard 15-HP units in service by 1980. This would be equivalent to about 830,000 conventional units but will actually

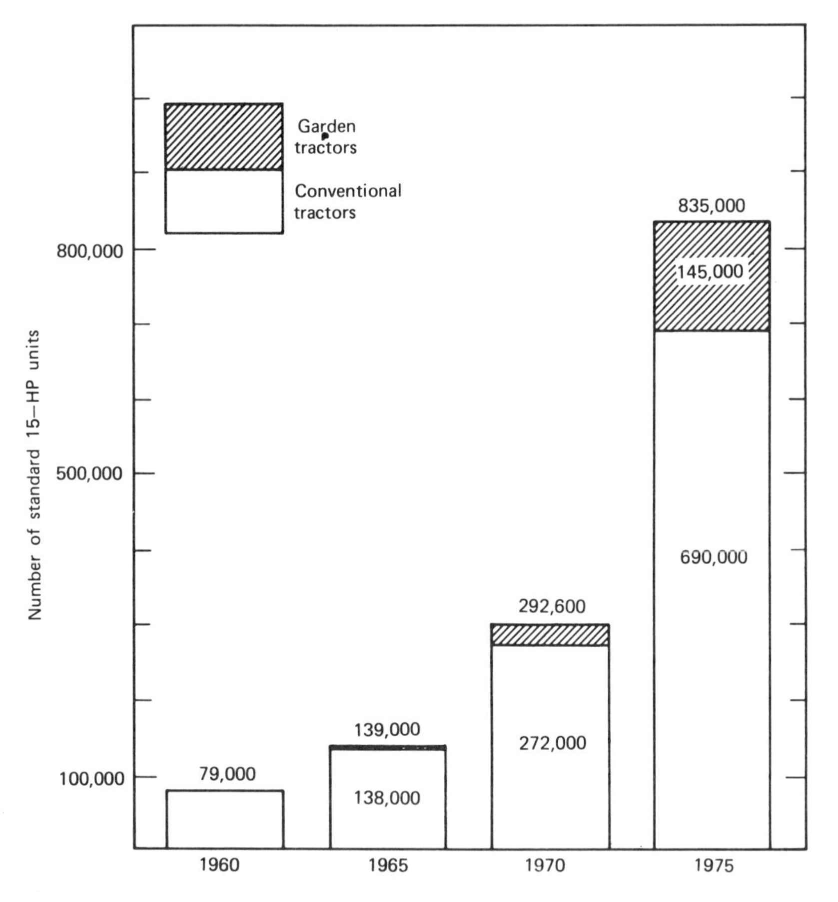

***Figure 12.4** Growth in conventional and garden tractor inventory.* (**Source:** *Derived from data in Joint Economic Committee study, "China: A Reassessment of the Economy," U.S. Congress, July 1975 and* **U.S. Government Research Aid,** *"Production of Machinery and Equipment in the People's Republic of China,"* A(ER)75-63 *May 1975.*)

consist of 570,000 conventional tractors and 2,400,000 garden tractors. Only a serious shortage of steel or electric power would slow down China's tractor production rate which in 1976 was 53,600 conventional units down from a high of 58,300 units in 1975.

The small garden tractors are particularly suitable for cultivation in China, where many fields and rice paddies are relatively small. The more powerful 12-HP Dong Feng model is also a versatile power unit used for local transport. Garden tractors can be fitted with antiskid iron wheels for working in wet fields, and several implements can be attached. These include ploughs, harrows, cultivators, and trailers. They are also used as stationary power takeoff units for operating agricultural machinery or irrigation pumps. These tractors use single-cylinder four-stroke diesel engines, which consume up to 9 kgs of fuel per hectare. About 0.12 to 0.15 hectares can be ploughed in an hour. Used for transport the tractor can pull 15.3 ton-km/h.

Almost thirty different tractor models have been identified in China. At least six of these are small two-wheel garden tractors ranging from 2 HP to 12 HP. China also makes at least seven track-laying tractors up to 120 HP, and there are reports of a 250-HP tractor made in China and even exported abroad (see Table 12.5).

Of the thirty or so tractor-manufacturing plants in China, the largest is the Tung Fang Hung Factory in Loyang in Honan Province. It was one of the thirty-two major projects built with Soviet assistance to form the backbone of the machine-building industry. The original annual output capacity was 15,000 45-HP tractors. Today it manufactures at least ten different models, ranging from 7-HP garden tractors to 80-HP track-laying units (see Table 12.6).

The tractor-manufacturing plant in Tientsin is the second largest factory, with potential capacity to produce 11,000 Tieh Niu 45-HP tractors and, since 1971, 55-HP tractors. The Yenchou County tractor works in Shantung Province is a new facility, in operation since 1976. It is designed to produce 10,000 Taishan brand 25-HP tractors per year. Other better-known plants include the Changchou Tractors Works and Wuhsi County Walking Tractor Works in Kiangsu Province, Anshan Hungchi Tractor Plant in Liaoning, and Hangchou Tractor Plant in Chekiang.

Other Agricultural Machinery Production

China manufactures numerous types of agricultural equipment. Besides tractors, equipment for irrigation and drainage, pumps, and combine/

Table 12.5 Basic characteristics and prices of Chinese tractors

Rated Power in HP	*Tractor Model*	*Type of Tractor*	*Traction in Tons*	*Basic Price in Yuans*
2	Hand tractor	2 wheels	—	2,000 (1973)
5	Hand tractor	2 wheels	—	4,000 (1973)
7	Iron Ox	2 wheels	—	10,000 (1973)
7	Kung Nung	2 wheels	—	
11	Kung Nung	2 wheels	—	
12	Dong feng	2 wheels	—	
20	Yao Ching	4 wheels	—	
20	Tung Fang Hung	4 wheels	0.5	
27	Feng Shou	4 wheels	—	
28	Tung Fang Hung	4 wheels	—	18,000 (1973)
30	Tung Fang Hung	4 wheels	1.0	
40	Chi Tsai		—	
40	Tieh Niu	4 wheels	—	
40	Tung Fang Hung	4 wheels	1.0	
45	Tieh Niu	Tracks	—	
50	Chi Tsai	4 wheels	—	
50	Tung Feng	4 wheels	1.2	
54	Tung Fang Hung	4 wheels	—	21,000 (1959)
55	Tieh Niu	4 wheels	1.4	13,000 (1974)
55	TN-55	Tracks	—	
75	Tung Fang Hung	Tracks	—	30,000 (1973)
80	Tung Fang Hung	Tracks	3.0	
80	Hsing Shu Kuan	4 wheels	2.8	
80	Hung Chi	Tracks	—	
100	Hung Chi	Tracks	—	
120	Hung Chi	Tracks	6.0	

Source: Based on descriptions given in JPRS #63091; *China Machinery Import and Export Corporation Catalogue, 1973; U.S.–China Trade Council Special Report on Prices of Chinese Commodities,* No. 11, 1975.

harvesters are the most important units. Total horsepower of irrigation systems was estimated at 43,000,000 HP in 1975, over twice the 1970 capacity. Machinery over 20 HP and engines over four cylinders are manufactured in large state plants for national distribution. Smaller units are made at local county and city plants.

Large production increases were also reported at internal-combustion engine plants manufacturing diesels and in production of processing equipment, sowers, rice transplanters, plows, and rural vehicles. Shortages are still believed to exist in rice-transplanting machines, power threshers, power pumps, and trucks.

Table 12.6 Major tractor-manufacturing plants

Plant Name and Location	*Types of Tractor Manufactured*
Anshan Hungchi Tractor Plant	one of the four basic plants built in 1958–1962
Canton Tractor Factory	2000 medium tractors per year
Changchiang Tractor Works	tractors and diesel engines
Changchou Tractor Works	largest walking tractor plant in China
Changchun Tractor Plant	Tung Fang Hung 28-HP tractors
Chengtu Hungchi Tractor Works	under construction in 1975
Chingchiang Tractor Works	5000 tractors per year (established 1975)
Hanchou Tractor Plant	Feng-Shou 27-HP all-purpose tractors
Hsinhui Agricultural Machinery Works	10,000 hand-guided tractors per year
Hunan Tractor Works	5000 30-HP tractors per year (established 1970)
Kiangsi Tractor Works	5000 Feng-Shou 27-HP tractors per year
Liuchou Farm Machinery Works	Liuchou tractors
Loyang Tung Fang Hung Plant	largest plant; ten tractor models
Nanning Hand Tractor Plant	4000 in 1974; Kung Nung 12 tractors,
Pengpu Tractor Parts Plant	fully mechanized, partly automated
Shanghai 1st July Tractor Plant	Bumper Harvest 35-HP tractors
Shanghai Feng-Shou Tractor Plant	10,000 units/year; mechanized and automated
Shanghai Kung Nung Motor Plant	hand-guided; Kung Nung 11 tractors
Shanghai Tung Fang Hung Tractor Plant	trial produced own design tractors, 1969–1971
Shenyan Small Tractor Plant	8-HP and 12-HP walking tractors
Shaokuan Tractor Plant	hand tractors
Sunchiang Tractor Plant	Chitsai 50, multipurpose tractors
Taiho Tractor Manufacturing & Repair Plant	Tung Feng 12, small rice harvesters
Tangtung Tractor Accessories Plant	parts for export
Tientsin Tractor Manufacturing Plant	11,000 Iron Bull 56-HP tractors per year
Wuhan Tractor Plant	1500 hand-guided tractors in July 1975
Wuhsi County Walking Tractor Works	50,000 units in 1975 (established 1969)
Yangchian Tractor Plant	Kung Nung 12, walking tractors
Yenchou County Tractor Works	10,000 Taishan, 25-HP tractors per year capacity

Source: Compiled by 21st Century Research from various articles, papers, studies, and publications listed at the end of this chapter.

Imports of Agricultural Machinery

Despite intensive agricultural mechanization campaigns, there is little likelihood that imports of agricultural machinery will increase significantly. China seems to be able to manufacture most agricultural equip-

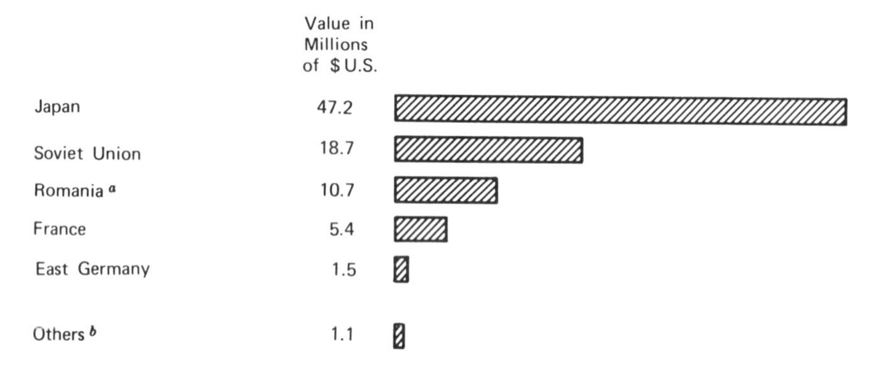

***Figure 12.5** Estimated market shares of major agricultural machinery suppliers identified during 1970–1975.* (Source: *OECD Commodity Trade Statistics Series B for SITC code 712,* Romanian Foreign Trade Yearbook 1974, *and United Nations bulletins of statistics on world trade in engineering products.*) [[a] *Data for 1970–1973 only.* [b] *Others include United States (0.5), United Kingdom (0.4), West German (0.1), Czechoslovakia (0.1), Italy (0.1).*]

ment it requires, and only serious shortages of steel for large tractor plants could slow down production.

Average agricultural machinery imports were under $20 million annually during 1970–1975. In 1975 purchases of agricultural machinery totalled $19.6 million, of which $14.4 million represented tractors purchased from Japan and France. Japan appears to be the largest supplier in recent years, but the Soviet Union is second, probably as a supplier of spare parts for a large inventory of Soviet designed equipment operating in China. Romania, France, and East Germany are the other suppliers. Sales of equipment from other countries probably supplied China with new ideas and solutions incorporated into that equipment (see Fig. 12.5).

REFERENCES

Cheng, Chu Y. *The Machine Building Industry in Communist China,* The Edinburgh University Press, 1972.

China Achieves Grain Self-Sufficiency. China Council for the Promotion of International Trade, Peking, 1974.

China Letter, The. Special Report, "China's Quiet Rural Revolution," Issue No. 64, The Asia Letter Ltd., 1977, Hong Kong.

Crook, F. W. "The Commune System in the People's Republic of China," in *China: A Reassessment of the Economy. Part II. Rural and Agricultural*

Development, Joint Economic Committee, U.S. Congress, July 1975, p. 366.

Erisman, Alva L. "China: Agriculture in the 1970's," in *China: A Reassessment of the Economy. Part II. Rural and Agricultural Development,* Joint Economic Committee, U.S. Congress, July 1975, p. 324.

Felix, Fremont. *World Markets of Tomorrow,* Harper & Row, London, 1972.

Field, Robert M. "Recent Chinese Grain Claims," *The China Quarterly,* March 1976, pp. 96–97.

GUS, Glowny Urzad Statystyczny. *Rocznik Statystyczny, 1967,* Part II, Warsaw, Poland.

Khan, Amir V. "Agricultural Mechanization in China," *Agricultural Engineering,* April 1976.

Khan, Amir V. "Agricultural Mechanization and Machinery Production in the People's Republic of China," *U.S.–China Business Review,* November–December 1976.

Le Monde. "La Presse de Pekin Fait Etat d'Importants Succes dans l'Agriculture Chinoise," Le Monde de l'Economie, p. 14, July 29, 1975.

McCarron, Geoffrey. "Agricultural Machinery," in William W. Whitson, ed., *Doing Business with China,* Praeger, New York, 1974.

Novosti Press, "Grozit-li Kitayu Perenasyelennost?," Moscow, 1971.

Perkins, Dwight H. "Constraints Influencing China's Agricultural Performance," in *China: A Reassessment of the Economy. Part III. Rural and Agricultural Development,* Joint Economic Committee, U.S. Congress, July 1975, p. 350.

Pomeranz, Y. "Food and Food Products in the People's Republic of China," *Food Technology,* March 1977, pp. 32–41.

Rinfret, Pierre A. "Agriculture–Yes, Agriculture Just May Be on the Verge of a Great Leap Forward," *Institutional Investor,* February 1973.

Sigurdson, Jon. "Rural Industrialization in China," in *China: A Reassessment of the Economy. Part II. Rural and Agricultural Development,* Joint Economic Committee, U.S. Congress, July 1975, p. 411.

United Nations. *Bulletin of Statistics on World Trade in Engineering Products, 1973, 1974, 1975,* New York.

U.S.–China Business Review. "China Market Prospects–the Next Five Years, Agriculture, Agricultural Mechanization," July–August 1976.

U.S. Department of Agriculture, Economic Research Service. "The Agricultural Situation in the People's Republic of China and Other Communist Asian Countries Review of 1975 and Outlook for 1976," *Foreign Agricultural Economic Report,* No. 124, *August 1976.*

U.S. Government. "Potential Implications of Trends in World Population Food Production and Climate," OPR-401, August 1974.

U.S. Government. "China Agricultural Performance in 1975," *Research Aid,* ER 76-10149, March 1976.

13

Transportation Systems and Equipment

Major transportation systems of China include railroads, highways, and coastal and inland waterways. Railroads constitute the predominant form of transportation, accounting for 59 percent of all freight carried. Highway transportation by truck is the second most important method, accounting for 28 percent of all the freight carried in China. Water transport is of importance in coastal shipping and on inland rivers to locations not readily accessible by railroad or highway. Air travel plays a very small role in domestic transportation.

Private car ownership is not permitted, and most passenger transport is by bus or by train. The first subway system in China was put into operation in 1969 in Peking. Bicycles are the basic personal transportation means for local travel for the majority of people.

China is fourth in manufacture of locomotives and freight cars in the world, ninth in production of trucks, and fifteenth in tonnage of merchant shipping launched each year. It has a considerable licensed production of small piston-engine civil planes and jet military aircraft, but there is no production of modern jet transports or jet helicopters. Passenger automobiles are manufactured in very small quantities, primarily for official use.

Because of inadequate domestic production China imports large quantities of transportation equipment, including ships, trucks, locomotives, tank cars, jet aircraft, helicopters, and jet engines. Japan, the Soviet Union, and eastern Europe as well as Western countries are equally important suppliers of all types of transportation equipment to China.

TRENDS IN CHINESE TRANSPORT PERFORMANCE

Total freight carried in China reached 1598 million metric tons in 1975, which is over twice the amount transported in 1965. Growth in Chinese transport was estimated at an average 8 percent annually since 1965, although it has been accelerating in recent years, averaging about 9 percent between 1970 and 1975.

Transport growth was much more rapid during the 1950s, with an average increase of 21 percent per year between 1952 and 1957. At that time railroads accounted for a much larger share of total freight carried, but their share decreased to present levels because of somewhat faster growth in highway transportation, which now accounts for 28 percent of all the freight carried.

Of the total 1598 million metric tons transported within China in 1975, 59 percent was carried by the railroads, which continue to be the largest freight carrier in modern China. Measured in ton-kilometers, railroads are believed to account for 75 percent of the total because average hauls by rail are almost 500 km, compared with only 35 km by truck (see Fig. 13.1).

Chinese transport performance first reached a high of 864 million metric tons in 1959, after a very rapid growth during the 1950s, but declined in the following year as a result of economic slowdowns of the "Great Leap Forward" period. Eventually transport performance declined to a low of 565 million metric tons in 1961, after which it began to recover. The traffic picked up and reached about 873 million metric tons in 1966, only to decline once again this time due to the disruptions of the Cultural Revolution, and it continued to decline until 1968, when a new low of 730 million metric tons was reached. Freight traffic recovered to 827 million metric tons in 1969 and increased rapidly to over 1 billion metric tons in 1970 and has been growing at an annual average of 9 percent since then.

RAILROADS, ROLLING STOCK, AND IMPORTS

Railroads constitute the predominant form of transportation in China and account for about 60 percent of all the freight carried, more than half of which is coal as fuel for China's industries. Grains, cotton, and pig iron are the other commodities conveyed by the railroads. Thus the railroads are critical to the growth of the economy and are managed and operated as a quasimilitary industrial organization supervised by a special ministry.

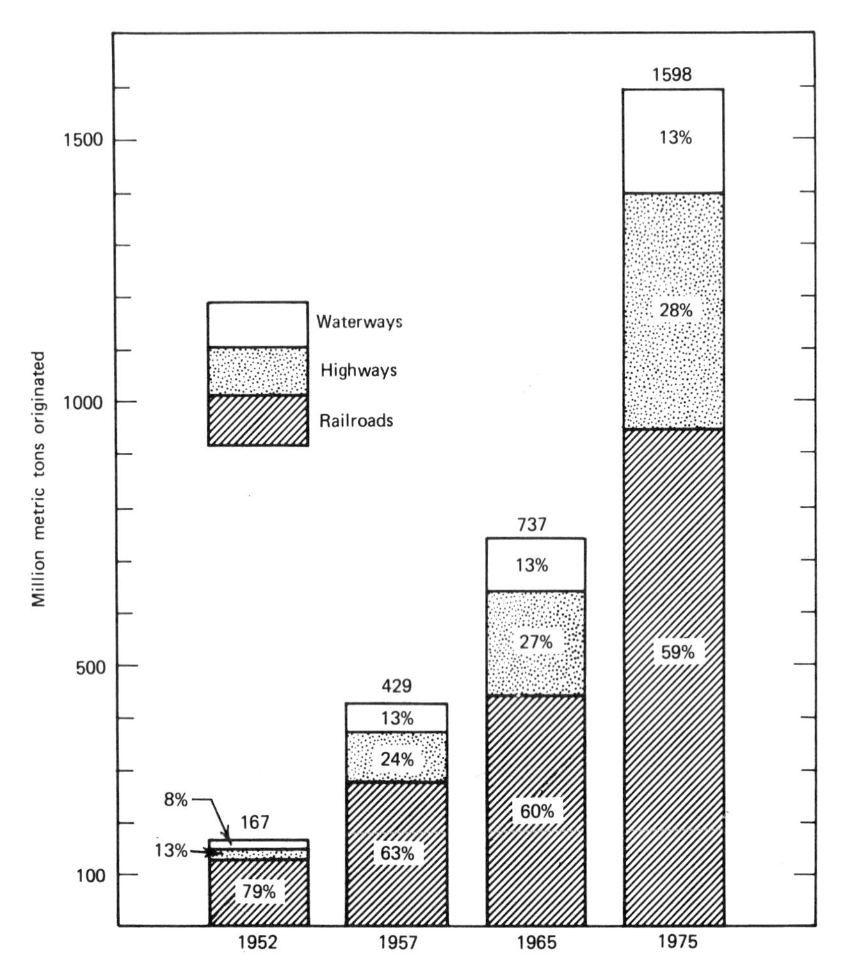

Figure 13.1 Trends in modern transport performance. **(Source: U.S. Government Handbook of Economic Indicators for People's Republic of China, August 1976.)**

Total railroad network was estimated at 47,500 km of mostly single track, which is comparable to the length of the railroads that existed in the United States almost a century ago. The populous interior province of Szechwan, believed to contain over 11 percent of the population of China, was not connected to the rest of the country by railroad until 1952.

Although it is the fifth largest railroad network in the world, it is still relatively modest when the population and the large area of China are taken into account. The Chinese network is less than one fifth that of the United States and a little over one third that of the Soviet Union. It is about 20 percent larger than the railroad network of Australia and about 30 percent smaller than that of India (see Table 13.1).

Table 13.1 Railroad length and freight carried in selected countries of the world in 1975

Country	*Freight Carried in Billion Metric Ton-km*	*Length of Railroad Network in km*	*Freight Density in Million Ton-km per km*
Soviet Union	3,236	138,362	23.4
United States	1,255	277,686	4.5
China	458	47,500	9.6
Canada	193	70,904	2.7
India	137	61,313	2.2
Poland	129	26,597	4.8
South Africa	67	19,902	3.4
France	64	36,720	1.7
Czechoslovakia	63	13,208	4.8
West Germany	55	33,453	1.6
East Germany	51	14,252	3.6
Japan	47	28,912	1.6
North Korea	24	4,535	5.3
United Kingdom	21	18,500	1.1
Argentina	12	38,971	0.3
Spain	12[a]	16,438	0.7
Mongolia	2.3	1,516	1.5

Source: U.S. Government Handbook of Economic Statistics, 1976.

[a] Data for 1974.

The present network is about twice that existing in 1949, and on the average China has been adding about 1000 km of new track every year. By 1980 total track laid is expected to increase to about 58,000 km. New lines known to be in planning or construction stages will extend the railroad network to Lhassa in Tibet and to western Sinkiang toward the Pakistan, Afghanistan, and Soviet Union borders. Other lines are planned in southern regions of China from Burma along Vietnam border areas and in southeastern China.

Most tracks are of standard 1.435 meter gauge, but there are about 9500 km of industrial lines with gauges of 0.59 to 1.453 m and 600 km of track with 1.0 meter gauge. In 1975 over 40 percent of all railroad lines were believed suitable for 90-km/h travel speeds.

About 8500 km of the total are double-track lines running from Harbin in Heilungkiang Province to Dairen in Liaoning, and from Shenyang to Peking, and Wuhan and Changsha in Hunan, as well as to Shanghai in the east. Double tracking continues in heavily traveled industrial areas as one method to improve the throughput of Chinese railroads. Other substitute procedures such as two-way radio and modern blocking methods are also reported in use.

At least 2300 km of lines are electrified, notably in the mountainous region between Paochi and Chengtu in Szechwan, and the Peking–Tatung–Paotu and Tatung–Tungkuan lines. Electrified lines use a 27,500-V (volt), single-phase AC system, and French specialists assisted China in developing the first electrified Paochi–Chengtu line.

Utilization of railroad lines has grown faster than their length, and some industry observers believe that China will soon develop container systems to further improve the efficiency of railways and provide better transshipment methods between rail and highway transportation.

In 1975 Chinese railroads originated 945 million metric tons of freight, making China the third largest rail-freight-transporting country in the world after the Soviet Union and the United States. In fact, this was equivalent to all rail freight originated that year by West Germany, France, the United Kingdom, and Australia combined.

Rail-freight turnover amounted to 458 billion metric ton-km, or about three and a half times the turnover of India, whose railroad network is 30 percent larger. By 1980 China's rail-freight turnover is expected to reach about 600 billion ton-km per year, or about 50 percent of that of the United States.

The traffic density on Chinese railroads is on the average over 9 million ton-km per route kilometer, which means that the railways are very heavily used. Among the large countries only the Soviet Union has a higher average freight density, with 23.4 million ton-km per route kilometer estimated for 1975. But some key Chinese lines between Shenyang, Peking, and Wuhan are reported to have traffic densities in excess of 25 million ton-km per route kilometer. On the other hand, lighter traffic lines may carry only 3 million ton-km per route kilometer, which is more comparable to railway traffic densities of Canada, India, South Africa, and the United States (see Table 13.1).

The average freight-train loading has increased from 1900 tons in 1958 to about 2300 tons in 1975, and the turnaround time for freight cars was 3.4 days in 1977. Dieselization of the locomotive inventory, which at present still consists of 75 percent steam engines, is further expected to improve this performance in the coming years.

Passenger traffic on China's railroads is extremely heavy because it is practically the only long-distance mode of transportation readily accessible to the ordinary people. It is estimated that 450 to 500 million people travel by railway in China every year. This is inevitable in the absence of individual automobile ownership and lack of long-distance bus services, similar, for example, to those of Mexico.

The subway era began in China in 1969, when the Peking subway first opened its doors to the public. This 24-km line, with seventeen stations operating between 6 A.M. and 9 P.M., is supervised by the Peking Underground Railways department, which is controlled by the Peking Municipality. It is a 750-V DC system using self-propelled electric cars operating at 60 to 80 km/h. By the end of 1976 the Peking subway carried no more than 80,000 passengers per day, but undoubtedly it will be expanded and similar systems may be built in other large cities, at least sixteen of which are believed to have populations over one million.

The railroads are supervised by a Ministry of Railways headed by a new minister, Tuan Chun-yi, who was appointed in 1977. It consists of four basic departments, which include operations, construction of lines, manufacture of equipment, and management of the labor force. Specialized departments of the ministry are concerned with planning, accounting, statistics, signaling, payments, international transactions, and railroad security.

The Ministry of Railways administers the operations by a system of twenty major railroad bureaus, each with several subdivisions. Individual railroad bureaus are large organizations employing tens of thousands of workers each. Each bureau operates its own locomotive and engineering plants, and maintains the track in its region of responsibility. The Peking bureau is probably the largest in the country; it was reported to handle 25 percent of the total volume of rail freight in China during the early 1970s.

Production of Locomotives and Rolling Stock

China is a major manufacturer of locomotives and freight cars. It is the third largest producer of all types of locomotives after the Soviet Union and the United States. China is also the largest, if not the sole, producer of steam locomotives in the world, with an estimated production of about 250 steam engines per year. As a producer of diesel and electric locomotives China is among the ten leading manufacturing countries in the world. It is also the fifth largest producer of freight cars after the Soviet Union, United States, West Germany, and Poland (see Table 13.2).

Production of all types of locomotives is estimated to have reached a total of 530 units in 1975, of which 275 were diesel and about five electric. By 1980 China should be able to increase its annual locomotive production to over 600 units per year with a diminishing share of steam

Table 13.2 Production of locomotives and freight cars by major manufacturing countries in 1975 in units

Country	*Locomotives*			*Freight Cars*
	Diesel	*Electric*	*Steam*	
Soviet Union	1400	395	—	69,900
United States	984			72,370
Poland[b]	373	72		17,850
Czechoslovakia[b]	361	87		5,130
Romania	334	44[b]		13,380[b]
China	275	5	250	18,500
West Germany	275	68		67,000
Japan	243[a]	172[a]		3,188
East Germany[b]	143	26		4,850
France	48	84		10,120
India				11,980
North Korea		55		9,100

Source: *U.S. Government Handbook of Economic Statistics 1976* and various United Nations statistical sources.

[a] Data for 1973. [b] Data for 1974.

locomotives. It is interesting to note that China already produced as many as 600 steam locomotives as far back as 1960, when it was also estimated to have made 23,000 freight cars, both of which levels have not been surpassed since. Disruptions of the Great Leap Forward and the Cultural Revolution are probably only partially responsible for this decrease in production. A much smaller railroad network in a relatively poorer state than today may not have permitted the utilization of all available equipment aside from any drop in freight turnover due to political disruptions (see Table 13.3).

At least thirty major locomotive and rolling-stock plants have been identified in China, several of which are very large enterprises by world standards. These produced over thirty models of steam, diesel hydraulic, diesel electric, and electric locomotives as well as several types of freight cars ranging from thirteen tin two-axle units to 120-ton flat cars (see Tables 13.4 and 13.5).

Production and imports of tank cars have had considerable priority in recent years because China lacks an extensive system of long-distance pipelines and has a need to carry chemicals and special fuels as its chemical and petrochemical industries expand rapidly.

Manufacture of locomotives and rolling stock is under the control of the Bureau of Locomotives and Vehicles of the Ministry of Railways, but not all the plants are controlled by this ministry. Those that produce equipment for narrow-gauge industrial and mining lines are believed to be supervised by the First Ministry of Machine Building.

Table 13.3 Estimated production and cumulative inventory of mainline locomotives and freight cars in China since 1952

	Mainline Locomotives		*Freight Cars*	
Year	*Production*	*Inventory*	*Production*	*Inventory*
1952	20	3,300	4,501	58,000
1960	600	5,900	23,000	132,000
1965	60	5,400	6,600	143,000
1970	435	6,400	12,000	175,000
1971	455	6,700	14,000	185,000
1972	475	7,100	15,000	197,000
1973	495	7,500	16,000	209,000
1974	505	7,900	16,800	222,000
1975	530	8,300	18,500	237,000

Source: Adapted from data presented in *U.S.–China Business Review*, March–April 1977 and *U.S. Government Handbook of Economic Statistics, 1976.*

China began building steam locomotives in the 1950s, when most of the world's railways were changing to diesels. This was a practical policy to follow at the time because though coal was plentiful, China had to import petroleum products. However, in the late 1950s and early 1960s prototypes of diesel and electric locomotives were built. During the 1960s China also became self-sufficient in oil products. As a result of these developments, China can be expected to eventually replace the bulk of its locomotives with diesel units because it is one of the easiest ways to improve railroad carrying capacity without expensive and time-consuming investment in new lines or double tracking of existing lines.

The Dairen Locomotive and Rolling Stock Works in Liaoning Province is the most important locomotive-manufacturing plant. Chinese 4000-HP "Dong Feng 4" diesel locomotives are built at this plant, which began serial production of diesel locomotives in 1965, producing several models in the 600-HP, 1200-HP, and 2000-HP categories as well. It may also be the most modern locomotive plant in the country, claiming use of numerically controlled machine tools and computer controls for diesel-engine manufacture.

The Szufang Locomotive and Rolling Stock Works in Tsingtao in Shantung Province is regarded as the second largest locomotive plant in China. In 1946 it made the first Chinese locomotive, the "Mao Tse-tung," rebuilding it from used foreign parts. In 1952 the plant began serial production of the "Construction" 2-8-2 steam locomotive similar to the Japanese MK-1 Mikado model. In 1959 the plant began serial production of the first 2000-HP diesel locomotive. The 5000-HP diesel hydraulic locomotive, "Dong Fang Hong," was first produced in 1969 at this plant. Smaller locomotives and passenger and sleeping cars are also manufac-

Table 13.4 Major models of locomotives manufactured in China since 1952

Rated Power in HP	*Type of Locomotive and Date First Produced*	*Manufacturing Location*
60	Industrial diesel engine (1965)	Changchou
80	Diesel hydraulic engine for mining use	Dairen
120	Diesel hydraulic	Dairen
120	Industrial diesel engine (1965)	Changchou
180	Diesel hydraulic for mining use	Dairen
234	"Rocket" small steam locomotive (1960)	Changchun
240	Narrow-gauge engine	Dairen
380	Narrow-gauge internal combustion engine (1975)	Canton
600	"Construction" steam locomotive (1958)	Peking
600	Diesel electric	Dairen
1000	"Satellite" diesel hydraulic (1959)	Tsingtao
1000	Diesel hydraulic	Chishuyen
1200	Diesel electric	Dairen
1500	Steam passenger T-6 locomotive (1960)	Tsingtao
1545	"Liberation" steam freight locomotive 2-8-2 (1956)	Dairen
1600	"Victory" steam passenger locomotive 4-6-2	
1860	"Dong Feng" diesel	Dairen
1900	"Renmin" steam locomotive 4-6-2 (pre-1956)	
2000	"Satellite" diesel hydraulic (1960)	Tsingtao
2000	Diesel electric (1959)	Dairen
2000	"Dong Fang Hong" diesel hydraulic (1968)	Tsingtao
2000	Diesel electric (1958)	Chishuyen
2270	"Construction" steam locomotive 2-8-2 (1952)	Tsingtao
2780	"Peace" steam locomotive 2-10-2 (1956)	Changchun
	"Red Flag" steam freight locomotive (1958)	Dairen
2980	"Forward" steam locomotive 2-10-2 (1965)	Tatung
3000	"Peking" diesel hydraulic (1976)	Peking
3000	"Chang-Chung" gas turbine locomotive (1970)	Tatung
3000	"Dong Fang Hong 4" diesel hydraulic (1976)	Chishuyen
3000+	"Giant Dragon" diesel electric (1962)	Dairen
3154	"Peace" steam locomotive development 2-10-2 (1959)	Changchun
4000	"Dong Feng 4" diesel electric (1974)	Dairen
4900	Electric locomotive based on Soviet N-60 model	Hsiangtan
5000	Diesel hydraulic	Tsingtao
5200	"Shaoshan No. 1" electric locomotive (1969)	Chuchou
6000	Diesel hydraulic (1975)	Peking

Source: "La Vie du Rail Outre-Mer," May–June 1975, JPRS #66459 and *U.S.–China Business Review,* March–April 1977.

tured here. The original plant capacity was estimated at twenty-five locomotives, fifty passenger coaches, and 900 freight cars per year.

The largest 6000-HP diesel locomotives have been produced at the Peking "February 7th" Locomotive and Rolling Stock Plant since 1975. This plant claims a work force of over 10,000 as of May 1976 but the factory area population is several times higher, with wives of the work-

Table 13.5 Major locomotive, rolling stock, and railroad equipment plants identified in China

Name of Manufacturing Plant	*Most Advanced Equipment Manufactured*
Canton Motive Power Machinery	narrow-gauge locomotive engines
Chishuyen Rolling Stock Plant	5000-HP diesel locomotive
Changchou Diesel Locomotive Plant	60-HP and 120-HP industrial locomotives
Chengtu Rolling Stock Works	repairs diesel and steam locomotives
Changchun Rolling Stock Plant	deluxe and sleeping railway coaches
Changchun Locomotive Works	self-propelled subway cars for Peking
Chuchou Rolling Stock Plant (Tienhsin)	5200-HP electric locomotive
Dairen Locomotive & Rolling Stock	4000-HP Tungfeng-4 diesel locomotives
Dairen Industrial/Mining Rolling Stock	80 to 240-HP diesel hydraulic engines
Fuchin Rolling Stock Plant	passenger and freight cars; parts
Harbin Rolling Stock Plant	100-ton automatic dumping wagon
Henyang Rail Machinery Plant	railroad construction equipment
Hsuanhua Railway Wheels Plant	wheels for 100,000 freight cars per year
Hsiangtan Locomotive & Rolling Stock	first electric 4900-HP locomotive
Lanchou Locomotive Works	locomotives for desert use
Lanchou Rolling Stock Plant	wheels for railway vehicles
Peking "February 7" Plant	6000-HP diesel hydraulic locomotives
Nankou Rolling Stock Parts Plant	railway equipment parts
Paochi Rolling Stock Plant	press to assemble wheels and axles
Shenyang Locomotive & Rolling Stock	railroad-track-building machines
Sian Railway Carriage Factory	oil-tank, freight, and passenger cars
Tangshan Locomotive & Rolling Stock	small steam and diesel locomotives
Tatung Locomotive Works	3000-HP gas-turbine locomotive
Tientsin Locomotive & Rolling Stock	internal-combustion locomotives
Tsingtao Locomotive & Rolling Stock	5000-HP diesel hydraulic locomotives
Tsinan Locomotive Works	no details
Tsitsihar Railway Carriage Works	370-ton flat car; 100-ton railway crane
Wuchang Rolling Stock Plant	refrigerated, ventilated, and stock cars
Wuhan Rolling Stock Works	large freight-car production

Source: Compiled by 21st Century Research from several reports, papers, articles, and publications listed at the end of this chapter.

ers providing additional output in manufacture of smaller parts from scrap metal left over from main production. The 6000-HP locomotive has a total weight of 184 tons and a tractive force of 60,700 kg. Its level-grade traction at 80 km/h is 3900 tons, and on 1.2 percent grade at 29 km/h it drops to 2400 tons. It is powered by two V-12 supercharged diesel engines of 3000 HP each.

The largest electric locomotive made in China is the 5200-HP "Shaoshan" unit developed in 1969 at the Tienhsin Locomotive and Rolling Stock Works in Hunan Province. It was first designed in 1958 and is used on the Paochi–Chengtu route, which is well known for its numerous steep-grade sections.

Other major locomotive plants that developed and produced large numbers of steam, diesel–electric, and hydraulic locomotives are located in Changchou, Changchun, Chishuyen, Tatung, and Peking. Electric locomotives are manufactured at Hsiangtan and Tienhsin (see Table 13.5).

Locomotive Imports and Markets

China is an importer as well as exporter of locomotives and an importer of railway tank cars, though its trade patterns in this equipment are irregular and somewhat unusual. In comparison with imports of other transportation equipment, imports of locomotives are relatively small, averaging less than $20 million annually during 1965–1975 and reaching a peak of $57 million in 1972. Probably no more than 12 percent of the total locomotive inventory of China comes from imports, which include a total of 235 modern mainline diesel and electric units. These account for only 10 percent of all such units in China, and their share is declining rapidly as domestic production of diesel locomotives increases.

The French Societe Alsthom is the largest supplier of locomotives to China in recent years and as of 1977 delivered a total of 115 units of electric and diesel–electric locomotives, including the 7300-HP and 6000-HP electric units, which are the most powerful in China. French specialists also participated in initial electrification of some Chinese lines, and France was the first Western supplier of locomotives, shipping twenty-five electric units as long ago as 1960. Henschel of West Germany is the other Western supplier, shipping a total of ninety large diesel locomotives to China since 1967.

Romania and East Germany are the other major suppliers of railroad equipment to China. Romania, the sixth largest diesel locomotive manufacturer in the world and a major manufacturer of tank-cars, shipped twenty units of diesel electric locomotives to China in 1975 and in recent years was shipping about 500 tank-car units per year, or about 10 percent of the total Romanian tank-car production. East Germany, also a major locomotive producer, shipped thirty electric locomotives to China in 1975.

Japan is a steady supplier of small industrial locomotives, but its annual shipments are in the range of only a few million dollars. However, Japanese companies also participated in electrification of Chinese lines, and a visit of a Chinese Railways Mission to Japan in November 1976 gave rise to speculation that China may order large quantities of rolling stock and marshaling yard technology, presumably from Japan.

The Soviet Union has not been involved in railroad equipment sales to China since 1964 but before then shipped a total of almost $100 million of railroad equipment to China, of which $74.5 million was shipped in 1959 alone. Most of this was 1050 units of used Soviet "Felix Dzherzhinsky" class steam locomotives, which were being replaced by diesels throughout the Soviet Union.

China acted as an exporter of locomotives and railway cars primarily during the construction of the TANZAM railroad, which was undertaken by China as its largest foreign-aid project in eastern Africa. The program involved the construction of 1156 miles of railroad and included the supply of 102 diesel–hydraulic locomotives in the 1000 to 2000-HP range, as well as 2100 freight cars and 100 passenger coaches, steel for rails, and a cement plant for the production of sleepers and other sections for the railroad. As many as 16,000 Chinese Nationals worked in eastern Africa during peak construction period, and PLA Railway Construction Corps reportedly provided up to 15,000 technicians of that total.

China will continue to rely on its railways as the primary transportation system for many years to come, and this raises the possibility of future import markets for railroad equipment as industrial and railroad expansion continues. So far imports have been relatively small, basically because there is a large locomotive and freight-car manufacturing industry in China. Because this industry was already able to manufacture at output levels higher than at present, it is clearly capable of supplying most of the equipment required by the growth of the Chinese railways.

If total tonnage carried by the railways increases by an average 10 percent annually in the near future, there will be a need for about 500,000 freight cars in the Chinese inventory to handle that traffic by 1985. Assuming about thirty freight cars per single locomotive in a typical freight train, this would suggest a need for about 16,000 locomotives by then, which is twice the present inventory.

Based on these assumptions, imports may be expected if China is unable to increase its production to about 1000 units of locomotives and 35,000 freight cars per year by 1985, even when only 2 percent retirement rate is taken into consideration. Although these production levels do not appear impossible for an already large industry, the limitation in output may result from a shortage of steel, which is already imported in large quantities. But even if 10 percent of China's demand for such equipment is imported, the total volume of this trade will most likely remain below $100 million per year.

Whatever the volume of this trade becomes, Western suppliers may not be able to obtain a more significant market share in the future because production of locomotives and even freight cars has been declining for several years in several Western countries. In contrast, production of such equipment is on the increase in COMECON countries (see Fig. 13.2).

Even small communist countries like Poland, Czechoslovakia, and Romania each have a larger production of diesel locomotives than West Germany, Japan, or France, and even China as yet. As their production increases these countries may be in a better position to meet Chinese demand for such equipment, offering shorter delivery time and politically attractive agreements, which China may want to exploit.

Most eastern European countries depend on the Soviet Union for their oil and petroleum supplies, but with the predicted oil-production slowdowns in the Soviet Union they may find themselves looking for alternate sources of energy. China, with its increasing oil production, presents an attractive trade partner because crude and petroleum products could be acquired from China in exchange for such equipment as locomotives and freight cars without the need to spend hard currencies in the Middle East or other oil-supplying countries outside of the COMECON.

Supplying eastern Europe with oil products would be most desirable for China politically because those countries are being supplied by the Soviet Union at present. If the Soviet Union increases the price or reduces the supply of oil to eastern Europe, China will be in a position to offer a solution. Although this will reduce the amount of hard currency earned by China, such trade would save the need to import additional steel products and possibly the same type of equipment from the West, where its production is in decline. At the same time such trade

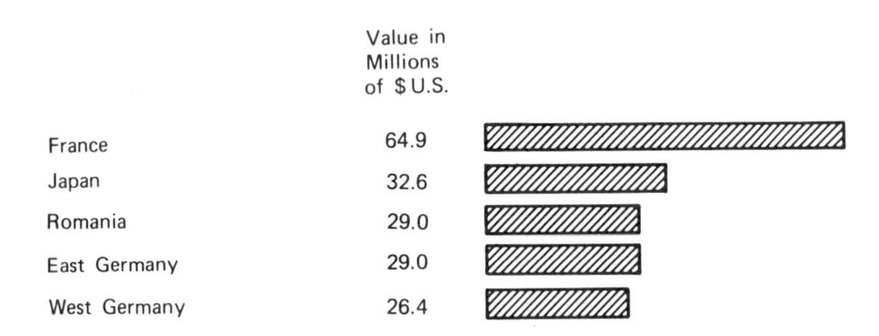

Figure 13.2** Market shares of major suppliers of locomotives and railroad equipment during 1965–1975.* (**Source:** ***Based on export statistics of individual countries reported under SITC commodity code 731 in United Nations publications and foreign-trade yearbooks for individual countries.)

would have highly desirable political effects, benefiting China in its struggle for influence in the communist world, which it may find very difficult or even impossible to resist.

Table 13.6 Major imports of locomotives and rolling-stock equipment to China since 1958

Year	*Type of Equipment Imported*	*Quantity*	*Supplier and Country*	*Value in Millions of $U.S.*
1959	Steam locomotives FD 2-10-2	1050	Soviet Union	70.0
1960	6000-HP Electric locomotives	25	Alsthom-MTE, France	
1965	Small electric locomotives	31	Japan	
1966	Small electric locomotives	17	Sweden	
1967	4000-HP Diesel hydraulic	4	Henschel, West Germany	
1970	Tank cars	298	Romania	3.42
1971	Tank cars	505	Romania	5.83
1972	4300-HP Diesel hydraulic	30	Henschel, West Germany	
	Tank cars	503	Romania	6.35
1973	7300-HP Electric locomotives	40	Alsthom-MTE, France	
	5500-HP Diesel locomotives	36	Henschel, West Germany	8.00
1974	4000-HP Diesel electric	50	Alsthom-MTE, France	
	Small diesel locomotives	6	Nippon Sheryo Seizo, Japan	0.5
1975	2100-HP Diesel electric	20	Electropuntere, Romania	
	EL 2 Electric locomotives	30	Hans Beimler, East Germany	

Source: Compiled by 21st Century Research from reports, articles, and publications listed at the end of this chapter.

HIGHWAYS AND MOTOR FREIGHT

China now has about 840,000 km of roads of all types, of which almost half (350,000 km) are unimproved earth roads and tracks. About 190,000 km is improved earth roads, 2 to 5 m in width, but the condition of these roads is generally considered poor to fair. The remaining 240,000 km include the principal highways, of which over 110,000 km is paved with asphalt.

The present Chinese highway network is probably the ninth or tenth largest road network in the world but is only 60 percent as large as that of Brazil or India, which is the nearest comparable country in size and population. The Soviet highway network is about 50 percent larger than that of China, but a larger percentage of it is believed to be improved and surfaced roads. The total length of Chinese roads is of the same magnitude as that of Canada and Australia but improved and surfaced roads comprise only 13 percent of the total network.

Chinese reports indicate that 35,000 km of new roads was built in 1976 and 15,000 km was surfaced with asphalt. This suggests a growth in highway network of about 4.7 percent, considerably below the 11 percent increase in 1975 but somewhat better than the average annual growth of 4.3 percent during 1965–1975 as a whole.

Total freight originated in highways in 1975 was estimated at 445 million metric tons, which was about 10.1 percent over the previous year. Freight turnover was 15.6 billion metric ton-km, also about 10.6 percent above the previous year. Measured in tons of freight originated, highways accounted for 28 percent of all the freight in China, only a slight increase in the decade since 1965, when it accounted for 27 percent of all the freight originated in the country at that time.

The Highways Administration of the Ministry of Communications is responsible for construction and maintenance of highways and operation of road-transport systems as well as the supply of trucks, spare parts, and transport services. Because highway transport is predominantly of short-haul nature the operation of trucks and buses is probably controlled by units at the provincial or local level. Highway construction is the responsibility of Provincial Highway Bureaus, which also operate highway-maintenance crews of eight to ten persons to keep road surfaces in reasonable condition.

General highway transport is provided by Provincial Transport Companies, which operate fleets of trucks, mechanical-repair workshops, and gasoline stations throughout various district bureaus. The Highways Administration also appears to act as a central equipment and spare-parts distributor for all motor vehicles in China.

Although its importance is growing, the highway network is traditionally a system of feeder roads to and from the railroad lines that constitute the main transportation system of China. However, in western China new long-distance highways are the only means of transportation between such regions as Szechwan, Tsinghai or Singkiang, and Tibet, and a strategic highway along China's southern and western borders provides an additional measure of control of those areas. Because of very rugged terrain, remoteness, and sparse populations, some of those highways are not economically justifiable and require the use of most modern and reliable equipment to keep the transportation links open.

Total truck inventory in China has been estimated at 1,044,000 units in 1976, making China about the eleventh largest truck-operating country in the world.

China's truck inventory is well over twice as large as that of India. It is somewhat larger than that of Mexico or Argentina and approximately comparable to that of Spain or Brazil. Soviet truck fleets are six times larger, and total inventory of trucks in the United States is about twenty-five times as large as that of China (see Table 13.7). The current density of trucks on Chinese roads is estimated at about 116 units per 100 km, which is nearly four times higher than in India but almost four times lower than in the Soviet Union. Because of very high population densities Chinese roads are probably unable to provide efficient road transport thoroughfares at much higher truck densities. Even though private automobile ownership is not permitted, roads in China are constantly in use by large numbers of pedestrians, bicycles, animal-drawn transport, and slow-moving agricultural machinery, which results in very low travel speeds for motor transport. In less populated regions, road surfaces do not permit higher speeds for efficient transportation.

Table 13.7 Civilian truck inventory, length of highway networks, and truck operation densities in selected countries of the world in 1975

Country	*Number of Trucks*	*Freight Carried in Billion Ton-km*	*Length of Highway Network in km*	*Truck Density Units per 100 km*
United States	22,250,000	723.0	6,059,200[a]	367
Japan	10,315,000	130.8[b]	1,059,100	974
Soviet Union	5,800,000	383.0	1,425,200	406
France	2,134,000	90.0[b]	795,920	268
Canada	2,027,600		829,308	244
United Kingdom	1,871,600	57.6[b]	366,000	511
West Germany	1,340,800	44.8[b]	636,040	210
Australia	1,200,300		863,767[b]	138
Spain	1,040,100	71.5[b]	138,560	750
Brazil	1,001,000[b]		1,312,700[b]	76
China	914,000	15.6	800,000	116
Mexico	887,900			
Argentina	879,800[b]		290,200	303
South Africa	800,300		332,500	240
East Germany	534,000	15.2	174,300	306
India	434,400		1,327,000	33
North Korea	113,000		20,278	557

Source: United Nations Statistical Yearbook, 1976; U.S. Government Handbook of Economic Statistics, 1976; U.S. Government "China Economic Indicators," A Reference Aid, October 1977.

[a] Data for 1972. [b] Data for 1974.

PRODUCTION AND IMPORTS OF MOTOR VEHICLES

China produced an estimated 146,000 trucks in 1976 and is the ninth largest truck manufacturer in the world in comparison with commercial vehicle production of other countries. Its motor-vehicle production, which consists mainly of trucks, is about 30 percent higher than that of Italy and equal to half the commercial vehicle production in West Germany. China produced a very small number of passenger cars, primarily for official use since private car ownership and operation are not permitted in China (see Table 13.8).

Table 13.8 Production of commercial vehicles in selected countries of the world in 1975

Country	*Number of Commercial Vehicles Produced*	*Percent of World Total*	*Country*	*Number of Commercial Vehicles Produced*	*Percent of World Total*
World	8,000,000	100.0	China[c]	133,000	1.66
			Italy	110,100	1.37
Japan	2,379,800	29.7	South Africa[a]	104,500	1.31
United States	2,269,600	28.7	Mexico	100,100	1.25
Soviet Union	764,900	9.56	Spain	100,600	1.25
Canada	379,200	4.74	Indonesia[a, b]	45,000	0.6
Brazil	369,800	4.6	Romania	39,300	0.5
France	346,500	4.5	India	38,000	0.47
West Germany	285,900	3.57	Nigeria[a]	17,200	0.2

Source: Adapted from *United Nations Statistical Yearbook, 1976.*

[a] Refers to number of vehicles whose components were assembled not originally manufactured in that country. [b] Data for 1974. [c] *U.S. Government Handbook of Economic Statistics, 1977;* Research Aid ER 77-1053, September 1977.

Almost thirty motor-vehicle plants with some serial production of trucks or other vehicles have been identified in China, but only four of those were believed to have production capacity of over 10,000 units per year in the early 1970s. Although China claims that there is motor-vehicle production in every province of the country, many of those plants produce only a few thousand or a few hundred vehicles per year, and some are collective undertakings by several motor-vehicle repair workshops and parts manufacturing machine shops in an area without benefit of modern serial-line production (see Table 13.9).

The largest motor-vehicle plant in China is the Changchun No. 1 Motor Vehicle Plant in Kirin Province, which is believed to have an annual capacity of 70,000 vehicles and could account for 50 percent of all truck production in China. It is an old plant, built in 1956 with Soviet aid and designed for a potential capacity of 35,000 trucks per year.

The Changchun plant manufactures mainly the 4.0-ton or 4.5-ton "Liberation"-brand trucks based on the Soviet ZIS-150 model, which in

Table 13.9 Automotive manufacturing plants identified in China in the 1970s

Plant Name and Location	*Estimated Capacity of Plant*	*Types of Vehicles Manufactured*
Canton Motor Vehicle Plant	5,000	3.5-ton trucks
Changchun No. 1. Plant	70,000	4.0-ton and 4.5-ton trucks
Chengchou Motor Vehicle Plant		makes "Chengchou" trucks
Chinghai No. 2. Motor Vehicle Plant		4.0-ton "Chinghai Lake" trucks
Chingkangshan Motor Vehicle Plant	3,800	2.5-ton and 1.5-ton trucks
Fuchou Motor Vehicle Plant		2.5-ton "Fukien 130" trucks
Hofei Bus Factory		JT-661 long-distance buses
Harbin Truck Plant		12.0-ton "Aurora" trucks
Hohukan Motor Vehicle Plant	3,000	4.0-ton "CA-10 Liberation" trucks
Kunming Motor Vehicle Works		4.5-ton "Kunming" trucks
Liuchou Machinery Plant		"Liuchan 130" trucks, motorcycles
Nanching Motor Vehicle Plant	18,000	2.5-ton "Leap Forward" trucks
Nanching Chiang Huai Plant		3.0-ton HF 140 trucks
Nanchang Motor Vehicle Plant	6,000	2.5-ton trucks
Nanyang Motor Vehicle Plant		7.0-ton "Nanyang 351" dump truck
Peking Motor Vehicle Plant	10,000	sedans, medium/heavy trucks, vans
Peking Changcheng Plant		12.0-ton and 10.0-ton trucks
Peking Tung Fang Hung Plant		all-terrain jeep-type vehicle
Peking Erhlikou Motor Vehicle Plant	3,500	2.0-ton trucks and BJ-130 jeeps
Shanghai Motor Vehicle Plant	3,000	2.0-ton and 4.0-ton trucks, sedans
Shanghai Small Car Plant		"Sea Swallow SWH 600" cars
Shanghai Cargo Vehicle Plant	2,000	32.0-ton, 15.0-ton and 4.0-ton dumpers
Tsinan Motor Vehicle Plant	5,000	8.0-ton "Yellow River" trucks
Tientsin Motor Vehicle Plant		5.0-ton trucks, coaches, vans
Taiyuan Motor Vehicle Plant		4.5-ton and 2.5-ton trucks
Tsingtao Motor Vehicle Plant		7.0-ton "Yellow River" trucks
Wuhan Motor Vehicle Plant	30,000	2.5-ton "Wuhan" WH-130 trucks

Source: Compiled by 21st Century Research from reports, articles, and publications listed at the end of this chapter.

turn was based on Ford trucks of World War II vintage. The plant also manufactures small numbers of "Red Flag" passenger sedans, cross-country vans, jeeps, and auto parts. It is equipped with the Soviet version of the vital Gleason axle-gear cutting machinery and is a major center for training technicians and supply of machine tools and molds for manufacturing motor vehicles in other plants.

Wuhan Motor Vehicle Plant in Hopeh Province may be the second largest truck manufacturer, with an estimated capacity of 30,000 units annually. It has been producing unknown quantities of the 2.5-ton WH-130 "Wuhan" trucks since 1970. Because of a large expansion and modernization of Wuhan Steel Rolling Mills in recent years it is believed that Wuhan may become the largest and most modern vehicle-manufacturing center in China. Its central location, removed from the proximity of Soviet borders, also has a strategic importance to China.

Nanching Motor Vehicle Plant in Kiangsu Province is known to have produced about 11,000 units in 1972, and as such it has been sometimes regarded as the second largest motor vehicle plant in China, though its maximum capacity is estimated at only 18,000 units per year. It manufactures 1.5-ton and 2.5-ton "Leap Forward" trucks as well as some passenger cars and engines. It is an important plant, also equipped with the Soviet version of the Gleason axle-gear cutting machinery.

Several motor vehicle plants have been identified in Peking, of which the Peking Motor Vehicle Plant is estimated to have 10,000 units of annual output. It produces small sedans, medium and heavy trucks, jeeps, vans, and auto parts. It is also regarded as the largest auto-accessory manufacturer in China.

Another Peking plant, the Changcheng Motor Vehicle Plant, produces 10 to 12-ton "Long March" trucks designed with the assistance of the TATRA Truck Manufacturing Enterprise from Czechoslovakia. The Peking Erhlikou Motor Vehicle Plant produced an estimated 3500 units of 2-ton trucks for both domestic use and export.

Shanghai is also a major motor-vehicle manufacturing center, with at least five plants in operation. The Shanghai Motive Power Company produces an estimated 3000 units per year, including the "Phoenix" sedan, a 2-ton truck, a 4-ton truck, passenger cars, jeeps, three-wheel trucks, and busses. The Shanghai Small Passenger Car Plant manufactures micro-trucks, while the Cargo Vehicle Plant is believed to have a 2000-unit annual capacity for production of dumper trucks in the 4-, 15,- and 32-ton range. This plant depended heavily on Soviet technology from the BELAZ plant in Minsk, which in turn depended on French Berliet 32-ton dumper truck technology of the 1960s. Characteristics of Chinese motor vehicles are listed in Table 13.10.

Truck production increased by 9.8 percent in 1976 and has been growing at the rate of about 10 percent since 1973. If production of trucks continues to grow at similar rates, China would have an output of 250,000 units by 1981, and by 1985 its truck production would be comparable to that of Canada or the United Kingdom in 1975 (see Table 13.8).

If truck production is to expand at such a rate, plant capacity will have to triple by 1985, and there have been some indications of plans for additional large motor-vehicle plants. It is not clear whether China will import an integrated modern plant or develop its own production. At present China appears to continue to obtain specialized machinery and technology licenses to selectively improve its ability to manufacture ma-

Table 13.10 Characteristics of Chinese motor vehicles

Model	*Brand Name*	*Maximum Horsepower*	*Cylinders and Engine Type*		*Maximum Payload in kg*[a]
SWH-600	Sea Swallow	20	2	gasoline	500
BJ-130	Peking	75	4	gasoline	2,000
BJ-212	Peking Jeep	75	4	gasoline	425
TJ-210C	Tientsin Jeep	70	4	gasoline	540
TJ-500	Wuhan Jeep	80	4	gasoline	500
SH-130	Shanghai	75	4	gasoline	2,000
NJ-130	Leap Forward	79	6	gasoline	2,000
WH-130	Wuhan	79	6	gasoline	2,500
WH-27	Ching-Kang	79	6	gasoline	2,000
NJ-230	Leap Forward	89	6	gasoline	1,500
CA-30A	Liberation	110	6	gasoline	2,500
CA-340	Liberation	95	6	gasoline	3,500
CA-10B	Liberation	95	6	gasoline	4,000
HF-140	Chiang-Huai	120	6	gasoline	3,000
GZ-140	Red Guard	120	6	gasoline	3,500
SH-141	Communication	90	6	gasoline	4,000
QX-10X	Ching Hai Lake	95	6	gasoline	4,000
QD-351	Yellow River	160	6	diesel	7,000
QD-351	Nan-Yang	160	6	diesel	7,000
JN-150	Yellow River	160	6	diesel	8,000
JN-151	Yellow River	160	6	diesel	8,000
XD-250	Long March	170	V-12	diesel	10,000
XD-160	Long March	180	V-8	diesel	12,000
SH-361	Communications	220	6	diesel	15,000
SH-380	Shanghai	400	V-12	diesel	32,000
XD-980	Long March Tractor	170	V-12	diesel	Tow 100,000

Source: "Technical Handbook for PRC Motor Vehicles," published by People's Transportation Agency, Peking; People's Republic of China (transl. by JPRS #60262), October 12, 1972.

[a] Payload figures indicate maximum payload on paved highways. This figure is estimated to be 30 to 50 percent less on country roads in China.

jor automotive components, including roller bearings, shell bearings, clutch and brake linings, fuel pumps, injectors, crankshafts, and transmission elements.

Imports of Motor Vehicles

Since 1970 China has been importing over $100 million worth of motor vehicles per year. In 1974 imports of motor vehicles and parts reached an all-time high of over $260 million but dropped to $204 million in 1975. Imports probably amount to about 20 percent of the value of domestic truck production, but their share in the future may decrease further if domestic manufacture continues to grow considerably faster than the construction of new highways, leading to very high and inefficient truck densities in some parts of China.

Speculation also continues as to whether China will decide to import a complete integrated automotive plant; previous experiences of the Soviet Union and eastern European countries would suggest that this will eventually happen. China is actually judged to possess sufficient ability to manufacture custom-made and hand-machined products and vehicles that are very satisfactory for the purpose for which they have been designed. However, it is the modern continuous flow production know-how that includes precise automation technology, as well as organizational and management experience, which is lacking.

Fiat, which provides advanced motor-vehicle production technology to the Soviet Union and some eastern European countries, already held high-level talks with Chinese authorities. Fiat's president visited Peking in 1975, and in March 1977 a group of Fiat executives went to Peking for further discussions.

Toyota of Japan discussed with China several times the importation of a modern production plant that would give China a technological jump greater than that realized by the Soviet Union from its Fiat-Togliatti auto plant a few years ago.

China also held discussions with Volkswagen about the possible acquisition of a plant to manufacture "VW Safari" vehicles and even purchased a number of such units manufactured at the Volkswagen plant in Mexico. There are indications that China also approached Chrysler and possibly other auto manufacturers about plant technology. On the other hand, China turned down an offer of a complete Leyland car plant from Australia in 1974, indicating that it has no interest in import of automobiles.

Recently the NHK Spring Company of Japan held talks in Peking that may lead to large Chinese purchases of flat springs for trucks. Japanese industry sources now appear to believe that China will import engine parts, interior fittings, and other parts for assembly in China. This would suggest that China will proceed to develop its own modern truck plants, incorporating innovations in production of various components while continuing to import large quantities of finished trucks until its own production begins to meet the demand. China also exports small quantities of its trucks to Third World countries primarily as part of its foreign-aid programs.

In 1970 half of all the 600,000 trucks then operating in China were

Chinese made, about 30 percent were imports from the Soviet Union and eastern Europe, and the rest were imports from Japan, France, and other Western countries. By 1975 probably as much as 65 percent of all the trucks from domestic production and this percentage will increase as Chinese production grows. On the other hand, it is Chinese practice to use a motor vehicle for 30 to 35 years before replacement, which creates a significant spare-parts aftermarket of very long duration.

Because of a large percentage of Soviet and eastern European trucks in the original Chinese fleet, imports of such trucks and their spare parts are expected to continue well into the 1980s. Romania also is a major supplier of new trucks to China, and during the early 1970s shipped 3000 to 5000 new trucks per year. During 1970–1975 total imports of trucks and parts were valued at almost $1 billion, of which about 40 percent were imports from the Soviet Union, Romania, Czechoslovakia, and East Germany, with smaller quantities from Hungary and Poland (see Fig. 13.3). In 1976 imports of trucks from the Soviet Union increased to 968 valued at $22 million, up from 398 units in 1975.

Japan, however, became the major truck supplier in 1970, and by 1974 its truck exports to China reached $117 million, amounting to about 45 percent of all Chinese truck imports. In 1976 truck imports from Japan were $44 million. Major Japanese suppliers include Isuzu, Toyota, Hino Motors, Mitsubishi, and Nissan. In 1976 Japan told a total of 4000 trucks to China, somewhat less than in previous years, but expects to ship again about 5000 units in 1977. Since 1965 Hino Motors alone supplied about 8000 trucks, including 1000 medium-sized trucks in 1976.

France is the second largest Western supplier of trucks to China, shipping a total of $161 million during 1970–1975 and approximately another $60 million during 1965–1970. In 1965 Berliet of France negotiated an agreement for the construction of a truck plant in China and by 1972 licensed four types of its trucks for manufacture in China. By mid-1975 it was estimated that a total of 11,000 Berliet trucks were operating in China, supplied since 1965 from French, Algerian, and Moroccan plants of the company. In 1976 France did not export any trucks to China any more.

Romania is probably the third largest truck supplier to China, and in 1972 it was the largest supplier, having shipped $29 million worth of trucks that year, almost 30 percent of all Chinese imports. But Romanian truck production is only about 39,000 units per year, and it is probably unable to supply larger quantities without disrupting its own domestic and European markets. Nevertheless, Romanian sales to China are comparable to those of the Soviet Union and Czechoslovakia, both of which have considerably larger truck-production capabilities.

In Chinese foreign trade, imports of trucks are the next largest

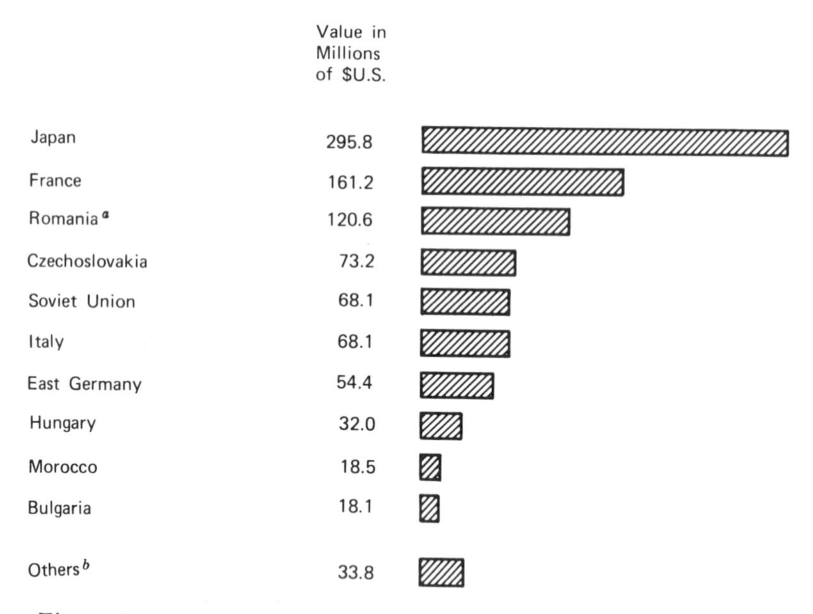

Figure 13.3 Relative market shares of suppliers of trucks, automobiles, and automotive parts to China during 1970–1975. (Source: OECD Commodity Trade Statistics *for SITC code 732 and United Nations international trade statistical publications and foreign-trade yearbooks of individual COMECON countries.*) [[a] *Includes authors' estimates for 1974 and 1975 comparable to earlier years.* [b] *Includes West Germany (11.9), United Kingdom (10.4), United States (5.3), Poland (3.3), Sweden (3.5), Finland (2.9), as well as very small imports in some years from Norway, Denmark, Belgium, Yugoslavia, and Spain.*]

machinery imports after imports of ships and thus it is reasonable to expect that China will make significant efforts to increase domestic production to reduce such imports. This is particularly more likely in the case of trucks rather than ships or aircraft because production of trucks uses relatively less steel and requires relatively less sophisticated technology for manufacture.

On the other hand, even if imports of trucks are progressively reduced from about 20 percent to only 10 percent of domestic production by 1985, the market may still remain at a level of about $200 million per year. This may happen as a result of a continuing need to import heavy trucks for mining, industrial construction, and off-highway use. These trucks are not yet manufactured in China in significant quantities and require more advanced materials and heavy-duty performance technology, not readily available in China.

MERCHANT MARINE OF CHINA

The Chinese merchant fleet in 1976 consisted of almost 500 ships, totaling over 7 million deadweight tons. It accounted for about 1 percent of total global carrying capacity, making the Chinese fleet the fifteenth largest in the world. It includes a sizable and growing tanker fleet, which in 1975 consisted of eighty tankers with a total carrying capacity of almost $2 million deadweight tons, making this the twelfth largest tanker fleet in the world (see Table 13.11).

Table 13.11 World significance of Chinese merchant and oil-tanker fleets in 1975

	Merchant Fleet		*Tanker Fleet*	
Country	*Units*	*Th. DWT*[a]	*Units*	*Th. DWT*[a]
World total	22,391	556,697	5,183	301,852
Liberia	2,546	132,694	1,014	89,470
Japan	2,051	63,238	531	33,950
United Kingdom	1,576	54,913	459	32,896
Panama	1,556	22,112	238	10,224
United States	857	17,694	267	9,711
Soviet Union	1,660	15,353	286	4,981
Malaysia/Singapore	429	8,118	48[b]	
India	308	6,943	25	1,510
China	495	6,082	80	1,916
Brazil	260	4,546	53	944

Source: U.S. Maritime Administration and *U.S. Government Handbook of Economic Statistics, 1976.*

[a] Th. DWT = thousand deadweight tons. [b] Data for Singapore for 1974 only.

The Chinese fleet grew at an average rate of 21 percent per year during 1970–1975, and it is estimated that it could reach as much as 17 million deadweight tons of oceangoing-capacity by 1985.

China owns an estimated 130 ships, totaling about 1.7 million deadweight tons, registered under flags of convenience. Until October 1976 most were registered under the Somalian flag, and a few were registered in Hong Kong. Since 1976 most vessels under the Somalian registry were transferred by China to the registry of Panama, a move believed to have been prompted by increasing Soviet–Somali relations.

Chinese merchant fleet-carrying capacity is comparable to that of India, but its ships are generally smaller averaging 12,000 deadweight tons, compared with India's average 19,000 deadweight tons and the world average of 24,000 deadweight tons. It is also about two and a half times smaller than the Soviet merchant fleet-carrying capacity, but the Soviet Union operates over 1600 ships, its average ship size is only 9200 deadweight tons, about 25 percent smaller than Chinese ships. China's tankers average 24,000 deadweight tons and are considerably smaller than the world average of about 58,000 deadweight tons for that type of ship.

Freighters account for about 65 percent of the Chinese fleet, tankers make up about 16 percent, and bulk carriers comprise 12 percent of the total. The remaining units are made up of reefers, container and semi-container ships, and about twenty-five passenger ships.

In comparison with fleets of other countries, Chinese ships are old, with an average vintage of 20 years, whereas the average age of ships of the total world fleet is only 12 years. Indian and Soviet fleets average only 11 years in comparison. This is so because a large part of the Chinese fleet consists of second-hand ships purchased in the world markets when prices are low. Only one-sixth of the fleet is believed to have been supplied by the Chinese shipyards so far.

The Chinese Merchant Marine is under the supervision of the Ministry of Communications, headed by Minister Yeh Fei, through an administrative structure that controls inland, coastal, and ocean transportation. In 1975 a total of 208 million metric tons, or 13 percent of all the freight, was carried on inland and coastal waterways. About 160 of China's oceangoing ships are believed to operate on coastal routes and on rivers that have adequate facilities for oceangoing vessels. These are mostly freighters or combination passenger/cargo ships, but there are also hundreds of small vessels until 1000 gross tons involved in domestic water transport.

Domestic shipping is under the control of three administrations. The Shanghai Maritime Administration controls the largest part of domestic fleet, and its jurisdiction extends from Wanchou in Chekiang Province all the way to the North Korean border. The Canton Maritime Administration is responsible for the area from Swatow south and west to the Vietnam border. The Yangtze River Navigation Authority controls shipping on inland waters as well as river systems, which are primarily centered on the Yangtze River and its tributaries.

That part of the Chinese fleet that is engaged in foreign trade is administered by China Ocean Shipping Company (COSCO), believed to control over 200 oceangoing vessels and carry over 30 percent of Chinese foreign trade. The China Merchant Steam Navigation Company is an agency in Hong Kong through which COSCO controls several steamship companies in Hong Kong and Macao, which in turn operate their ships under flags of convenience. The China Ocean Shipping Company also participates in joint shipping operations such as the Sino-Albanian Shipping Company, the Sino-Polish Shipbrokers Company, and the Sino-Tanzanian Shipping Company.

China has maritime agreements with seventeen countries, including Belgium, Brazil, Bulgaria, Chile, Denmark, France, West Germany, Greece, Italy, Japan, Mexico, the Netherlands, North Korea, Sri Lanka, Yugoslavia, and Zaire. In many cases such agreements merely formalize existing maritime relations between various countries, but in some cases these contain exemption clauses permitting ships of foreign countries to enter Chinese ports without paying the 3 percent tax that is normally levied.

In general the merchant fleet in China is not adequate to meet all transport needs, and China charters additional ships from owners in various countries. Over the years a large percentage of Greek-owned vessels registered under Greek, Liberian, and Cypriot flags have been chartered to China.

Containerization is in its infancy in China, but Japanese shipping lines have already begun containership operations between Japan and China in 1973. It is a logical expansion area for the Chinese merchant marine if it plans to continue to carry a large percentage of China's trade in its own vessels and compete in the international shipping market.

PORTS AND WATERWAYS

There are seventeen principal ports in China, administered locally by Harbor Superintendent Bureaus of the Ministry of Communications. The Ministry also conducts port inspection and cooperates closely with the Ministries of Public Security and Public Health. In previous years the Ministry of Communications also controlled at least seven large shipyards, which are now all believed to be under the supervision of the Sixth Ministry of Machine Building, which is specifically responsible for production of naval and merchant-marine equipment.

All Chinese harbors are relatively shallow with depths of 30 feet or less at high tide, which limits the size of ships that can be handled. Ports are also very congested, and one estimate in 1976 put the number of ships in Chinese ports at 200 per day and increasing rapidly. Ships with cargo bound for China often report delays of weeks and months before they are unloaded. Significant imports of dredgers from Japan and western Europe are believed to be the result of Chinese decisions to deepen the channels in Chinese ports and improve cargo handling and throughput capacity. Another reason for delays and congestion in ports is the rapid increase in foreign trade, coupled with the fact that roads and railroad facilities to and from the ports were not originally designed to handle the intense cargo traffic of today.

In 1973 China began a program of port modernization and expansion and by the year 1976 claimed to have completed forty new deepwater berths capable of handling ships of 10,000 tons or over. Some facilities designed to handle 25,000-ton and 50,000-ton tankers were also developed in recent years. In 1976 China also announced that the first containership-handling terminal will be developed in the port of Tientsin. Due to poor highway networks and lack of transshipment facilities, containers are not believed to be an immediate answer to China's shipping congestion until container-handling facilities on Chinese railroads are developed in ports and at major destination points.

Shanghai is the largest of Chinese ports and has over fifty berths, of which at least twelve are capable of handling 10,000-ton freighters. In 1974 local authorities claimed unloading time for a general cargo vessel to be 82 h and for a 10,000-ton collier, 68 h, with over 75 percent of general cargo handling to be mechanized. The port of Shanghai already handles small containerships, but larger vessels draw over 44 feet of water while Shanghai channel can only accommodate a maximum of 25 feet of draught (see Table 13.12).

Tientsin is believed to be the second largest port in China, and at least eleven of its berths can handle ships of over 10,000 tons. The ports' importance is underscored by the plans for China's first containership terminal. Ports of Whampoa and Chanchiang near Canton are also of great importance; several new 10,000-ton berths have been built in Whampoa, and the first wharf for tankers up to 50,000 tons was built in Chanchiang in 1975, providing more efficient means of transporting large quantities of oil to energy-short southern China areas. Chinghuangtao is connected by a 1000-km pipeline with the Taching Oilfield in northeastern China and is the main port handling export of oil. Tsingtao and Lienyungkiang are major coal ports.

SHIPBUILDING AND IMPORTS OF SHIPS

Total deadweight tonnage of merchant vessels launched in China during 1975 was estimated at 494,000 metric tons, which is about 1.4 percent of

Table 13.12 Shipping and cargo-handling capacities of Chinese ports

Location	*Maximum Draft in m*	*Ships of Foreign Registry per Year*	*Largest Berth Size in Tons*	*Maximum Lifting Capacity in Tons*	*Total Available Storage in m²*	*Largest Tugs in HP*
Amoy	8.0	—	3,000			
Canton	5.0	400		15	200,000	400
Chanchiang	11.0	200	10,000	40	240,000	1,200
Chinghuantao	10.6	1,000+	10,000		250,000	1,080
Dairen	12.0	1,500	10,000+	100	1,000,000	1,200
Foochow	6.0	—	—	—	—	—
Haikou	5.0	30+		20	16,000	400
Lienyunkiang	8.0	400	10,000	23	70,000	1,080
Paso	9.0	200	—	23	40,000	650
Peihai	9.0	30	—	15	20,000+	400
Shanghai	9.0	2,900	—	100	1,600,000	1,080
Swatow	6.0	500	4,000	15	100,000	300
Tientsin	10.0	1,600+	10,000	150	300,000	1,200
Tsingtao	12.0	900	10,000	50	200,000	1,080
Whampoa	9.0	1,000	20,000	30	250,000	1,080
Yenchou	6.0	—	—	—	—	300
Yentai	8.0	—	5,000	16	70,000	400

Source: *General Condition of China's Seaports,* published in China and translated by JPRS #62776 on August 21, 1974, Washington, D.C. 20402.

the world's total. This is equivalent to about 335,900 metric tons of actual ship weight which is more indicative of material consumption and cost of production. Chinese merchant ship production level is comparable to that of Poland or South Korea and is at least ten times larger than tonnage launched by India. However, China's merchant ship production is only a fraction of the tonnage launched by major shipbuilding countries every year (see Table 13.13).

Merchant vessel construction is not the best measure of China's total shipbuilding capabilities because China now has what is believed to be the third largest navy in the world, measured in terms of number of ships, many of which have also been built in Chinese shipyards. Over 700 navy vessels are deployed along the China coast and inland waterways, but many are very small patrol boats, though larger units, including a nuclear-powered submarine are also in construction. Since China is also building freighters of 25,000 tons there is now speculation that it may attempt to develop its own aircraft carrier to provide more flexibility to the Chinese Naval Air Force, which already operates about 700 land-based fighters, bombers, and reconnaissance planes.

There was a dramatic increase in merchant ship construction since

Table 13.13 Merchant vessels launched by various countries during 1975 in Gross Registered Tons (GRT)

Country	*Tonnage Launched in GRT*[a]	*Percent of World Total*	*Country*	*Tonnage Launched in GRT*[a]	*Percent of World Total*
World	35,898,000	100.0	Denmark	961,000	2.6
			Netherlands	951,000	2.6
Japan	17,978,000	50.0	Italy	847,000	2.3
West Germany	2,549,000	7.0	Yugoslavia	639,000	1.8
Sweden	2,460,000	6.8	Poland	608,000	1.7
Spain	1,638,000	4.6	China[c]	494,000	1.4
Soviet Union[b]	1,432,000	3.9	South Korea	441,000	1.2
United Kingdom	1,304,000	3.6	Brazil	389,000	1.1
France	1,301,000	3.6	East Germany	338,000	0.9
Norway	1,029,000	2.8	India	45,000	0.1
United States	1,004,000	2.7			

Source: *United Nations Statistical Yearbook 1976.*

[a] Gross registered tons represents total volume of all enclosed space in vessels launched at 1 ton = 100 ft^3 or 2.83 m^3. [b] Data represent additions to Soviet merchant fleet during 1975–1976 in net registered tons and include imports from abroad. [c] Data in deadweight tons indicate weight of cargo the ships can handle.

the end of the Cultural Revolution, and although industrial production only doubled between 1965 and 1974, total tonnage launched increased six times in the same time period while major ship tonnage in 1974 was 14 times larger than in 1965 (see Fig. 13.4).

This is the result of construction of increasingly larger units by the Chinese shipyards. During the 1960s ships of 10,000 tons were still experimental, but by the 1970s these entered into mass production. In 1975 total tonnage of major ships launched was estimated at a minimum 308,000 tons, or about 60 percent of all merchant-ship output. Minor vessels built and launched in China include river craft, ferries, fishing boats, barges, and research and observation vessels. By 1976 Chinese shipyards were building freighters up to 16,000 tons, bulk carriers up to 24,000 tons, and at least one tanker of 50,000-ton capacity was reported in construction at the Hunchi Shipyard in Dairen (see Table 13.14).

Some insight into Chinese shipbuilding technology can be gained from reports that shipyards in Dairen and Shanghai are involved in submarine construction. China is known to have already built a G-class submarine equipped with three missile tubes and other reports of Albacore-type hull construction suggest that a nuclear-powered submarine may be in construction.

About 110 shipyards are believed to be in operation in China, but only ten appear to be capable of constructing ships in the 10,000-ton class,

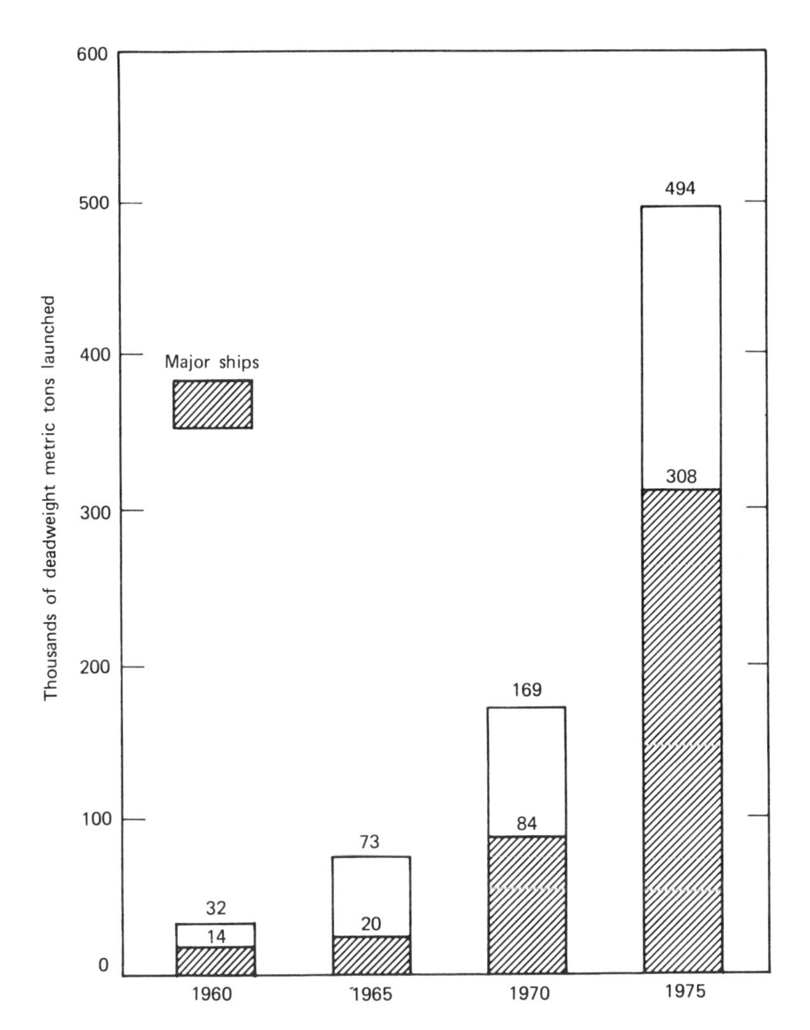

Figure 13.4 Growth of merchant ship production. **(Source:** ***Adapted from*** **U.S. Government Research Aid,** ***"Chinese Merchant Ship Production," ER 76-10193, March 1976.*****)**

and only four of those have capabilities in the 25,000-ton category. Although there is little doubt that China will be building larger ships in the future, industry observers believe such expansion now depends on construction of larger slipways, deepening of Chinese harbors, and provision of larger berthing facilities.

The largest shipbuilding complex in China is operated by the Shanghai Shipbuilding Company, which operates six shipyards in the

Table 13.14 Major shipyards identified in China in recent years

Shipyard Name and Location	*Largest Ships Launched and Types Built*
Chekiang Shipyard in Hangchou	1000-ton coastal freighters
Whampoa Shipyard near Canton	20,300-ton ship built in 1975; gunboats
Chiangmen Pearl River Shipyard	600-ton class steel and wooden ships
Canton Fishing Boat Works	800-ton cold storage vessels
Chehyang County Shipyard	1400-ton concrete seagoing vessel
Chiuchiang Experimental Boatyard	500-ton passenger/cargo concrete ship
Hunchi Shipyard in Dairen	50,000-ton oil tanker in construction
Tungfeng Shipyard in Szechwan	2000-HP marine diesel engines
Canton Diesel Engine Plant	1200-HP marine diesel engine
Lutung Shipyard	1000-ton tankers
Chiangnan Shipyard in Shanghai	20,000-ton freighter launched in 1971
Hutung Shipyard in Shanghai	25,000-ton freighters; 12,000-HP diesel
Shanghai Ship Repair Yard	25,000-ton freighter, tugs, ferries
Chungkua Shipyard in Shanghai	10,000-ton freighters and dredging vessels
Tunghai Shipyard in Shanghai	small hydrographic vessels
Chiu Hsin Shipyard in Shanghai	3200-ton vessels, icebreaker in 1969
Hsinko Shipyard in Tientsin	500-ton floating crane, dredgers
Hsinkang Shipyard in Tientsin	10,000-ton freighters since 1969
Hunghsing Shipyard in Tsingtao	10,000-ton freighters; 5000-ton tankers
Weihaiwei Shipyard in Shantung Prov.	passenger/cargo diesel vessels
Wuchou Shipyard	cement and steel ships
Wuchang Shipyard in Wuhan	one of largest in the country
Ningpo Shipyard in Chekiang Prov.	1000-ton cold storage vessels
Chinling Shipyard in Kiangsu Prov.	10,000-ton ships and floating cranes
Canton Shipyard	10,000-ton passenger/cargo vessel
Wenfang Shipyard near Canton	2000-ton passenger/cargo vessels
Wenchung Shipyard near Canton	3000-ton passenger/cargo vessels

Source: Compiled by 21st Century Research from articles, reports, and publications listed in reference section to this chapter.

Shanghai area and probably also runs the Shanghai Marine Diesel Plant, which produces 10,000-HP diesel engines for ships. Its Hutung shipyard built the first 25,000-ton freighter in China in 1971 as well as the first 12,000-HP marine diesel engine, which may not be in serial production. In 1974 the "Kantan No. 1" deep-sea oil-drilling rig was completed here. Since the mid-1960s over 400 ships have been launched and over 100 large diesel engines totaling 500,000-HP have been produced. Other shipyards in the Shanghai area also built floating docks, research and geological survey ships, icebreakers, dredgers, and submarines.

Several large shipyards are operated by the Canton Shipbuilding Industry, which controls six shipyards and a marine diesel-engine plant. Other major shipyards are located in Tientsin, Tsingtao, and Wuhan (see Table 13.14).

Imports of Ships

During 1976 Hong Kong sources reported that Shanghai shipyards were taking orders for exports of 3000-ton freighters and tankers, quoting an 8 month delivery time, which led some industry observers to conclude that as soon as expansion of Chinese shipyards is complete, China will compete with major shipbuilding countries for export orders.

However, despite rapid growth of the Chinese shipbuilding industry the supply of larger and more modern ships does not appear to be adequate, and China has been ordering ships in foreign yards. Imports of ships amounted to almost $300 million in 1974 and at least $377 million in 1975, which represents about 5 percent of all Chinese imports and constitutes the fourth largest single import category in Chinese foreign trade. In 1976 preliminary results show that only $48 million in ships was imported by China from Japan and Singapore, although import data from East European countries were not available at the time of this writing.

China may be importing ships equivalent to about 50 percent of its production, which is not only limited by the capacity of Chinese shipyards but also by availability of steel. China already imports over $1 billion worth of iron and steel per year, and imports of ships may be regarded as a way to save steel imports and at the same time obtain the latest shipbuilding technology.

Assuming that the demand for ships grows at about 10 percent per year, which is considerably lower than the recent Chinese shipbuilding growth rate, there may simply not be enough steel to sustain such growth domestically. Even if imports of ships decline with increasing production to only 25 percent of the total output by 1985, this will still mean about $300 to $400 million of ship imports per year.

Although China purchases many second-hand ships when shipping prices are depressed, many orders have been placed for large modern vessels with Japanese, Norwegian, Dutch, Yugoslav, German, and Danish shipyards in recent years. Previously China was also buying small freighters from Bulgaria, Romania, and Poland (see Table 13.15).

Recent orders placed in Yugoslavia included a 45,000-ton freighter from the 3 Maj Shipyard in Rijeka, which is also producing 12,000-HP diesel engines for China. Orders for 40,000 to 50,000-ton tankers were also placed with Japanese shipyards. East Germany is believed to be the supplier of a new 13,000-ton containership, and Norway and Netherlands are suppliers of specialized ships for offshore oil and dredging operations.

Prior to 1970 China was a minor purchaser of ships, averaging about $20 million per year. Poland, East Germany, France, and the United

Table 13.15 Typical imports of ships to China identified since 1972

Year	*Type of Vessel Imported*	*Quantity*	*Supplier and Country*	*Order Value in Millions of $U.S.*
1972	13,000-ton freighters	6	Yugoslavia	
	dredgers	2	IHC, Netherlands	31.0
	dredgers	6	Marubeni, Japan	13.0
	14,300-ton cargo ships	2	Hitachi, Japan	17.0
1973	tugboats and barges		Malta Drydocks, Malta	2.4
	refrigerator ships	10	Kanazashi Zosen, Japan	18.0
	large crane ships	2	Mitsui, Japan	6.7
	cargo ships	4	3 Maj Shipyard, Yugoslavia	
	20,000-ton freighter	1	Japan	
	800-ton supply ships	4	Solstad Rederi, Norway	8.7
	supply ships	2	Norse Petroleum, Norway	
	tugboats		Hitachi, Japan	17.4
	31,825-ton freighter	1	Olsen Daughter, Norway	
	12,000-ton freighters	3	Yamashita Shinnihon, Japan	
	1800-ton freezer ship	1	Tokyo Hroichi Shoji, Japan	1.1
	offshore supply ships	8	Aarhus Flydedok, Denmark	41.8
	3200-ton cargo/passenger	1	Shikoku Dockyard, Japan	
	passenger ship	1	France	
	fishing boats	9	Nisii Dockyard, Japan	
	suction dredgers	16	IHC, Netherlands	
	dredgers	10	IHC, Netherlands	52.0
	self-propelled dredgers	8	Nippon Kokan, Japan	56.6
	45,000-ton ships	2	3 Maj Shipyard, Yugoslavia	
1974	55,000-ton freighter	1	Sweden	
	33,822-ton tanker	1	Norway	14.5
	800-ton supply ships	2	Fred Olsen, Norway	
	1500-ton supply ships	2	Fred Olsen, Norway	
1975	3100-ton salvage ship	1	Sumitomo, Japan	21.7
1976	12,000-HP marine diesels	4	3 Maj Shipyard, Yugoslavia	
	14,800-ton ship	1	Primo Transport, Liberia	
	salvage tugs	2	Hitachi Zosen, Japan	
1977	39,000-ton and 37,000-ton ships	2	second-hand	10.5

Source: Compiled by 21st Century Research from various reports, articles, and publications. Order values are approximate converted at exchange rates in effect during the years in question.

Kingdom were then major suppliers, and Poland alone accounted for at least $25 million worth of ships during 1965–1970.

Since 1970 purchases of ships grew rapidly, increasing tenfold by 1974 and reaching a total of $376 million in 1975. All this occurred in parallel with an equally unprecedented growth in Chinese shipbuilding. Many new ships were ordered during that period, but China again began buying second-hand ships in 1977, when $10.5 million was spent on

30,000-ton and 37,000-ton ships of the 1967 and 1969 vintage, respectively.

Japan now became the leading supplier of ships to China and in 1975 accounted for at least 46 percent of the value of all ships imported by China. During 1970–1975 Norway, the Netherlands, and Yugoslavia each sold over $100 million worth of ships to China and displaced the supplier countries of the 1960s as major ship exporters to China (see Fig. 13.5). East Germany, however, continues to export ships to China, but Poland's exports dwindled practically to zero by 1973, though Polish sources claim this to be the result of unavailable capacity in the Polish shipyards at present.

Because China will increasingly depend on foreign trade and exports of oil it may be a large future market for oil tankers in the 100,000-ton class, which are not being built in China. The present limitation on ships of that size is the lack of docking facilities in China, but modernization of ports and expansion of the merchant fleet are progressing and should create additional opportunities for exports of new and second-hand ships to China in the future.

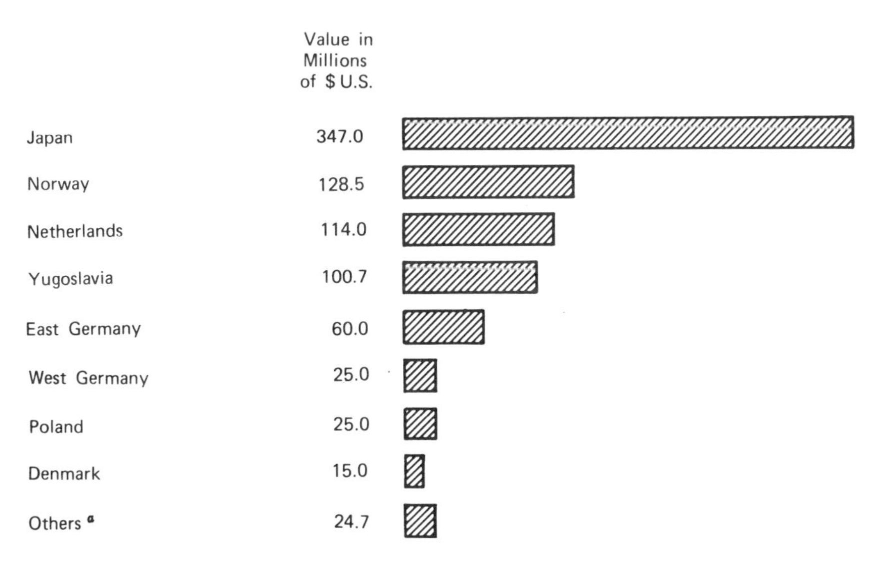

Figure 13.5 Market shares of major suppliers of ships to China during 1970–1975. (Source: OECD Commodity Trade Statistics *and United Nations trade statistics as well as some individual country foreign-trade yearbooks.*) [[a] *Others include Finland (8.8), Romania (7.7), Singapore (2.5), Italy (2.2), France (2.3), United Kingdom (0.6), Sweden (0.7), United States (0.2), Canada (0.2).*]

CIVIL AVIATION ADMINISTRATION OF CHINA

This organization (CAAC) exists since 1962 as a special agency of the State Council and combines the functions of a national airline, airport operator, and central aeronautical authority. Some of the priority functions of the CAAC are special aerial services such as application of fertilizers and pesticides, geological surveys, timberland patrols, and emergency flights. In 1964 special-purpose services occupied 33 percent of total CAAC flying time. In general, the organization is similar in many respects to Aeroflot of the Soviet Union. Operation of aircraft outside of CAAC or the military services is forbidden, though gliding probably exists as organized sport.

Civil Aviation Administration of China operations are organized geographically under six regional directorates in Canton, Lanchou, Peking, Shanghai, Chengtu, and Shenyang and are supervised by Director General Liu Tsun-hsih in Peking. The International Bureau is responsible for CAAC activities outside China and is headed by Hsiao-Feng-pu. In 1974 China began participating in the International Civil Aviation Organization (ICAO) and is one of thirty member states with a seat in the ICAO Council. China is not yet a member of the IATA organization, but discussions began in 1976.

In comparison with many other large countries, CAAC operations are relatively modest and in 1976 consisted of only 328 flights per week. This is several times fewer flights than in India, which has a comparable population and a smaller area and handles as many as 600 flights on any single day as well as operating a large international network.

During the 1950s there were two organizations providing aerial services in China. One was a Sino-Soviet Joint Stock Company SKAOGA (Sovietsko-Kitaysko Aktsionereu Obschestvo Grazhdanskoi Aviatsii), which operated on northern China, and the other was Chinese Civil Aviation Corporation (CCAC) operating in southern China. Soviet equipment and training were critical to operation of both. In 1955 the two organizations were merged into a single China Civil Aviation Bureau, which became the present CAAC in 1962.

Peking is the main aviation hub of China, and thirty-six of seventy-seven interprovince flights originate there. Shanghai is the next most important aviation center, with seventeen such flights originating there. Total domestic unduplicated routes were estimated at over 45,000 miles in 1973 and have been expanding at almost 1500 miles per year since 1950. However, it is now believed that China will first develop a more extensive international network before significantly expanding domestic services. In fact, domestic services were down from 128 in 1975 to 119 in 1976, but so far international services did not increase significantly, except for a new biweekly flight between Peking and Phnom Penh.

Air agreements exist between China and over thirty countries, but only a few are actually exploited and in most cases by the foreign airline making flights to China. International routes of importance to China are through: (1) Pakistan, Iran, Romania, Albania, Italy, and France into western Europe, (2) to the Soviet Union, neighboring communist countries, and Burma, (3) the link through Japan to Canada, which could be expanded to South America, and (4) the eastern African route through Ethiopia.

At present the CAAC operates international services only along routes to: (1) Paris and Tirana, (2) Moscow, Hanoi, Pyonyang, and Rangoon, and (3) Tokyo. Expansion of international service to and from China could in fact proceed, and the CAAC as well as the foreign air carriers appear to be limited primarily by economic considerations, mainly due to lack of sufficient traffic. In fact, during 1976 Air France reduced its flights from two to one per week. Despite existing air agreements, many other Western airlines never initiated any flights to China.

Civil Aircraft Fleet

The civil aircraft fleet includes over 600 aircraft, half of which are Soviet Antonov An-2 designs built under license in China. About 200 of these single-engine aircraft are believed to be adapted for agricultural use, and over 100 are employed in passenger operations. The An-2 can accommodate up to thirteen passengers, or 1500 km of cargo and requires only 300 m of runway. It is particularly well suited to the small Chinese airfields, and because it is so widely used it is perhaps representative of the state of Chinese aviation (see Table 13.16).

The mainstay of China's international fleet are the ten Boeing 707 aircraft, which operate to Tokyo and Tirana via Bucharest and Paris. British Tridents are used on flights to Rangoon, Hanoi, and Pyongyang, and the Soviet Il-62 long-range jetliners fly to Moscow.

The total CAAC modern jet fleet appears to consist of about fifty-five jet aircraft, thirty-eight of which are the medium-range Tridents. In this respect CAAC is comparable to the fleet of National or Western Airlines in the United States. It is also slightly larger than the jet fleets of Air India and Indian Airlines and is slightly smaller than the combined jet fleets of Varig, Trans-Brazil, and VASP airlines of Brazil.

Table 13.16 Composition of the Chinese civil aviation fleet circa 1976

Aircraft Type	*Number*	*Maximum Passenger Capacity*	*Range in Miles*	*Year Acquired*
An-2 Passenger (P.R.C.)	100+	13	560[b]	1950s
An-2 Agricultural (P.R.C.)	200+	—	560[b]	1950s
An-12 Cargo (U.S.S.R.)	—	14	2110[a]	—
An-24 Turboprop (U.S.S.R.)	25+	44	1490	1965–1973
Aero 45 Transport (Czech.)	—	5	930	1950s
Alouette III Helicopters (France)	15	6	335	1967
Boeing 707-320B (U.S.A.)	4	189	3925	1973–1974
Boeing 707-320C (U.S.A.)	6	pass./cargo	3925	1973–1974
Capital No. 1 (P.R.C.)	10–15	5–10	—	late 1950s
Concorde (France/U.K.)	3	108–128	3896	options to buy
HP Herald (U.K.)	2	56	700	—
HS Trident 1E (U.K.)	3	109	1100	1970
HS Trident 2E (U.K.)	33	149	1100	1972–1974
HS Trident 3B (U.K.)	2	179	1785	1973–1974
Ilyushin Il-12 (U.S.S.R.)	5	27	1240	1950s
Ilyushin Il-14 (U.S.S.R./P.R.C.)	58	32	920	1950s
Ilyushin Il-18 (U.S.S.R.)	14	84–110	2300	1960–1967
Ilyushin Il-62 (U.S.S.R.)	5	186	4160	1971–1973
Lisunov Li-2 (U.S.S.R. DC-3 copy)	28	25	1500	1950s
Messerschmitt BO-105 (German)	10+	5	720	1976
Mil-1 Helicopter (U.S.S.R.)	—	—	—	—
Mil-2 Helicopter (U.S.S.R.)	—	6–8	360	—
Mil-4 Helicopter (U.S.S.R./P.R.C.)	15+	—	—	—
Mil-6 Helicopter (U.S.S.R.)	5	65	652	1965
Nomad STOL (Australia)	15	15	840	1973
Peking No. 1 (P.R.C. Yak-16 like)	50	8–10	620[a]	1958–1960
Super Frelon Helicopter (France)	13	27–37	633	1974
Tupolev Tu-124 (U.S.S.R.)	2	60	1550	1970
Twin Otter (Canada)[c]	3	13–20	1103	1977
Vickers Viscount 843 (U.K.)	6	53+	1100	1961
Whirlwind Helicopter (P.R.C.)	—	—	—	—
Yakovlev Yak-12 (U.S.S.R.)	—	3	—	—
Yakovlev Yak-18 Trainer (U.S.S.R.)	—	2	560	—

Source: Compiled from various reports, articles, and publications listed at the end of this chapter.

[a] Data for Soviet design. [b] Data for Polish version. [c] To be delivered.

Airports

Only three airports in China, including those in Peking, Shanghai, and Canton, which have runways exceeding 10,000 ft in length, are suitable for largest jet aircraft. Navigational facilities at these and other airports are believed to be rather limited, and there are reports of planes being grounded by adverse atmospheric conditions. There is also apparently very little flying during the night.

A major program now underway in China appears to be modernization of airports and Plessey Company of the United Kingdom already received orders for air-traffic control and surveillance radar equipment for Peking and Shanghai airports. British, U.S., and French firms involved in airport design and construction have been approached for discussions about runway, ramps, buildings, and procedures of airport construction. Ultimately it is estimated that China will have to upgrade and modernize up to fifty airports to develop a modern air-transportation system.

Lack of navigational facilities and inadequate airports are probably the reason for a low utilization rate of aircraft now in the CAAC fleet. Observers who visited China in 1975 and later reported that the CAAC was managing only 2 h per month on its fleet of Boeing 707-320s, Tridents, and Il-62 aircraft.

One of the recent winners in airport-modernization program appears to be the French company Thompson-C.S.F., which was reported during 1977 to have received a contract for $36.6 million worth of navigational aids. This equipment is to include radars, information-processing systems, display terminals, and other support equipment for aerial navigation control.

AIRCRAFT PRODUCTION AND IMPORTS

China's technological lag is particularly noticeable in the aircraft-manufacturing industry, which at present appears to be in the process of extensive modernization. Some analysts suggest that faced with limited advanced technological resources in earlier years China concentrated on development of nuclear and missile industries and licensed production of military aircraft to the detriment of its aircraft industry as a whole and the transport aviation sector in particular.

Despite its "self-sufficiency," policies China has so far failed to produce its own jet transport airliner or jet engine, although several piston engine aircraft have been designed and built by the Chinese. This does not mean that China does not have jet-aircraft manufacturing experience and know-how, and large numbers of obsolescent jet fighters and jet engines have been produced under Soviet licenses. A supersonic twin jet fighter aircraft has also been developed by China in recent years, but advanced jet-engine technology is still lacking, and this deficiency is being met by imports and licenses for Rolls Royce turbofan jet engines.

Aircraft Manufacturing

During 1958–1960 a piston-engine twin passenger aircraft similar to the Yakovlev Yak-16 but designed by the Peking Institute of Aeronautical Engineering was developed in China and subsequently went into production. It is known as "Peking No. 1" and is capable of carrying eight to ten persons, and as many as fifty units are believed to be in service. This would make it the most popular Chinese designed and built aircraft in the civil fleet.

Another original Chinese design also developed in Peking at the Aviation College was known as the "Red Banner No. 1," a training aircraft, but no details of its manufacture are available. Other Chinese aircraft designs include the "Heilungkiang No. 1" prototype developed at the Harbin Engineering College, which resembles the Soviet Yakovlev Yak-12 aircraft; the "Capital No. 1" twin-engine transports based on Soviet Antonov An-14 design, of which ten to fifteen units are believed to be in service; the "Shenyang No. 1" aircraft built by the Shanyang Aeronautical Institute; the "Yen-an No. 1" plane designed at the Sian Northwestern Industrial University; the "Flying Dragon Seaplane"; and the "Sungari No. 1" aircraft, about which no details are known.

Aircraft design and production in China are believed to be under the control of the Seventh Ministry of Machine Building headed by Minister Wang Yang. This ministry is a major element in the military–industrial complex of China and is probably also responsible for construction of missiles and satellites. Its military orientation is well reflected in the fact that most of the aircraft production is concentrated on jet fighters and some light and medium bombers based on obsolescent Soviet designs of the 1950s (see Table 13.17).

The magnitude of Chinese aircraft industry production effort can best be judged by analysis of Chinese air force equipment. The International Institute of Strategic Studies in London estimates a total of 5650 aircraft operated by the Chinese air forces in 1976 and believes that most of the 600 aircraft of the Civil Aviation Administration of China are probably available to the military in emergencies as an auxiliary transport fleet.

The largest and best known of Chinese aircraft factories is the Shenyang State Aircraft Factory in Liaoning Province. This plant was originally the Manchu Aeroplane Company built by the Japanese in 1938 and since the 1950s has produced several Soviet aircraft types under license for use in China as well as for export to Albania, Cambodia, Vietnam, Tanzania, and Pakistan.

Table 13.17 Types of aircraft manufactured in China since 1950

Aircraft Type	*Basic Design Source*	*Manufacturing Location*
Capital No. 1	An-12-based twin	Peking
Chinko No. 1	Yak-12 license	Shenyang
Feng-Shou No. 2	An-2 licenses	Shenyang
F-2 Fighter Trainer	MiG-15 UT license	Shenyang
F-4 Fighter	MiG-17 license	Shenyang
F-6 Fighter	MiG-19 license	Shenyang
F-8 Fighter	MiG-21 license	Shenyang
F-9 Mach 2 Fighter	Chinese twin jet design	Shenyang
Flying Dragon Seaplane	Chinese design	unknown
Heilungkiang No. 1	Similar to Yak-12	Harbin
Il-28 Light bomber	Soviet Il-28 license	Shenyang
Mi-4 Helicopter	Mi-4 license	unknown
Peking No. 1	Chinese twin transport	Peking
Red Banner No. 1	Chinese trainer design	Peking
Sungari No. 1	Chinese design	unknown
Super Aero 45	Czech Aero 45 license	Harbin
Shenyang No. 1	Chinese prototype	Shenyang
Tu-16 Medium bomber	Soviet Tu-16 copy	Sian
Yak-18 Trainer	Yak-18 license	Shenyang
Yin-an No. 1	Chinese prototype	Sian
Whirlwind Helicopter	Chinese design	Shenyang
Agricultural aircraft	New Chinese design 1977	unknown

Source: Compiled by 21st Century Research from various reports and publications listed at the end of this chapter.

About 4500 of all the aircraft in China are fighter planes produced under Soviet licenses for the Chinese Air Force and the Chinese Naval Air Force, which operates about 700 ground-based aircraft in support of the Navy. Over 1500 F-6 fighters based on Soviet MiG-19 design and over 1000 F-4 planes based on the MiG-17 have been built, representing the largest aircraft production runs in China. In addition, several hundred F-2 and F-8 aircraft based on MiG-15UT trainer and MiG-21 fighter, respectively, have also been produced in Shenyang. The latest fighter plane, designated F-9, is believed to be an original Chinese design capable of Mach 2 speeds and may become powered with the Rolls-Royce Spey jet engines for which licenses were purchased in 1975 (see Table 11.10).

Another aircraft built in large quantities by the Shenyang plant is the Antonov An-2 transport, apparently known as "Feng-Shou No. 2" in its Chinese version. About 400 An-2 may have been produced for use as civilian and military transports as well as agricultural aircraft. The Chinese air forces also include several hundred of the Ilyushin Il-18 light bombers, most of which have been built under Soviet licenses at the Shenyang plant.

A large aircraft-manufacturing center also exists in Sian in Shensi Province, where the Chinese version of the Soviet Tupolev Tu-16 medium bombers was reported in production at the rate of six aircraft per month in 1973. Licensed Rolls-Royce Spey jet engines will be manufactured at this plant.

Harbin Aircraft and Aeroengine Plants are another major aircraft-design center believed to have produced the Super Aero 45 twin transports and its engines under licenses from Czechoslovakia. This plant also built the "Heilungkiang No. 1" aircraft similar to the Soviet Yak-12 design. There is also an aircraft and aeroengine plant at Chungking and a glider-manufacturing factory in Chengtu in Szechwan Province.

The aircraft industry is critically dependent on availability of modern jet engines for production of more advanced aircraft. In this respect the Chinese themselves admitted that they may have been 20 years behind in technology in comparison to Western aeroengine manufacturers. This is why the purchase of licenses to manufacture Rolls-Royce Spey turbofan jet engines is viewed as an important step in modernization of the Chinese aircraft industry and its air force as well. The Spey engines are used in transport and military fighter aircraft in the West, and their manufacture in China will reduce the technology gap in Chinese jet-engine capabilities to about 10 years behind the West depending how soon their production gets under way.

The other prominent gap in Chinese aircraft industry is the lack of large jet-transport manufacturing capability. Such aircraft are vital to modern military forces as well as to efficient air-transportation systems. Domestic production of more advanced helicopters is also believed to be receiving priority in China's aircraft-industry modernization.

Aircraft Imports and Markets

During the last decade China spent over $600 million for imports of aircraft and associated aviation equipment from the Soviet Union, the United Kingdom, the United States, France, and to a lesser degree West Germany and Australia. During 1965–1975 imports of aviation equipment from the Soviet Union alone amounted to over 50 percent of all such imports, and the Soviet Union consistently remained the largest or second largest supplier. In 1976 China imported a total of $137 million of aircraft, of which $60 million came from the Soviet Union and another $57 million represented Trident deliveries from the United Kingdom.

During 1975 aircraft shipments from the United Kingdom were leading with $60 million also as a result of deliveries of many of the Trident jet transports ordered in previous years. In 1973 the United States was momentarily the largest supplier when Boeing 707 transports were being shipped to China, but even then the Soviet Union shipped $55.8 million of equipment, only $3 million less than Boeing shipments that year. In 1974 the Soviet Union was once again in the lead with $67.4 million of shipments (see Fig. 13.6).

China first purchased aircraft in the West in 1961, when it acquired six Vickers Viscounts 843 turboprop transports. During the 1960s interest was shown in the purchase of jet transports such as the French Caravelle, the British BAC-111, Vickers VC-10, and the De Havilland Comet. But all such sales were opposed by the United States through the COCOM export controls committee. As a result of this policy China did not obtain jet-transport aircraft until 1970. There are unconfirmed reports that two Tupolev Tu-134 medium-sized jet transports were ordered in 1965 from the Soviet Union, but it is not clear whether these aircraft were ever delivered to China. Disruptions of the Cultural Revolution may have led to the cancellation of those orders, but a refusal to ship the aircraft to China by the Soviets cannot be ruled out. There are also reports that the Soviets refused to sell China supersonic fighter planes such as the MiG-25 and certain Sukhoi models.

In 1970 China placed an order with Aviaexport of the Soviet Union for five Ilyushin Il-62 long-range jet transports and also purchased four used Hawker Siddeley Tridents from Pakistan International Airways. Although the Soviet Union has been the leading, if not the sole, aircraft

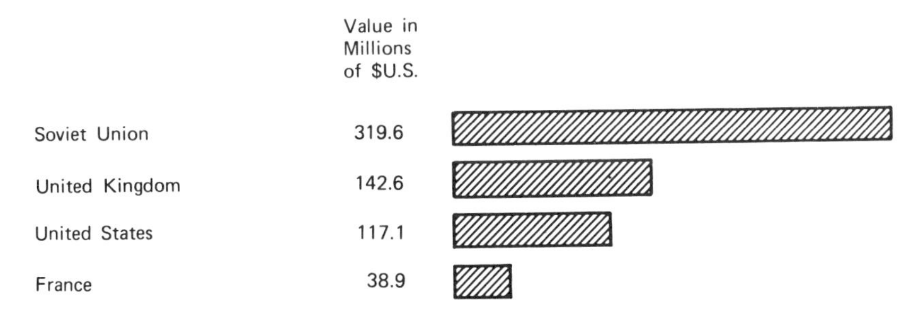

***Figure 13.6** Relative market shares of suppliers of aircraft equipment and facilities to China during 1965–1975.* (**Source:** *Based on export statistics for individual countries reported under SITC commodity code 734 or equivalent for the Soviet Union.*)

Table 13.18 Major sales and orders of aircraft and aviation equipment to China identified since 1961

Year	*Type of Equipment Sold*	*Quantity*	*Supplier and Country*	*Contract Value in Millions of $U.S.*
1961	Viscount 843 turboprops	6	Vickers Aircraft, U.K.	
1965	Two An-24, two Tu-124, five Mil-6	9	Aviaexport, U.S.S.R.	
			Sud Aviation, France	
1967	Alouette III helicopters	15		
1970	Ilyushin Il-62 jet transports	5	Aviaexport, U.S.S.R.	
	Trident 1 jet transports	4	PAI, Pakistan[a]	
	Radio altimeters		T.R.T., France	
1971	Trident 2E jet transports	6	Hawker Siddeley, U.K.	46.0
1972	Concorde (preliminary)	2	Aerospatiale, France	70.0
	Concorde (preliminary)	1	British Aircraft Corp., U.K.	
	Trident 2E jet transports	6	Hawker Siddeley, U.K.	57.0
	Boeing 707 jet transports	10	Boeing Airplane, U.S.A.	125.0
	JT3D turbofan jet engines	40	Pratt & Whitney, U.S.A.	21.0
	Trident jet transports	8	Hawker Siddeley, U.K.	65.0
	Boeing 707 & Trident simulators		Redifon Ltd., U.K.	4.8
1973	Trident support systems		Lucas Aerospace Ltd., U.K.	6.8
	An-24 transports	30	Aviaexport, U.S.S.R	
	Nomad STOL aircraft	15	GAF, Australia	
	Aircraft finishing machines	7	Ratier-Forest, France	4.2
	Trident 2E jet transports	15	Hawker Siddeley, U.K.	116.2
	JT15D jet engines	16	United Aircraft, Canada	2.0
	Inertial navigation systems	8	Litton Industries, U.S.A.	1.0
1974	Super Frelon helicopters	13	Aerospatiale, France	1.9
	Sets Trident equipment	15	Smith Industries, U.K.	6.8
	Viscount aircraft servicing		Hong Kong Aircraft Eng.	0.8
	Aircraft cables for Tridents		BICC, U.K.	
	ILS Landing systems	11	Plessey Navaids, U.K.	2.04
1975	Spey jet engines and license		Rolls-Royce, U.K.	150.0
	Spey engine reheat systems		Dowty Fuel Systems, U.K.	
1976	BO-105 Helicopters	4	Messerschmitt, West Germany	
	Computerized navigation system		S.F.I.M., France	4.2
1977	Air navigation control equip.		Thomson-CSF, France	36.6
	Allison 250 jet engines		HDH, Australia	0.3
	Aircraft wheel-testing systems		Vickers, U.K.	9.8
	Twin Otter survey planes	3	De Havilland, Canada	6.0

Source: Compiled by 21st Century Research from reports, articles, and publications listed at the end of this chapter.

[a] These believed to have been in exchange for a squadron of Chinese-built MiGs.

supplier to China for many years, its shipments dropped to only about $5 million per year in 1969 and 1970, but with the latter order they recovered quickly, to almost $34 million already by 1971.

It is difficult to determine whether the availability of Soviet jet transports in 1970 played a role in relaxation of Western export controls policies, because China was always importing and manufacturing licensed Soviet aircraft and aeroengines of many types. However, in 1971 China was able to order six Trident 2E jet transports directly from Hawker Siddeley in the United Kingdom for $46 million. In 1972 an additional six Tridents and parts were purchased for $57 million, and later that year eight more Tridents, valued at $65 million, were purchased. In 1973 China ordered fifteen more Trident 2Es for another $116.2 million, for a total of thirty-five modern medium jet transports. (See Table 13.18).

The most spectacular Chinese aircraft purchase came in the summer of 1972, when China placed preliminary orders for three Concordes with Aerospatiale of France and British Aircraft Corporation of the United Kingdom for a total of $70 million. China still retains its options to buy the Concordes, although the traffic on its long-distance international routes cannot justify such an aircraft. Some analysts believe that China will purchase the Concordes eventually and adapt the supersonic transports as manned strategic-weapons delivery systems as an alternative to its intermediate-range ballistic missiles and the intercontinental ballistic missile under development. This is a plausible argument because China's strategic bombers, which form the mainstay of its strategic-weapons delivery system, are Soviet designed Tu-16 medium-range bombers with neither the range nor the speed advantages offered by an aircraft like the Concorde.

In September 1972 China again placed an order in the West with Boeing Aeroplane Company of the United States for ten Boeing 707 jet transports for a total of $125 million and also ordered 100 percent of spare Pratt & Whitney JT3D turbofan jet engines from United Technologies for another $20 million. This in itself may be an indication that long service life is projected for the Boeing 707 transports, with few prospects for additional sales of similar aircraft in the future (see Table 13.19).

By the end of 1973 China ordered fifteen Nomad STOL (short takeoff and landing) aircraft manufactured by the Government Aircraft Factory (GAF) in Australia. This twin-engine transport has conversion capa-

Table 13.19 Market shares of major suppliers of aircraft engines SITC Commodity Code 711.4 values in millions of current U.S. dollars

Country	*1973*	*1974*	*1975*	*1973–1975*
United Kingdom	11.4	10.4	17.4	39.2
United States	3.9	16.2	—	20.1
Soviet Union	—	—	6.2	6.2
France	—	0.8	1.4	2.2
Canada	—	0.1	1.0	1.1

Source: United Nations Bulletin of Statistics on World Trade in Engineering Products for 1973, 1974 and 1975.

bility from civil to military roles, and its purchase by China is curious because a similar twin-engine feeder transport known as the Norman-Britten Islander is manufactured under license from the United Kingdom in Romania, with whom China maintains extensive trade relations and where it could purchase a similar aircraft without expenditure of hard currencies.

In modernizing its aircraft industry China is clearly seeking equipment with military applications, and chances are that further sales will be made if strategic trade export controls are not strictly enforced against China. There are already indications that sales of equipment with military potential to China are treated more leniently by western European countries and the United States than they would be if such sales were proposed to the Soviet Union or eastern Europe.

Recent sales of French and West German helicopters are also believed to be of military value to China, although ostensibly both transactions involved aircraft for civilian and off-shore oil production applications. Aerospatiale of France sold thirteen Super Frelons helicopters in 1974 for $1.9 million, and unspecified aircraft imports from France amounted to $28 million in 1975 and $16 million in 1976. Messerschmitt-Boelkow-Blohm of West Germany sold four BO-105 helicopters in 1976. According to the Strategic Survey report for 1976 issued by the International Institute for Strategic Studies in London, China also placed an order for an additional hundred BO-105 helicopters equipped for military missions, which was approved by the West German Bundestag Defense Committee. A British manufacturer, Westland Aircraft, was permitted to show its 606 helicopter in Peking during a British National Exhibition in China, also clearly with sales of this aircraft in mind. As for Moscow's reaction to these sales, one Soviet broadcast in 1974 stated that the United States had already sold China a complete helicopter-assembly plant.

Chinese interest in military aircraft has been further demonstrated on several occasions by specific inquiries in recent years. One of the most publicized was the interest shown during 1974 and again more recently in the Hawker Siddeley Harrier VTOL (vertical takeoff and landing) fighter plane manufactured in the United Kingdom but also used by the U.S. Marines. Despite long-standing embargoes on sales of such equipment to any communist country, neither Whitehall nor the U.S. export-control agencies have made any major comments on such proposals. Serious interest was also shown by the Chinese in the Lockheed C-141 military transport plane and a Japanese-made aintisubmarine patrol aircraft.

Because of a very low utilization factor and a relatively significant fleet of imported jet transports, China does not appear to be an immediate market for additional aircraft of this type. However, the exception may be the wide-body jet transport, which may be of interest as a possible convertible cargo–military transport that could contribute immediately to the modernization of China's military forces by providing additional mobility to large numbers of manpower and military equipment.

Neither the present air traffic or the state of Chinese airports can justify the use of "jumbo" jet transports in civilian aviation alone. In this respect the European Airbus 300 aircraft produced jointly by France and West Germany is probably most interesting to the Chinese. The possibility of subassembly of such an aircraft in China should not be ruled out.

Interest was also shown by the Chinese in the VWF-614 medium transport manufactured by the Netherlands and West Germany and in the Fletcher FU-24 agricultural aircraft manufactured under American license by Aerospace Industries of New Zealand. This latter interest may signal a need to upgrade the relatively old An-2 design manufactured under Soviet licenses in China since the 1950s, which performs almost all the agricultural flying in China today. During 1977, however, Chinese media indicated the construction in China of a new agricultural aircraft prototype.

The most recent aviation sales to China were made by De Havilland of Canada and by Hawker De Havilland of Australia in 1977. The Canadian firm sold three Twin Otter survey planes whereas the Australian company received a contract to supply two Allison 250 gas-turbine jet engines valued at $318,000 for the BO-105 helicopters previously purchased from Germany for offshore oil-rig work. Hawker De Havilland holds options for sixteen additional units and proposed to set up an Allison engine-overhaul facility in Peking. Servicing foreign aircraft in or for China may in fact develop into an additional business for Western suppliers as China's air fleets expand in the future. Hong Kong Aircraft Engineering already won a contract in 1974 worth $800,000 to service the venerable Vickers Viscounts transports flying with CAAC since 1961.

REFERENCES

Aviation Week. "China Reviews Aerospace Requirements," June 2, 1975, p. 275.

Baranson, Jack. "The Automotive Industry," in William W. Whitson, ed., *Doing Business with China,* Chapter 13, Praeger, New York, 1974.

BBC Summary of World Broadcasts. "Yugoslavia Shipbuilding," FE/W943/A/19, August 24, 1977.

Centre Francais du Commerce Exterieur. *Nouvelles Commerciales,* DRGS 2eme Trimestre 1977, Paris.

Chambers, David. "Civil Aviation in the PRC," *Current Scene,* August 1974, Hong Kong.

Chiang, Huai. "Communist Activities in Communications," June 10, 1972 (transl. by JPRS #57106), Taipei, Taiwan.

Chiang, Huai. "The Situation in Communist Communications," June 10, 1972 (transl. by JPRS #57106), Taipei, Taiwan.

China Briefing. "China's Automotive Industry," October 22, 1974.

China Trade Report. "Extending China's Ports," March 1976, p. 6, Hong Kong.

China Trade Report. "China's Tanker Fleet: Growing," January 1977, p. 6, Hong Kong.

China Trade Report. "Business Pipeline–Aviation," February 1977, p. 3, Hong Kong.

Current Scene. "PRC Highway Network," March 1977, p. 23, Hong Kong.

Current Scene. "CAAC Air Routes," April–May 1977, p. 22, Hong Kong.

Europa Yearbook 1977. "PRC, Transport and Tourism," London.

Forbes Magazine. "US Arms for Red China?," June 1, 1976.

Green, Stephanie R. "When Will Your Ship Come In?," Part I, *U.S.–China Business Review,* November–December 1975, p. 14.

Green, Stephanie R. "When Will Your Ship Come In?," Part II, *U.S.–China Business Review,* July–August 1976, p. 26.

Heine, Irwin Millard. "China's Merchant Marine," *U.S.–China Business Review,* March–April, 1976, p. 7.

Heyman, Hans. "The Air Transport Industry" in William W. Whitman, ed., *Doing Business with China,* Chapter 12, Praeger, New York, 1974.

Huenemann, Ralph W. and Nicholas H. Ludlow. "China's Railroads," *U.S.–China Business Review,* March–April 1977, p. 27.

International Institute for Strategic Studies. *"Military Balance 1976–1977,"* p. 50, London.

Jane's "All the World's Aircraft," 1966–1977 yearbooks, London.

Ministry of Foreign Trade, *Commertul Exterior al Republicii Socialiste Romania 1974,* Central Statistical Department, Bucharest, Romania.

Pillsbury, Michael P. "Future Sino-American Security Ties," *International Security,* Spring 1977.

Porch, E. Harriet. *"CAAC–the Airline of Communist China,"* Rand Corporation Report #P-4184, September 1969.

U.S.–China Business Review. "Transportation and Communications Equipment Sold to China," September–October 1974, pp. 8–15.

U.S.–China Business Review. "International Air Routes to Peking," Map, January–February 1975, p. 77.

U.S. Government. "Value Added by Work Brigades in Railroad and Highway Construction in China in 1952–1957," *Research Aid,* A(ER)75-74, November 1975.

U.S. Government. "Chinese Merchant Ship Production," *Research Aid,* ER 76-10193, March 1976.

U.S. Government. "China: International Trade, 1976–77," *A Research Paper ER77-10674,* November 1977.

Vetterling, Philip W. and James J. Wagy. "China: The Transportation Sector, 1950–1971," in *PRC: an Economic Assessment,* Joint Economic Committee, U.S. Congress, May 1972.

Wettern, Desmond. "China to Buy First British War Material," *Sunday Telegraph,* July 13, 1975, London.

Wilson, Dick. "Modernising China's Transport and Communications," *Financial Times,* November 1, 1974, London.

14

Communications

One of the priorities of the fourth Five Year Plan (1971–1975) was the development of a modern telecommunications system to support China's increasing role on the international scene. It is also aimed at ironing out many inefficiencies within a rapidly growing and industrializing economy.

Control of the means of communications is vital to political power and in China the government and the military establishment supervise all such facilities under the guidance of the leadership of the Chinese Communist Party. The responsibility for the operation of most facilities is delegated to the Ministry of Posts and Telecommunications, headed by Minister Chung Fu-hsiang, a former PLA signal-corps general. The ministry operates the mails as well as telephone, telegraph, telex, and facsimile services. It also operates international cables and through INTELSAT, the communications satellite services.

There are additional specialized telecommunications networks operated by the Ministry of National Defense and the Ministry of Foreign Affairs, and by weather, shipping, and news organizations. Special agencies and bureaus of the State Council are responsible for radio and television broadcasting, news gathering and dissemination, and all publishing activity in China.

In comparison with other large and populous countries, China's communication systems are limited. Except for private radio receivers all other systems are primarily intended to meet the needs of the state and industry. With an estimated 5 million telephones China has the same number as Sweden, Australia, or the Netherlands. Relative to its large population, however, China has only one telephone for 200 inhabitants. This is still considerably better than India with 330, or Indonesia with 400, inhabitants per telephone (see Table 14.1).

Use and production of radios increased rapidly during the 1970s.

Table 14.1 Comparative use of electronic communications equipment by selected countries in 1976 unless otherwise noted

Country	*Telephones in Use (Millions)*	*Radios in Use (Millions)*	*Radio Stations, AM + FM*	*Television Sets in Use (Millions)*	*Television Stations*	*Earth Satellite Antennas*
China	5.0	45.0	251	0.35	105	3
United States	147.0[a]	402.0[b]	7549	121.1[b]	940[b]	14
Soviet Union	17.7[b]	57.1[b]	3034[c]	52.5[b]	1466[c]	2[d]
Japan	41.9	76.0[e]	889[c]	25.5[b]	4991[c]	2
West Germany	20.4	21.5	219	19.5	2350	4
France	14.1	18.5	136	15.0	1473	5
United Kingdom	22.4	41.7	215	18.7	300	3
Brazil	3.45	32.0	494	11.0	79	5
Poland	2.18[b]	5.92[b]	51[c]	6.1[b]	52	1
India	1.82	14.1	172	0.275[b]	7	2
Mexico	2.96	19.0	683	4.9	163	1
Indonesia	0.305	5.0	151	0.3	13	2 + 40[f]
Nigeria	0.115	5.0	31	0.1	8	1 + 6[f]
Taiwan	1.1	3.0	117	2.9	3	2

Source: United Nations Statistical Yearbook, 1976; U.S. Government Basic Intelligence Factbook, July 1977.

[a] Data for 1975. [b] Data for 1974. [c] UNESCO data for 1971–1972 denoting number of transmitters. [d] Civilian use satellite stations only. [e] Data for 1973. [f] Domestic satellite communication systems earth stations.

Besides an estimated 45 million receivers about 140 million wired-speakers are deployed, particularly throughout rural China. Television is in its infancy and is growing rapidly. China may become a first major nation to move into color television on a massive scale without having developed black-and-white systems to any significant level.

Production of communications equipment is relatively well developed in China. Although the Ministry of Posts and Telecommunications operates research and manufacturing facilities, most important production comes from the radio equipment and electronics plants under the supervision of the Fourth Ministry of Machine Building. This is part of the military–industrial complex of China, and up to 66 percent of its output is still believed destined for military use.

Imports of communications equipment have been relatively small in comparison with imports of other industrial products. During 1965–1975 China imported only about $64 million of equipment identifiable as telecommunications products under SITC commodity code 724. Because of the importance of this industry to national defense, China is committed to development of its own equipment. Thus "prototype purchasing" for research and development purposes will continue from all countries with advanced electronic industries. Recent imports from the Soviet Union and particularly East Germany appear to be a new development in this market.

MINISTRY OF POSTS AND TELECOMMUNICATIONS

This ministry came into being in 1949. By 1952 it succeeded in restoring a telecommunications system in China, comparable to that existing in 1936. Since then it has been expanding a telephone network to serve numerous outlying construction sites and communes as well as a growing industrial demand.

In 1971 the ministry was abolished, and its function was taken over by a Telecommunications General Administration of China and the Postal General Administration of China, which was under joint military and civilian control. In 1973 the ministry was reestablished and now operates through the Directorate General of Telecommunications and a Directorate General of Posts. The ministry also has a Foreign Affairs Bureau headed by Chiao Wei-chung as director.

There is a Scientific and Technical Committee attached to the ministry, which presumably is concerned with research and development of the most advanced methods and technology. It probably supervises the work of several institutes and colleges involved in training and research. Peking Institute of Posts and Telecommunications Research, established in 1954, appears to be the leading institution and was instrumental in developing the original 12-year scientific development plan formulation for this industry. The institute also developed prototypes of facsimile machines, radio transmitters, telephone switchboards, and 24-channel microwave equipment.

The Peking College of Posts and Telecommunications is another institute that began operation of a television station in 1956 and developed some color television equipment as early as 1958. The Chengtu College of Telecommunications Engineering in Szechwan Province specializes in long-distance telecommunications, municipal communications facilities, and design of radio-measuring instruments. Two other Colleges of Posts and Telecommunications have been identified in Nanching in Kiangsu Province and Chungching in Szechwan Province.

The Ministry operates through offices and facilities at the province, special district, county, and municipality levels. Besides day-to-day operation of the postal and telecommunications services, it also produces, distributes and maintains the equipment, trains operators, and supplies the maintenance engineering support.

China was admitted to the International Telecommunications Union

(ITU) in 1972 after the expulsion of Taiwan and has taken an active role in ITU activities. It is also a major user of the International Telecommunications Satellite Organization (INTELSAT) but is not a member of this group, presumably because Taiwan retained its membership in that organization. In 1973 China also became a member of the Asian Broadcasting Union and the Universal Postal Union (UPU). China's membership in other international communications organizations is expected to follow according to its needs and the existence of an acceptable political setting.

TELEPHONE NETWORK

One of the most useful indicators of the state of modernization of an economy is its telephone network, which is indispensable to efficient operation of modern enterprises. Latest estimates for China are 5 million telephones, which are believed to provide a reasonably adequate service in urban and industrial areas for domestic and international communication. This means that there is only one telephone per 200 inhabitants, which is more than in India but considerably less than in Mongolia, South Korea, Brazil, or Taiwan. There appear to be thirteen times as many telephones in the Soviet Union for the same number of inhabitants and almost 140 times as many in the United States. On the other hand, the impact of the 5 million telephones on the Chinese economy is probably much greater than would appear at first sight because practically all of them are in official use and only the highest-ranking officials have telephones in their private residences (see Table 14.2).

The Chinese telephone system is believed to meet the basic needs of the government, industry, and armed forces, and the minimal need of

Table 14.2 Telephone usage in selected countries in 1975 in number of units per 100 inhabitants

Country	*No. Units*	*Country*	*No. Units*
China	0.5	Taiwan	6.9
		Soviet Union	6.6
United States	69.5	South Korea	4.0
Japan	40.5	Brazil	3.1
United Kingdom	37.9	Mongolia	2.1
West Germany	31.7	India	0.3
France	26.2	Indonesia	0.2
Poland	7.5	Nigeria[a]	0.2

Source: United Nations Statistical Yearbook, 1976.

[a] Data for 1974.

agricultural communities. More than 98 percent of the communes have been reported to have telephone service. Telephones are made available on a priority basis for party, military, government, industrial, and commune officials. In recent years an increasing number of public pay phones have been installed in public parks, post offices, shops, and on the streets.

The Chinese bureaucracy requires extensive communications, meetings, and exchanges of information. As a result of this demand a common feature of the telephone service is the telephone conference. It is preferred in place of time-consuming and costly meetings and is regarded as a means of saving other scarce personal resources such as travel space and accommodations.

In 1975 approximately 80 percent of China's cities were believed to be equipped with automatic dial systems. First long-distance direct dialing began in 1969 between Peking and Tientsin. Local services are generally rated as poor, but long-distance services are considered much better as a result of recent installation of modern transmission equipment. In small cities and rural areas there are still magneto-type manual exchanges in operation. Other exchanges manufactured in China are based on obsolete Soviet and eastern European designs.

At the end of 1974 China claimed a telephone network of over 4 million kilometers of long-distance wire telephone lines. Main trunk routes have carrier multiplex and can provide twelve two-way voice channels on a single wire pair. Secondary routes have three-channel and single-channel voice-carrier systems. The long-distance network is a "four-level converging radial system." Controls are at the inter- and intraprovince and inter- and intracounty levels.

Telephone switching is performed by a network of regional offices located at various administrative levels. In 1975 it was estimated that about 3000 local and toll telephone exchange offices existed in the cities. A minimum of 40,000 lower-level telephone exchanges were also believed to be in operation.

Chinese authorities appear to be aware that a modern telephone switching system is essential for efficient use of high-capacity telecommunications transmission networks. Since 1971 Chinese study groups have been looking at the least technology available in the West. There is speculation among industry observers as to whether the Chinese will decide on the crossbar or electronic switching equipment to modernize their system.

RADIO AND TELEVISION

China's broadcasting system consists of radio, television, and wired-speaker broadcast facilities. With 45 million receivers China is the fifth largest radio user in the world after the United States, Japan, the Soviet Union, and the United Kingdom. There are three times more radios in China than in India, and there are also an estimated 140 million wired speakers deployed throughout the rural areas of the country, which means practically one speaker for every Chinese peasant family.

Two hundred and fifty AM and two FM radio stations are operating in China. In 1975 the total power output of Chinese radio transmitters was estimated at over 5 MW. The most important radio station is the Central People's Broadcasting Station in Peking, which uses at least twenty-five shortwave and about thirty-five medium-wave transmitters for domestic broadcasting. Local radio stations receive Peking broadcasts and retransmit them through the AM facilities in their region. China also operates an international broadcasting service, using over sixty shortwave transmitters in the Peking area, some provincial and regional relay stations and at least four transmitters in Tirana, Albania, which relay Peking broadcasts to Europe, Africa, and the Americas.

China's radio production was relatively constant at about 1.5 million receivers throughout the 1960s. In 1970 it suddenly tripled in one year and doubled again to 8,064,000 units in 1973, coinciding with a large upsurge in electronics production in China. By 1975 China was producing an estimated 18,000,000 radio receivers and became the second largest radio producer in the world after the United States (see Fig. 14.1).

However, it must be remembered that since the early 1970s radio production in all Western countries, Japan, and even Bulgaria and Czechoslovakia has been declining, as the electronic industries of these countries expanded their production of television, computers, and other more sophisticated equipment. With a huge domestic market China will probably become the largest radio manufacturer in the world in a few years' time. Economies of scale available with such large production volume may also establish China as an extremely competitive exporter of transistor radios throughout the world.

The wired broadcasting network is designed to provide a direct radio link between Peking and the heavily populated rural areas. The 140 million speakers estimated to be deployed throughout rural China reach about 90 percent of all the production teams and brigades and approximately 65 percent of households in rural areas. Radio broadcasts reach primarily the heavily populated urban areas in China. The wired-speaker systems are operated at the county level and originate from the county broadcasting station. Relay stations at communes connect production brigades, teams, and households into the system. The development of this massive wire broadcasting system has been an impressive achievement of rural China. Most power comes from the 60,000 small hydroelectric power stations built by local units. During construction periods thousands

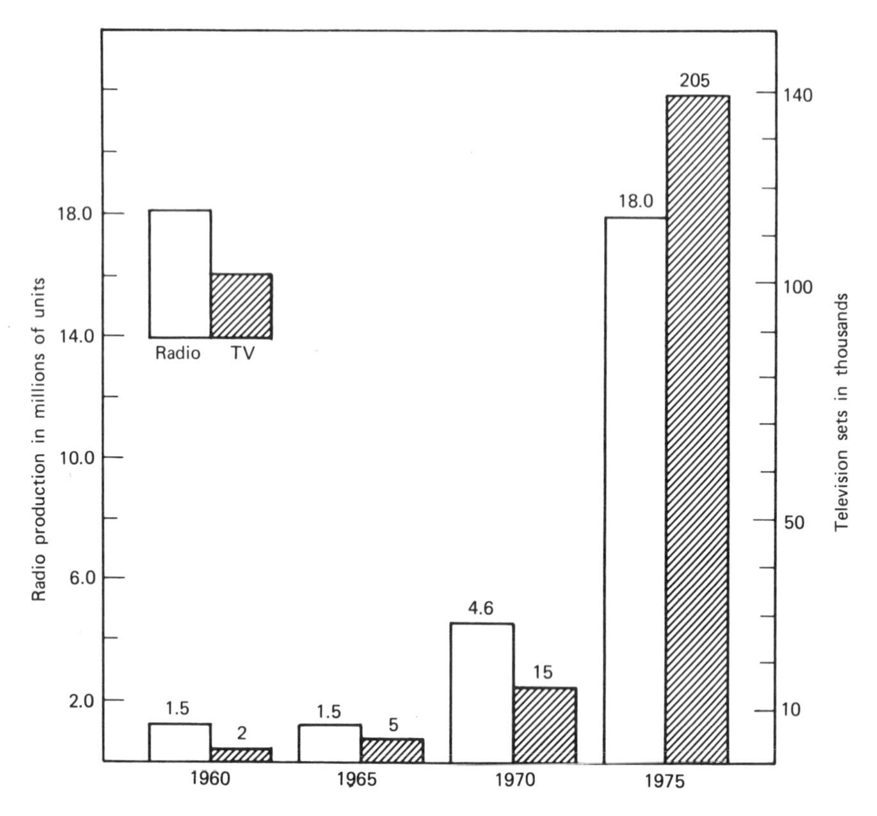

***Figure 14.1** Growth in production of radios and television.* (**Source:** *U.S. Government International Handbook of Economic Statistics, 1976 and 1977.*) ([a] *Projection from 115,000 in 1974 using growth rate similar to that of 1973–1974.*)

of peasants have been pressed into service to construct wire broadcasting lines, and in many areas manufacturing facilities were organized to produce loudspeakers and other radio-receiving equipment. At the height of development of this program in 1973 almost 39 million loudspeakers were produced in China.

In contrast, television is only in its infancy in China. There are an estimated 350,000 receivers, which is comparable to the number installed in India or Indonesia. The Soviet Union has about 150 times more television sets than China (see Table 14.1). On the other hand, production of television sets increased by 25 percent in 1973–1974 and is estimated to have reached about 205,000 units by 1975. This is already comparable to 50 percent of television production of Hungary or about one tenth the television production of United Kingdom (see Fig. 14.1). Current television sets production levels are estimated on the basis of an assumption that China has a single television tube plant with Soviet equipment with a capacity to produce 150,000 tubes per year. It is obvious that given the commitment to national television, China is expanding this capacity. Industry observers believe that the next television tube plant will produce color tubes and that China will not increase its production of black-and-white sets. A small production of color-television sets, estimated to be in the order of 5000 units per year, is already under way.

Since 1969 a national television network has been developed and by 1975 all provincial capitals except Lhassa in Tibet had broadcasting stations. Television in China primarily carries live programs from Peking, and only stations in Peking, Shanghai, Canton, and probably Tientsin are considered capable of transmitting programs to other stations through microwave systems. Television reception is being extended beyond the transmitter range of main provincial stations by numerous rebroadcasting stations.

Practically no television receivers are owned by private individuals. Most are installed in public meeting places, factories, military units, communes, and schools. Typically, at least 100 persons are watching such a television set, not unlike standing crowds in town or village squares in Mexico or other Latin American countries. High government officials and model workers are the first to receive private receivers, apparently as reward for unusual performance. Ability to obtain a color receiver may indeed become an important work incentive in the future, particularly among peasants in outlying areas who may even be illiterate.

Indications are that China will choose the PAL (phase alternation, line) color television broadcasting system developed by AEG-Telefunken in West Germany. Trial color broadcasts have begun at the Peking television station using the PAL process back in 1973 and later in Shanghai and Canton. Several color-television studio equipment items, relay vehicles, and other units have been purchased also from Japan, and the United Kingdom.

In September 1976 China claimed to have produced its first color videotape at the Wuhan Experimental Factory. Continuing developments in color-television equipment and transmission suggest that very much attention is being paid to all aspects of color-television industry. Observers believe that color television is indeed a priority item for China's electronic industry. They also expect that television-receiver production may reach 1 million units in 1980, which would make China one of the leading color-television manufacturing countries in the world. This would mean that a color-television receiver plant using domestic and imported color-television tubes is already operating. With the massive domestic market China may also obtain considerable economies of scale in production of color-television receivers and become a factor in the international color-television export market.

INTERNATIONAL RADIO, CABLE, AND SATELLITE COMMUNICATIONS

Since China's admission to the United Nations in 1971 many countries established diplomatic relations with Peking. Subsequent expansion of demand for communications with China and increasing foreign trade forced China to expand its international communications facilities.

Until 1972 high-frequency (HF) radio and wireless handled most of China's international traffic. Multiconductor cable and low-capacity microwave radio relays handled the remaining volume. Capacity of all those international links was limited, with the twelve-channel microwave radio-relay system between Hong Kong and Canton offering the largest international communications capacity. Radio telephone continues to play an important role in international communications. Direct service is available to Australia, Austria, Canada, Ethiopia, France, Hong Kong, Italy, New Zealand, the United Kingdom, and the United States.

In 1972 China acquired modern communications satellite earth stations from RCA and established wideband telecommunications links through INTELSAT. The Shanghai earth station provides sixty telephone channels and one television channel with potential to double its channel capacity. The Peking No. 1 and No. 2 earth stations are linked to Pacific and Indian Ocean satellites, respectively, and provide additional circuits for international communications. In 1970 China launched its first satellite, and since 1971 six more were placed in orbit. Those are believed to be primarily intelligence reconnaissance satellites and are not involved in international communications. They demonstrate, however, China's technological capabilities and future potential for operating its own satellite-based communications systems.

In recent years two major cable facilities were also established linking China with Hong Kong and Japan. The Hong Kong to Canton coaxial cable, valued at $1 million, was provided by Cable and Wireless Ltd. of Hong Kong and went into operation in 1974. Its capacity is 300 high-quality two-way telephone channels suitable for telegraph, telex, data, facsimile, and voice transmission.

Another 850-km submarine coaxial cable link was recently installed linking Shanghai with Kumamoto on Kuyushu in Papan. This cable has a capacity for 480 high-quality two-way telephone channels, and its $24 million cost was shared equally by China and Japan. The Hong Kong and Japan cable links provide access to the Trans-Pacific cable networks, which connect China to most countries in Southeast Asia, Hawaii, Australia, and the Soviet Union.

TELEGRAPH, TELEX, AND FACSIMILE TRANSMISSION

Telegraph in China appears to play a relatively smaller role than in other countries. The wired network and microwave radio provide only a limited transmission capacity. In addition, Chinese telegraph operation is hampered by the nature of written Chinese language. A numerical four-digit code is used to transmit Chinese characters (up to 9999), but it is a time-consuming method involving encoding and decoding of messages.

In 1969 a special electrostatic telegraph printer was developed that automatically converts the four-digit telegraphic code into Chinese character text at the speed of 1500 characters per minute. This is about seventy-five times faster than manual decoding. Most telegraph-transmission equipment, however, is considered obsolete. Teletypes include the Model 51 imported from East Germany in 1955, Model 55 manufactured in China, and Model 68 units imported from Siemens of West Germany in 1957.

Telex service that allows users to communicate directly with each other by means of a teletype is gaining in importance. It is, however, limited to international communications and is available only to government and press agencies. Telex use is expected to expand at Chinese organizations that are cleared to deal directly with foreign business contacts.

China introduced facsimile transmission to overcome the language problem in telegraph services during the 1950s. Imported equipment used at first, and initial facsimile service was available in 1957. It linked Peking with Shanghai, Canton, Wuhan, Moscow, Berlin, Warsaw, New Delhi, and Stockholm.

In subsequent years several facsimile transmission devices were developed in China. They were used to transmit full pages of the *People's Daily* for use in remote rural areas. The Tung Fang Hung No. 1 model announced in 1969 was reported capable of transmitting a complete newsprint page in 24 minutes. The Model No. 2, announced in 1972, was a faster version, transmitting a newsprint page in 3.7 min. Announcements of developments of other specialized facsimile-transmission units at provincial postal and telecommunications departments appeared in Chinese media fairly frequently. Despite facsimile-equipment developments in China, it is believed that such units are not mass produced, and China continues to rely on imports. In 1974 Fujitsu of Japan received an order for a facsimile system, and Matsushita Graphic sold a laser-beam color-printing facsimile system. However, although individual purchases will continue, these sales are very small in comparison with total electronic imports.

NEW CHINA NEWS AGENCY

This organization dates back to 1937 and in recent years was believed to have enjoyed the status of a Special Agency of the State Council designed to act as China's official news-dissemination organization and policy spokesman. The agency operates an international network for worldwide information collection and distribution. It is also known as the Hsinhua News Agency, and representatives of the organization abroad sometimes perform functions of a quasidiplomatic nature in countries where China has no diplomatic representation.

The New China News Agency maintains representatives in at least sixty-five countries. The largest foreign offices are maintained in Hong Kong, Tokyo, Cairo, and at the United Nations in New York. There is also a subsidiary agency known as China News Service, which specializes in information designed for overseas Chinese newspapers and magazines.

The agency is considered to be a part of China's foreign affairs establishment. It is directly under the control of the State Council but also receives some direction from the Politburo of the Central Committee of the Chinese Communist Party. It is headed by Director Chu Mu-chih, a member of the Chinese Communist Party Central Committee. Its International Department is managed by Hsieh Wen-ching.

The *Renmin Ribao* (*People's Daily*) is the official newspaper, controlled directly by the Politburo of the Central Committee of the Chinese Communist Party. The New China News Agency distributed major policy statements of the *People's Daily* throughout the world and supplies the bulk of the input for other major newspapers and broadcasting stations in China (see Table 14.3).

COMMUNICATIONS EQUIPMENT PRODUCTION AND IMPORTS

Manufacture of communications equipment is partially under the supervision of the Ministry of Posts and Telecommunications, but most of the

Table 14.3 Major newspapers published in People's Republic of China and their estimated daily circulations

Name of Newspaper	*Estimated Daily Circulation*	*Location*
Renmin Ribao (People's Daily)	3,400,000	Peking
Wen Hui Pao	900,000	Shanghai
Jiefang Ribao (Liberation Daily)	500,000	Shanghai
Chung-kuo	475,000	Peking
Tachung Daily	280,000	Peking
Sin Wain Ribao	132,000	Shanghai
Peking Daily News	130,000	Peking
Da Gong Ribao (Commercial, Financial)	125,000	Peking
Jiefang Jun Bao (Liberation Army Daily)	—	Peking
Kwangming Ribao	—	Peking
Kongren Ribao (Workers' Daily)	—	Peking

Source: Based on data in *Editor & Publisher International Yearbook, 1977* and *Europa Yearbooks,* London.

output such as radios and television is from factories that are probably controlled by the Fourth Ministry of Machine Building. There is also a Production Control Bureau at the First Ministry of Machine Building, believed to be responsible for production of radio, television, telecommunications, electronic components, and computers for civilian use.

Equipment Production

In 1972 it was estimated that at least sixty different plants in China were engaged in the manufacture of telecommunications equipment, with a total work force of 100,000 persons, believed to represent about 20% of all the employment in electronics. Most plants are small enterprises, employing 200 to 300 workers and producing a limited number of radios or telephone exchanges. Most are operating in major cities throughout the country, but the major telecommunications equipment-manufacturing centers are in Shanghai, Peking, Chengtu, Changchow, Canton, Hangchow, Nanking, Harbin, and Tientsin.

Manufacture under the direct control of the Ministry of Posts and Telecommunications is believed to include automatic switching equipment for long-distance telephone lines since 1969. Until recently, microwave equipment was manufactured primarily for military applications. In 1977, however, a 960-channel microwave telecommunications system was developed and is now providing trunklines linking Peking and at least twenty provinces, municipalities, and autonomous regions.

The industry also manufactures carrier-multiplexing equipment of twelve-channel capacity and telegraph multiplexes of sixteen-channel capacity on a single-voice grade line. Frequency-division multiplexes (FDM) with up to sixty channels are most common in China, but some time-division multiplexes (TDM) have also been noted in recent years.

One of the major basic electronic plants is the Chengtu Radio Factory in Szechwan Province. It was built during the first Five Year Plan with Soviet assistance. The area has become a major electronics-manufacturing region. The Peking Wire Communications Equipment Plant is also in operation since 1957, when it was completed with Soviet assistance; it became the first factory to produce telephone exchanges. Several major plants appear to specialize in particular products, but there is also considerable product mix in some factories and product change in others (see Table 14.4).

Table 14.4 Typical plants manufacturing telecommunications equipment in China

Name and Location of Plant	*Type of Product Manufactured*
Canton Broadcasting Equipment Plant	television receivers
Canton Wire Communications Plant	telephone equipment
Changchow Radio Factory	transistorized receivers
Changchow Electric Equipment Works	facsimile transmitters
Changchow Communications Supplies Plant	telephone equipment
Chengtu Radio Factory	basic electronics plant; Soviet aid
Hangchow Radio Factory	simultaneous translation equipment
Harbin Radio Factory	TV receivers; radio transmitters
(Hopei) Anhwei Broadcasting Appliance Works	large-screen TV receivers
Liuchou Electric Sound Equipment Plant	speakers and broadcasting equipment
Nanking Radio Equipment Plant	transistor-radio receivers
Peking Wire Communications Plant	telephone exchanges; basic plant
Peking Tungfeng Television Plant	color-TV and radio receivers
Peking Broadcasting Equipment Plant	closed-circuit TV units
Shanghai Marine Meter Plant	radar for ships
Shanghai People's Radio Equipment Plant	TV receivers; small plant
Shanghai Huaihai Electric Lamp Plant	CRT[a] tubes; 350 workers
Shanghai No. 4 Telecommunications Plant	electronic telephone exchanges
Shanghai Postal & Telecommunic. Equip. Plant	BD 055 teleprinter
Shanghai Sound Recording Equipment Plant	magnetic tape recorders
Shanghai Huatang Electric Switch Plant	remote telecommunications controls
Suchou Municipal Electronics Plant	TV and radio receivers
Taiyuan Telecommunications Equipment Plant	teletype equipment
Tzupo No. 5 Radio Plant	sonar equipment for surveys
Wuhu Kuanghua Glass Works	2.5″ and 5.0″ CRTs

Source: Compiled by 21st Century Research from various articles, papers, and publications listed in reference sections of this book.

[a] CRT = cathode ray tube.

Table 14.5 Recent telecommunications equipment sales to China

Year	*Type of Equipment Imported*	*Supplier and Country*	*Contract Value in Millions of $U.S.*
1972	Earth satellite station, Peking	RCA Global Communications, U.S.A.	3.7
	Earth satellite station, Shanghai	RCA Global Communications, U.S.A.	5.0
	Earth satellite station, Peking	Nippon Electric, Japan	1.0
	Earth satellite station, Peking	Western Union International, U.S.A.	3.8
	Television studio equipment	Central Dynamics Ltd., Canada	0.035
1973	Coaxial cable, 180km, 60 channels	Hong Kong Cable & Wireless Ltd.	1.0
	Submarine coaxial cable	Fujitsu/Nippon Electric, Japan	24.0
	Submarine repeaters and terminals	Fujitsu et al., Japan	3.9
	Coaxial cable with microwave system	Fujitsu et al., Japan	1.33
	Earth satellite station equipment	Comtech Laboratories, U.S.A.	0.381
	Satellite reception equipment	Nippon Electric, Japan	0.066
	Microwave network equipment	Nippon Electric, Japan	0.037
	Telephone cable, Hong Kong-Canton	Pirelli General Cable, U.K.	NA
	Color television studio equipment	Rank Organization, U.K.	0.47
	Color television studio equipment	Rank Cintel, U.K.	0.20
	Television camera	Marconi Communications, U.K.	NA
	Color television relay vehicles (PAL)	Toshiba, Japan	0.714
	Color television relay apparatus	Kannematsu Gosho, Japan	0.30
	Microwave telecommunications test eqpt.	Wiltron, U.S.A.	0.084
	Marine communications equipment	Marconi Marine Co. Ltd., U.K.	NA
1974	Color television broadcast vehicle	Pye TV Ltd., U.K.	0.57
	Multiplex radio equipment	Northern Electric Co.,	0.5
	Shipboard navigation, satellite rec. eqpt.	Magnavox Co., U.S.A.	NA
	Facsimile transmission system	Fujitsu Ltd., Japan	0.53
	Instrument landing system	Plessey Navaids, U.K.	1.9
	Facsimile transmission system	Matsushita Graphics, Japan	0.69
	Motion picture and television lenses	Rank Optics Ltd., U.K.	0.113
	Telephone switchboards	Japan	21.0+
1975	2000 color television tubes	Hitachi, Japan	NA
	Naval radar components	U.K.	NA
	Television studio equipment	unidentified	0.22
1976	Television broadcast and studio eqpt.	unidentified	0.451
	Magnetic tape recorders	U.S.A.	0.166
	Videotape recorders and TV studio eqpt.	International Video Corp., U.S.A.	0.200
	Non-video television studio equipment	unidentified	0.96
	Television broadcast transmitters	unidentified	0.35
1977	Television receivers	Hungary	NA
	Marine transmitter	Redifon Telecommunications, U.K.	NA

Source: Compiled by 21st Century Research from various reports, articles, and publications listed in references at the end of this chapter.

Telecommunications Equipment Imports

During 1965–1975 total imports of telecommunications equipment products was only $63.5 million. This is a relatively low volume of imports, in most cases consisting of single systems of color-television studio equipment, television cameras, relay vehicles, marine communications equipment, or facsimile transmission system (see Table 14.5).

Some telephone switchboards and television receivers have been imported in large numbers from Japan, but these are probably temporary measures designed to fill in the gaps while Chinese production catches up with the demand. In fact, a French observer of China trade suggests that China is already manufacturing some television receivers for Japanese companies under special contracts. If this activity expands further, China may find itself in the same role that Japan plays as a major supplier of television components and receivers to the United States and other markets. One cannot exclude the possibility that Japan may continue to

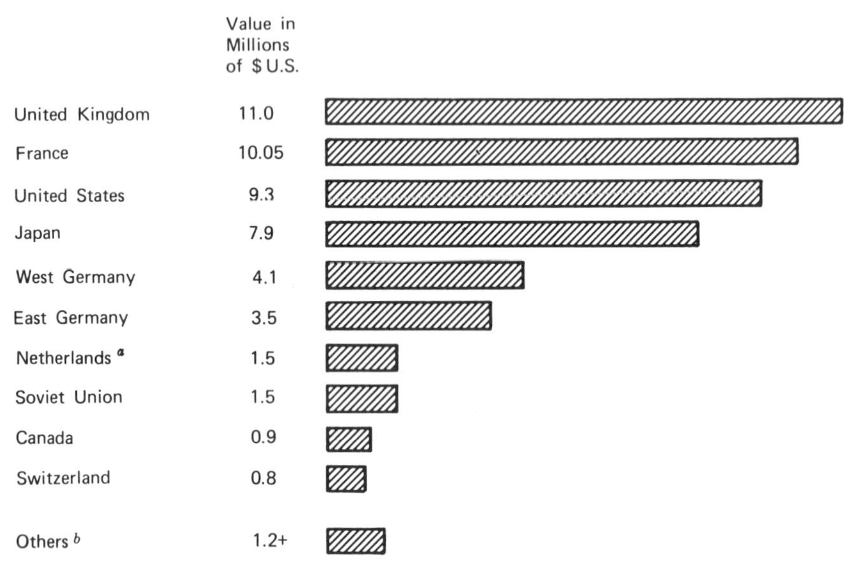

Figure 14.2 Estimated market shares of telecommunications equipment suppliers during 1969–1975. Data are for SITC commodity code 724. (**Source: OECD International Commodity Trade Statistics Series B** *and foreign-trade yearbooks of individual countries.*) [[a] *Data derived from SITC category 72 by subtracting totals for other subcategories in that group.* [b] *Includes Italy (0.5), Sweden (0.1), Belgium/Luxembourg (0.1), and amounts of less than $100,000 from Australia, Austria, Czechoslovakia, Denmark, Finland, Hungary, and Singapore.*]

supply its markets with television and radio receivers under Japanese brand names that will be manufactured in China. This is all the more likely because China's development of its own massive production and export capacity would eventually threaten Japanese domination of this market, anyway.

Until 1975 China imported telecommunications equipment almost exclusively from western Europe and Japan. In comparison with other imports, Japan does not have a market share significantly larger than other supplier countries. The United Kingdom and France are the leaders, although the United States still remains a significant supplier as a result of the satellite earth stations first sold to China during the Nixon visit to Peking (see Fig. 14.2).

Perhaps the most intriguing development in the trade are recent Chinese imports of telecommunications equipment from East Germany and the Soviet Union, from whom China did not import such products for years. China also agreed to import television sets from Hungary in its 1977 trade agreement. Perhaps this is a new policy designed to save hard currencies by buying comparable equipment from COMECON. It may also mean that plants originally developed with Soviet or eastern European assistance are being modernized and that COMECON equipment is the most suitable for the job.

REFERENCES

Books and Periodicals General Catalogue (Guozi Shudian), Peking, 1974.

Chinese Newspapers and Periodicals (Guozi Shudian), Peking, 1974.

Craig, Jack. "China: Domestic and International Telecommunications 1949–1974," in *China: A Reassessment of the Economy,* Part II, Urban and Industrial Development, Joint Economic Committee, U.S. Congress, July 1975, p. 289.

Editor & Publisher. International Yearbook 1977, New York.

Europa Yearbook, People's Republic of China, London, 1977.

McGurk, D. L. "Electronic Products," in William W. Whitman, ed., *Doing Business with China,* Praeger, New York, 1974.

Murphy, Charles H. "Mainland China's Evolving Nuclear Deterrent," *Bulletin of the Atomic Scientists,* January 1972.

Szuprowicz, Bohdan O. "Electronics in China," *U.S.–China Business Review,* May–June 1976, p. 21.

U.S. Government. "Production of Machinery and Equipment in the People's Republic of China," *Research Aid,* A(ER)75063, May 1975.

U.S. Government. "Directory of Officials of the People's Republic of China," *Reference Aid,* CR 77-15208, October 1977.

15

Construction and Building Materials Industry

During the last 20 years capital construction activity in China has grown to over four times its estimated level in 1957. Besides large modern industrial plant construction undertaken by central authorities under investment plans prepared by the State Capital Construction Commission, there is a considerable amount of rural construction performed by communes, production brigades, production teams, and individual households at the local level. Special units of the PLA also play a significant role in capital construction activity.

Construction priorities include projects that increase the productivity of agriculture and supporting industries and that will render operative major turnkey plants in petrochemicals, fertilizers, mining, metallurgy, and electric power generation that were imported from Japan, western Europe, and the United States during the mid-1970s.

Construction activity trends follow the general trends in industrial output but have been more sensitive to political disruptions of the Great Leap Forward and Cultural Revolution periods. Until 1957 construction activity increased at a higher rate than industrial output and the GNP of China. Then until 1965 construction activity grew practically in parallel with increases shown by industrial output as a whole. Since 1965 construction activity has been increasing at a slower rate than industrial output, although it still grew more rapidly than the GNP. Although construction activity shows rapid declines during periods of political unrest and economic downturns, it also shows much more rapid recovery during periods of economic upturns and relative political stability. Since

1970 construction activity is estimated to have grown at about 9 percent annually.

Actual construction activity in China is decentralized, and various government units operate design bureaus and construction companies at different levels of authority. China publicizes the need to construct its plants on wastelands rather than using land suitable for agriculture,

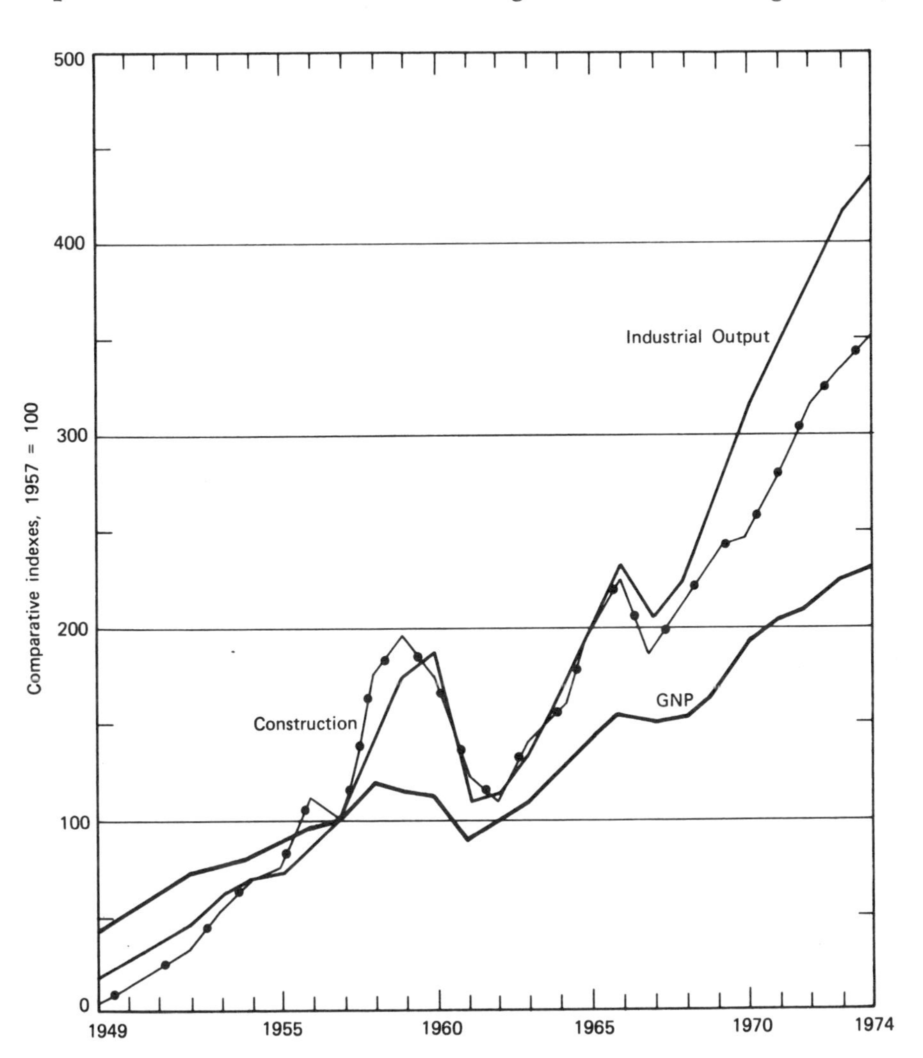

***Figure 15.1** Indexes of construction, industrial output, and GNP.* **(Source: U.S. Congress Joint Economic Committee,** ***Washington, D.C., July 1975.*)**

whenever possible building on slopes rather than level ground as well as in arid regions and other useless lands. Capital construction in rural areas, besides serving a strategic purpose for dispersing industrial production throughout the country, is believed to slow down the migration of people to the large and overpopulated urban centers.

China manufactures most building materials used in construction but has inadequate domestic production of steel and is critically short of timber. Iron and steel and particularly finished steel products are the largest single commodity imported by China, valued at $1.5 billion in 1975. Shortages of timber are periodically met by imports from the Soviet Union. Production of cement is adequate, and some is exported by China to nearby countries. Glass, bricks, tiles, fiberboard, and refractory and insulating materials are also domestically produced.

Most promising markets in the construction industry are for heavy construction equipment. China manufactures most standard machinery but lacks in serial production of large bulldozers, shovel loaders, hydraulic cranes, pipelayers, and heavy off-highway vehicles. Japan, the Soviet Union, France, Scandinavian and eastern European countries, and the United States have been the major construction equipment suppliers in recent years.

ROLE OF THE STATE CAPITAL CONSTRUCTION COMMISSION

The State Capital Construction Commission has overall responsibility for construction activity and determines the policy goals for major industry programs for the Five Year Plans. Its decisions are based on inputs received during discussions at operational levels in regions, provinces, and districts. During these discussions consideration is given to availability of investment funds as well as to the type and number of industries that require development.

The commission was established on October 12, 1958 but was abolished in January 1961, when it was merged with the State Planning Commission. It was reestablished in 1965 during another reorganization when an existing Ministry of Construction was divided into a Ministry of Construction and a Ministry of Building Materials. The two ministries were again combined into a single Ministry of Construction by 1972. The National People's Congress in 1975 confirmed the existence of the state Capital Construction Commission and elevated it to ministerial status with Ku Mu as the Minister-in-Charge. But that congress failed to reaffirm the existence of either the Ministry of Construction or the Ministry of Building Materials, although a State Building Materials Department was mentioned in a Chinese radio broadcast in 1976.

The State Capital Construction Commission is known to include a Building Construction Bureau as well as Building Materials Institute and a Building Research Institute. This suggests that the commission keeps track independently of the latest developments in construction and building-materials technologies.

There are several other research institutes specializing in building construction and materials research, some of which were previously known to have been operated by the Ministry of Construction. That ministry also operated factories and workshops that manufacture standard construction and roadbuilding equipment, but large construction machinery is produced by plants controlled by the First Ministry of Machine Building. Building materials and some ceramic and glass-manufacturing equipment were also previously under supervision of the Ministry of Construction. It is at present unclear which government units control that activity.

When annual targets for investment are determined the State Capital Construction Commission ensures that maximum utilization of domestic resources is given the utmost consideration. The commission authorizes foreign imports of construction equipment only after domestic capabilities and availability of hard currency have been investigated. As such it may be regarded as the final import approval authority, although actual purchases are usually made by TECHIMPORT or MACHIMPEX foreign-trade corporations.

CONSTRUCTION PRIORITIES

Current construction priorities are reflected by construction activity that features major projects designed to increase the productivity of agriculture and supporting industries. These include water management and irrigation schemes, electric power plants, transportation facilities, and above all, chemical fertilizer manufacture. Extensive purchases of turnkey plants during 1972–1974 from Japan, western Europe, and the United States also resulted in concentration of major construction activity in petrochemical, fertilizer, mining, metallurgy, and electric power industries (see Table 15.1).

Expansion of foreign trade and shipping bottlenecks in recent years led to priority in construction of port and harbor facilities. Construction of additional steel-finishing plants will continue to have high priority because its output is critical to growth of other vital sectors such as pro-

Table 15.1 Types of turnkey plants purchased by China during 1972–1975 indicating major construction priorities

Type of Plant	*Number of Units*	*Estimated Cost in Millions of $U.S.*
Iron and Steel plants		635
Rolling mills	5	
Iron works	1	
Power Generating Plants		303
Complete power stations	3	
Turbines and generators	46	
Petroleum Exploration & Extraction		127
Offshore drilling platforms	4	
Oil rigs	2	
Survey and supply vessels	33	
Petrochemical and Synthetic Fiber Plants		900
Intermediate products plants	33	
Synthetic fiber plants	11	
Chemical Fertilizer Plants		600
Ammonia plants	17	
Urea plants	15	
Other fertilizer plants	4	

Source: "China: A Reassessment of the Economy," U.S. Congress Joint Economic Committee, p. 701, July 1975.

duction of tractors and agricultural machinery. The petroleum and petrochemical industries command priority because of rapidly growing demand for more energy and fertilizer inputs to agriculture and also the export potential of the Chinese oil and petroleum products.

Since 1970 construction of small cement and fertilizer plants also received a certain amount of priority alongside the large modern units. However, construction of small plants, which are not centrally planned, often outpaces the availability of construction materials and energy, and recent reexaminations of these priorities resulted in renewed slowdowns of small plant construction.

Revision of investment priorities during late 1972 and 1973 also resulted in the announcement of the Five Principles of the 1973 Capital Construction Plan, which provide additional insight into China's construction industry problems and operation. In an obvious attempt to optimize capital construction resources priority was given first to "urgently needed" plants nearing completion within the 1971–1975 Five Year Plan. Particular attention was also called to projects under way without approved design documentation and unless approval of such projects was forthcoming shortly, construction was to be postponed.

Completion dates were also postponed for those projects for which building materials supplies and service connections could not be assured. Another principle stated that if increased output of existing plants and even some military plants could satisfy the demand that was to be met by the new plant under construction, such new plants should also be delayed unless construction had already progressed beyond the foundation stages. In other cases standards of construction were permitted to drop and equipment to be reduced if projects otherwise could not be completed sooner than originally scheduled.

During 1970–1975 capital construction activity as measured by an index based on estimated inputs of building materials was growing at an average annual rate of 9 percent. This was well over twice the 4-percent growth rate experienced by construction during the 1965–1970 period, which included the Cultural Revolution. During that time urban construction suffered the effects of political turmoil experiencing work disruptions and severe building-materials shortages. Many local investment decisions were not coordinated, and building materials and resources were wasted on inefficient small-scale plants.

Prior to the Cultural Revolution during the 1961–1965 period of readjustment and recovery from the Great Leap Forward, construction activity was also concentrated on major industrial projects, and many inefficient small plants were shut down. Chemical fertilizer, petroleum, electronics, and military plants received priority investment, and China began purchasing equipment and modern plants in the West. The policy of assigning more priority to agriculture and supporting industries also originated during those years.

In the Great Leap Forward years of 1958–1960 China attempted to speed up economic development by massive addition of human labor to the construction activity. Thousands of small plants producing iron, fertilizer, cement, and simple agricultural machines were started in those years, but severe building materials shortages developed rapidly and slowed down this activity to inefficient levels.

In the initial years beginning in 1949 most construction activity was directed toward rehabilitation and reconstruction. Only with the adoption of the first Five Year Plan (1953–1957) was new industry given priority, Soviet building specifications adopted, and significant Soviet technical assistance obtained in the construction of the "backbone" industrial plants for China. In that period construction activity was estimated to have been growing at an average 20 percent per year, but it has been slowing down ever since. Nevertheless, construction activity is estimated to have averaged an annual growth rate of 13 percent during 1949–1975, even when considerable slowdowns of the Great Leap Forward and the Cultural Revolution are taken into account.

ADMINISTRATION OF CONSTRUCTION ACTIVITY

Actual construction activity in China is decentralized, and responsibility for construction is diffused among various government organizations. Although the State Capital Construction Commission has overall responsibility for construction investment planning, considerable authority is exercised by industrial ministries that actually operate their own design bureaus and construction companies to undertake major industrial projects.

The No. 11 Metallurgical Construction Company, for example, employs 11,000 workers and operates five subsidiaries. These include an Earthwork Construction Company and an Installation Company, and they provide twenty-eight engineering teams to undertake construction of large projects. There are also Chemical Construction Companies, Electric Power Construction Companies, and Municipal Construction Companies, some of which also engage in the manufacture of their own specialized construction equipment.

Smaller projects are undertaken by provincial authorities who also operate Construction Bureaus to undertake construction work within their jurisdictions. The Anhwei Province Capital Construction Department, for example, was reported to have worked on 250 capital construction projects in 1976, including eight large undertakings. Fulfillment of capital construction investment plans is often reported on a provincial basis.

Railroad construction, which maintains the major transportation system of China, is the responsibility of the Ministry of Railways and the PLA Railway Engineering Corps as well as other civilian engineering bureaus. The corps plays a leading role in planning and design of large segments of Chinese railroads. During initial reconstruction years when operation of railways was vital to the consolidation of political power and stimulation of economic growth, the PLA Railway Engineering Corps supervised hundreds of thousands of peasants and youths recruited from farms and towns along the lines. Some laborers were inducted as noncombat troops and sent to construction sites as members of the corps.

In 1975 Chinese sources also revealed the existence of the PLA Capital Construction Engineering Corps, which includes cadres, soldiers, and technicians involved in construction of key industrial projects in metallurgy, coal mining, chemicals, transportation, communications, and geological exploration. This announcement came soon after the National People's Congress in January 1975, which did not mention the previously existing Ministries of Construction and Ministry of Building Materials. It is possible that some of the functions of those ministries have been taken over by the PLA Capital Construction Engineering Corps. People's Liberation Army units also regularly provide assistance in construction of schools, dams and reservoirs, irrigation systems, and in some regions such as Sinkiang virtually run the economy by building and operating most major industrial plants.

Small-scale plants throughout China are not centrally planned or controlled and depend on investment funds and construction work on local county authorities and smaller units such as communes, brigades, production teams, and even households. In 1974 about 500,000 small plants were estimated to be in operation, producing electric power, cement, fertilizers, coal, iron, agricultural machinery, and light industrial products.

Rural construction teams operate on a full-time basis under Provincial Agricultural Bureaus. These are involved in water irrigation, flood control, land leveling, slope terracing, and similar farmland construction projects. Tsinghai Province, for example, reported in recent years at least 966 full-time farmland capital construction teams operating at various levels and employing 64,000 members. In addition, large numbers of "volunteers" are often reported to participate in rural capital construction projects on a temporary basis.

There have been films and photographs of hundreds of thousands of Chinese people laboring on construction projects, digging soil by hand, loading with shovels on wheelbarrows and carts, and transporting and compacting also by manual effort. Sometimes soil is carted in baskets on shoulder poles in work on hillside terracing with rocks, reservoirs, dams, wells, and canals.

In recent years an estimated 3000 technicians and engineers from around the world have been involved in installation and follow-up work on major turnkey plants purchased from abroad, and some spent months and even years living on site in China. Their reports suggest that Chinese construction design criteria are not regarded as fluid and subject to change as would be the case in the West. Observance of the "self-sufficiency" rule also often leads to curious results. RCA engineers who went to Peking and Shanghai to help install satellite-communication earth stations reported that using identical equipment parts each of the two sites constructed a station completely different in design.

BUILDING-MATERIALS MANUFACTURING INDUSTRIES

Most building materials consumed by the construction industry in China are manufactured domestically, although some finished steel products and timber are imported. Production of cement, on the other hand, is estimated to be more than adequate and it is even exported to Hong Kong and other nearby countries alongside some hardboard, glass, and tiles made in China. Cement and other building-materials plants have been imported in a few cases, and it is believed these are primarily to serve as model plants for the construction of more sophisticated domestic manufacturing facilities.

Cement Production and Cement Plant Imports

China is the fourth largest cement-producing country in the world after the Soviet Union, Japan, and the United States. Production of cement rose from 37.3 million metric tons in 1973 to 46.9 million in 1975 and 49.1 million metric tons in 1976. The high rate of growth is attributed mainly to the expansion of the small-plant sector of this industry. In 1976 only 40 percent of the total national cement output was produced in large modern cement plants; 29.5 million metric tons, or 60 percent of the output, came from an estimated 3000 small, vertical-kiln plants operated by local units. China has been exporting cement since 1953, in amounts ranging from 300,000 to 1.5 million metric tons per year. Hong Kong is the primary market for Chinese cement, but China also exports some cement to Africa, Bangladesh, and Vietnam, much of it as aid to support projects such as TANZAM railroad.

There are some ninety modern cement plants with 200,000 to 300,000 metric tons annual capacity. Plants of at least 500,000 metric tons annual capacity are in operation in Canton, Chunching, Chiangshan, Fushun, Liuliho (Peking Municipality), Tangshan, Tatung, and Yungteng. The largest cement plant with an annual capacity over 1 million metric tons is located in Hantan in Hopeh Province. At least two other plants in the million-ton category are in operation at Huahs in Hupeh Province and at Yao Hsien in Shensi Province. The large modern plants mostly use horizontal rotary kilns and produce high-quality cement for priority domestic use and for export. At present it is believed that the modern cement plant sector is not utilizing its full capacity.

Small cement plants range in capacity from several thousand tons to a maximum of 32,000 tons per year. The largest of the small plants employ 16,000-ton standard vertical stationary kilns developed by a Chinese design institute in the late 1950s. Half of the small plant production is believed to come from 200 to 400 small but relatively modern plants.

There are also many small "native" kilns that produce as little as 100 tons per year and are not supplied with equipment by the central authorities but rely on improvised or second-hand equipment. These vary greatly in size and quality of output, which is mostly consumed in local agricultural projects.

In 1974 China negotiated with Ciments Le Farge SA of France for the purchase of a modern cement plant of 1.2 million metric tons annual capacity, valued at $4.5 million. Ten years previously China had purchased a foaming concrete plant of 150,000 cubic meter annual capacity from Sweden for $1.8 million. It is presumed that such plants are purchased primarily as models for developing more modern units in China.

Use of Steel in Construction Industry

At least 33 percent of the total supply of finished steel in China, which amounted to about 7 million metric tons in 1974, was used in construction. This is an estimate based on Chinese press statements about steel consumption for construction prior to 1958 and the assumption that there is a stable relationship between the amount of steel and cement used in construction.

The ratio of steel to cement in construction by 1958 is believed to have been 0.30, which is relatively high but justifiable because of concentration of steel-intensive projects such as industrial plants and railroad lines. This ratio for the Soviet Union in 1955 was also 0.30 but declined to 0.19 by 1969–1970. It was 0.14 for the United States during the 1950s and 1960s and as low as 0.09 in Japan in 1970–1971. Therefore, although the amount of steel used in construction has been increasing steadily, the share of steel supply used for this purpose has been dropping since 1957. This is a reasonable trend, and the growing total steel supplies have been consumed by the machine-building sector, which has been growing at a much faster rate than the rest of industrial production in China.

China imported 3.7 million metric tons of finished-steel products in 1974, primarily from Japan and West Germany, and in 1975 its imports of iron and steel amounted to over $1.5 billion, the largest single imported product category. West Germany overtook Japan as the largest iron and steel pipe supplier in 1975, but Japan remained the largest overall supplier of iron and steel products, totaling $837 million. France, Italy, Australia, and Luxembourg are also major iron and steel products suppliers to China. Besides steel pipe, universals, plates, thick steel sheet, heavy plate, and steel sheet less than 3 mm thick are the major steel products imported by China and may constitute a large proportion of all steel used in major construction.

Timber Production and Consumption

China is extremely short of timber as a result of many centuries of extensive exploitation of its forests. Timber consumption per capita in 1973 was estimated at 0.039 m^3 (cubic meters) per person, compared with 1.691 m^3 per person in the United States. Shortage of timber is also reflected in substitution of other materials such as concrete for railroad ties and utility posts, cement-filled bamboo pitprops in coal mines, and non-timber inputs for paper production.

Although extensive reforestation programs have been under way since 1949, not all of them have met with sufficient success so far to assure total self-sufficiency in timber in the face of a growing demand. It is believed that China will continue to be critically short of timber in the foreseeable future but may also use an increasingly large share of plastics and synthetic building materials as its petrochemical industries develop.

Production of timber in 1975 was estimated at over 36 million m^3. Between 1961 and 1967 and during the early 1970s domestic production was supplemented by timber imports from the Soviet Union, ranging in amounts from 120,000 to 1.5 million m^3 annually. In 1976 imports of sawn timber from the Soviet Union amounted to 586,200 m^3 valued at $23.2 million, up significantly from 161,500 m^3 imported in 1975.

The construction industry is the largest consumer of timber in China; it is estimated to use 41 percent of all the supply. Coal mining is the second largest user, accounting for about 23 percent of the total, and almost 11 percent is used for wood-pulp production in the paper industry. Railroad construction, an important consumer of timber for railroad ties, is nevertheless a relatively small user of timber by volume, accounting for no more than 3 percent of the total supply.

Other Building Materials

Practically no data are available on use for construction of such materials as plywood, plastics, precast concrete, or tiles, and only fragmentary data have been compiled on production and use of glass and bricks.

Production of bricks in 1974 was estimated in the 45 to 60-billion range, up almost 100 percent from the 30 billion reportedly produced in 1960. Chinese bricks are valued at 40 yuan (ca. $20) per 1000 and are believed to constitute about 17 to 21 percent of the total value of building materials used in construction.

Glass was estimated to represent only about 1.5 percent of the total value of building materials used in construction. Assuming a unit price of

1.29 yuan per square meter (m^2), it was estimated that about 130 million m^2 of flat glass were produced in 1974, twice the 66.5 million m^2 made in 1960. Some Chinese glass is available for export, which suggests an adequate supply to meet domestic demand.

At least twenty-five glass factories have been identified in various parts of China, but data about their production levels are not available. These include the Chuchou Rolled Glass Factory, Lanchou Plate Glass Factory, Peking Optical and Precision Glass Factory, Peking Quartz Works, Shanghai Tungfeng and Yaohua Glass Fiber Plants, Taiyuan High Grade Glass Factory, and Nanching Continuous Filament Glass Factory.

Additional building materials factories have been identified in China, manufacturing refractory materials, ceramic tile, fiberboard, roofing felt, insulating materials, and cast stone products. Only occasional plant imports occur in these product categories. In July 1975 China signed a contract with Ibigawa of Japan for a laminated board plant valued at $1 million. In April 1976 Nakajima Seiki and Kumiai Boeki of Japan also signed a contract for the supply of a wallpaper manufacturing plant valued at about $1 million. However, the low technology content of such manufacturing facilities precludes the expectation of any significant increase in imports of this type.

MARKETS FOR CONSTRUCTION EQUIPMENT

If projected rates of industrial growth up to 1985 are realized, China's demand for construction equipment should increase by 8 to 10 percent annually. Imports of construction equipment are expected to continue because even increasing domestic production is unlikely to be sufficient to meet the demand.

The U.S. Department of Commerce estimates that China's imports of construction machinery will also expand 8 to 10 percent annually, which means that by 1985 China will purchase $60 to $75 million worth of construction machinery, including trucks. However, the bulk of construction equipment in demand is manufactured in China, and only very large and sophisticated equipment items are being imported. On the other hand, many items are imported in large quantities, which suggests a lack of manufacturing capabilities of such equipment in China and raises the possibility of complete plant imports in the future (see Table 15.2).

Although China manufactures various types of construction machinery, ranging from cement mixers to road rollers and large cranes and dredgers, there is a significant technological gap between Chinese equipment and that of leading Western manufacturers. Very large ma-

Table 15.2 Typical imports of construction equipment by China since 1970

Year	*Quantity*	*Type of Equipment Imported*	*Supplier and Country*	*Value in Millions of $U.S.*
1972	1000	8-Ton dump trucks	C. Itoh & Co., Japan	6.6
	1192	16- to 33-ton bulldozers	Komatsu, Japan	43.8
	195	Air compressors	Mikuni, Japan	
	8	Air compressors	Tanabe (6), Hitachi (2), Japan	
	14	Hydraulic mobile cranes	Kato, Japan	
	6	Hydraulic mobile cranes	Tadano, Japan	
	63	Loaders	Toyo, Japan	
		Loaders BM 640/641 & 840/845	Volvo BM, Sweden	
	140	7-Ton dump trucks	Hino, Japan	0.5+
1973		Angledozers Model D80A-12	Komatsu, Japan	
		Bulldozers D 155A-1	Komatsu, Japan	
	87	Hydraulic 6-ton mobile cranes	Ransomes & Rapier, U.K.	2.4
	24	Truck cranes	Kobe Steel Works, Japan	
	146	Wheeled tractor loaders	Toyo, Japan	
		Trucks	Tatra, Czechoslovakia	0.75
	7	Power shovels and drills	Bucyrus-Erie, U.S.A.	20.0+
	1060	Heavy-duty trucks	UNIC-Fiat, Italy	26.1
	3020	8-Ton dump trucks	Isuzu, Japan	23.0
	400	12-Ton trucks	Hino, Japan	4.8
	700	20-Ton dump trucks	Perlini, Italy	
	500	Heavy-duty trucks (for timber)	SAAB-Scania, Sweden	12.5
	700	Cranes	Nichimen, Japan	
	250	Forklift trucks	Toyo Hauling, Japan	1.1
	769	Forklift trucks to 10 tons	Toyo Hauling, Japan	8.0
1974		Truck cranes	Kato, Japan	
	140	Hydraulic 12-ton mobile cranes	Coles, U.K.	4.6
	10	Loaders and dumpers	Volvo, Sweden	1.0
	12	Pipelayers, 70-ton lift	Komatsu, Japan	
		20-Ton tractors	Clark Equipment, U.S.A.	
		120-Ton off-highway trucks	WABCO, U.S.A.	7.0
1975	38	Pipelayers	Caterpillar, U.S.A.	3.8
	40	25-Ton and 36-ton mobile cranes	Raume-Repola, Finland	
	54	18-Ton dump trucks	Komatsu, Japan	
1976	20	Shovel loaders	Kawasaki, Japan	0.7

Source: Compiled by 21st Century Research from various publications.

chines are seldom produced serially, and China has yet to solve all the associated problems of hydraulics, metallurgy, and special machine tools before large construction equipment can be put into serial production.

Future demand is most likely expected to be for large equipment not yet manufactured in China. This includes bulldozers, shovel loaders, power shovels in the 8 to 15-yard category, hydraulic truck cranes, pipelayers, and off-highway trucks of 80 to 200-ton capacity. Some potential is also seen for sales of motorgraders, scrapers, road rollers, and spare parts for this equipment.

China imports construction machinery primarily from Japan, which accounts for about 75 percent of all the imports in this category, but the Soviet Union, Poland, France, West Germany, United Kingdom, Sweden, Finland, and more recently the United States also supplied some specialized construction equipment.

Although some individual orders have been relatively large, total imports in this category averaged only $20 million per year during 1970–1975. However, there are now signs that China may be changing its trade policies and may be more inclined to enter into long-term agreements with suppliers of specific equipment in demand.

The recent Sino-Japanese long-term trade agreement signed in 1977 covers the imports of construction equipment quite specifically. It specifies the import of shovels, cranes, bulldozers, and mining dump trucks, setting minimum Japanese export levels for such equipment to China at $15 million in 1977. This may be an indication of future Chinese willingness to enter into such long-term trade agreements with other supplier countries, particularly those that may be interested in purchases of Chinese crude and petroleum products.

REFERENCES

Albertson, Maurice L. "Impressions of the PRC," *ASME Mechanical Engineering,* March 1975; "Agricultural Engineering," *ASME Mechanical Engineering,* June 1975.

BBC, Summary of World Broadcasts. "Chinaghai Meeting on Farmland Capital Construction," FE/W935/A3, July 29, 1977; "Construction," FE/938/A2, July 20, 1977; "Capital construction in Jan–Jun," FE/W837/A3, July 30, 1975.

Chao, Kang. *The Construction Industry in Communist China,* Aldine, Chicago, 1968.

Cheng, Chu-yuan. *"The Machine Building Industry in Communist China,"* Aldine, Chicago, 1972, pp. 15, 16, 200.

Chien Yuan-heng. *"Study and Analysis of Chinese Communist Capital Construction work,"* May 1973, (transl. by JPRS #59528), Taipei, Taiwan.

CIA Economic Research. "People's Republic of China Timber Production and End Uses," ER 76-10493, October 1976.

Daily, James W. "Industry in a Planned Society," *ASME Mechanical Engineering,* July 1975.

Green, Stephanie R. "China's American Residents: US company technical personnel in China," *U.S.–China Business Review,* January–February 1977.

Lin, Pin. "Current Economic Trends of the Chinese Communists," May 1973, (transl. by JPRS #59528), Taipei, Taiwan.

McFarlane, Ian H. "Construction Trends in China 1949–1974," *China: A Reas-*

sessment of the economy, U.S. Congress Joint Economic Committee, July 1975.

Rouse, Hunter. "Impressions of the PRC," *ASME Mechanical Engineering,* February 1975.

U.S. Department of Commerce. "Construction Equipment: a Market Assessment for the People's Republic of China," Bureau of East–West Trade, March 1976.

U.S. Government Research Aid. "Directory of Officials in the PRC," April 1975.

U.S. Government National Foreign Assessment Center. "China's Cement Industry," *A Research Paper* ER-77-10704, November 1977.

Vneshnaya Torgovlya SSSR za 1976 God, Ministry of Foreign Trade of the USSR,Mezhdunarodnye Otnoshenya, Moscow, 1977.

16

Light Industries

Light industries include the textile industry, paper and pulp production, food processing, manufacture of consumer goods, and the handicraft industry. Light industry supplies over 70 percent of all the goods for China's domestic market, including such products as detergent, cotton yarn and cloth, food produce, and consumer appliances. It also provides some important products for export, particularly textiles, food products, and handicrafts.

In October 1977 there were some 120,000 light industrial enterprises reported in existence. However, that number includes numerous small and medium-sized plants run by districts, counties, communes, or even production brigades throughout China. The industry was originally concentrated solely in the coastal areas. Although many of the larger state-run enterprises have remained there, modern large-scale plants have been systematically constructed in the inland areas near raw mineral bases and the heavy industry plants that supply light-industry manufacturers with synthetic fiber, metals, and other materials. Raw-material supplies from heavy industry to light industries increased to 30 percent in 1976.

The industrial activity of plants at the county or commune level consists primarily of processing the agricultural and sideline produce, such as grain milling, oil pressing, and cotton ginning, but may also include production of basic necessities such as shoes, clothing, light bulbs, household utensils, and similar items, both to satisfy local needs and for sale to the state. Some commune factories produce porcelain or jewelry for export, and teams may set up workshops to produce handicraft items such as mats, woven baskets, and gunny sacks.

The Ministry of Light Industry was first established in 1949 but in 1958 became part of the Ministry of Food Industry, which in turn was divided in 1965 into the First and the Second Ministry of Light Industry.

In 1970 both these ministries were combined with the Ministry of Textile Industry to form the present Ministry of Light Industry, headed by Chien Chih-kuang. This ministry is responsible for production of consumer goods and operates several research institutes. It is also believed to operate factories manufacturing light-industry equipment such as textile machinery, brewing equipment, flour-mill equipment, and probably food-processing and packaging machinery.

In the early 1950s light industry was the largest sector of the Chinese industry, albeit the slowest-growing one. Its share in the industrial output declined from 56 percent in 1952 to 45 percent in 1957 and 28 percent in 1970. During 1957–1973 light industry grew at an average 6 percent per year, in contrast to the 12–13 percent average compound growth per year shown by the heavy industry. Because much of its production is directly affected by the agricultural output, the growth of light industry declined during the early 1960s following the agricultural crisis of 1959–1961 and did not recover its 1959 peak until 1967.

Since October 1976 there has been a new emphasis on light industry. Growing production and sales of consumer goods are now regarded as a major vehicle of capital formation for investment in other sectors. Hence light industry is now ranked second in priority after agriculture. With this government emphasis on the role of light industry as a capital generating vehicle, one can expect expansion of the production of the consumer goods, but it is highly unlikely that China will be interested in significant imports of such goods other than high-quality watches.

TEXTILES

The Chinese textile industry ranks among the largest in the world. China is one of the world leaders in production of cotton textiles and is also the most important world supplier of silk and a leading producer and exporter of cashmere wool. Production of man-made fibers is still relatively small, but considerable progress is being made and industry observers predict that in a few years China may become Asia's second largest producer of man-made fabrics after Japan.

Since textiles represent one of the basic necessities, the textile industry receives much attention in the economic plans. It is also one of China's main export industries. In 1976 exports of textile products totaled $1,655 million, with $1,235 million for textile yarn and fabrics and $420 million for clothing. Together, this represented about 54 percent of total manufactured goods exported that year and almost 23 percent of the total Chinese exports. Exports of textiles rose from less than 4 percent of total

Chinese exports in 1953 to an average 40 percent in 1961–1963, when textiles replaced foodstuffs as the leading export. Between 1966 and 1975 the textile share stabilized to an average 21 percent of total exports, ranging from 25 percent in 1969 and 1970 to 18 percent in 1973.

Cotton remains the main textile fiber in China, accounting for 80 percent of total Chinese textile production. China is the second largest producer of cotton in the world, and it also imports cotton to supplement its production (see Chapter 12). In the early 1970s there were an estimated 12 to 14 million cotton spindles in place, and the number of looms making cotton fabric was believed to be in the 300,000 to 350,000 range. It is also estimated that China is the largest exporter of cotton cloth, at about 600 million yards per year. Production of cotton cloth reached 7500 million m in 1970 and since 1973 has remained at the level of 7600 million m annually. Per capita production during 1966–1975 ranged from 7 to 9 m and averaged 8.2 m per year. In 1975, when China's production was reported as 7600 million m, India produced 7675 million m and the United States, 3956 million m of woven cotton fabrics (see Table 16.1).

Although 90 percent of all cotton production is allocated for domestic consumption, the cotton industry cannot meet the internal demand. Per capita consumption of textile fiber in China is 2.1 kg per year, which is about one-third of the world average. To maintain export levels, cotton cloth is rationed to 6 to 9 m per person per year, but had been as low as 2 m in 1961 and 1962. Wool, silk, linen, synthetic fibers, and knitted cottons are not rationed, but they represent only a small percent of textiles used in China and their domestic supplies are limited. Production of silk

Table 16.1 Production of cotton and wool cloth in selected countries in 1975

Country	*Unit of Measure*	*Cotton Cloth*	*Wool Cloth*
China	millions m	7600	65.2[a, d]
India	millions m	7675	38.0[b]
Soviet Union	millions m^2	6635	740.0
United States	millions m	3956	74.0
Japan	millions m^2	2115	355.0
Italy	thousands metric tons	113	132.0[d]
Poland	millions m	929	125.0
United Kingdom	millions m	400	151.0
Australia	millions m^2	—	7.0[c]
Brazil	millions m	864[b]	4.0[a]

Source: Unless otherwise stated, *Handbook of Economic Statistics, 1976,* U.S. Government, Washington, D.C.

[a] Data obtained from *United Nations Statistical Yearbook, 1976.* [b] Data obtained from *PRC Handbook of Economic Indicators, August 1976.* [c] Excluding worsted fabrics. [d] Data for 1974.

cloth rose from 50.16 million m to 401.28 million m in 1971. Most of the silk, however, is exported. Production of wool cloth rose from 5.4 million m in 1949 to 65.2 million m in 1974. China is the leading exporter of cashmere wool, but its overall production of wool is relatively small (see Table 16.1).

Imports of textiles and textile yarn as defined by SITC category 65 represented only 1 or 2 percent of total Chinese imports for 1966–1975. As measured by value in millions of $U.S., largest imports occurred in 1974 when China imported $150.8 million worth of textiles. In 1975 this figure dropped to $85 million and moved up again in 1976 to $125 million. Except for minor imports of cotton fabrics in 1971 and 1972 from Switzerland, the textile imports consisted mainly of synthetic yarn and thread and synthetic fibers fabrics. Japan is China's largest supplier of textile products. During 1970–1975 it supplied $285.1 million worth of textiles, including $125.9 million in 1974 and $59.0 million in 1975. The second largest supplier, Italy, exported only $34.1 million worth during the entire 1970–1975 period (see Fig. 16.1).

Japan is also the largest market for Chinese textiles. In 1975 its imports accounted for 10.5 percent of total exports of Chinese textiles and clothing and were valued at $139 million. Hong Kong ranked a close second with $136 million worth of textile and clothing imports from China.

Textile Machinery Production and Imports

Originally production of textile machinery was under supervision of the Manufacturing Bureau of Textile Machinery under the Ministry of Textile Industry, which became absorbed by the Ministry of Light Industry in 1970. It is believed that the Manufacturing Bureau of Textile Machinery now exists within that ministry.

Major textile machinery plants include the Chingwei plant in Yutzo, Shansi province. Built in 1953 with Soviet aid, it was one of the thirty-two major plants designed to form the backbone of the machine-building industry. It had a planned capacity of 0.5 million spindles annually and could turn out complete spindle frames. Six other major plants in existence at that time cooperated in production of parts for uniform machinery. Other key plants include Shanghai No. 7 Textile Machinery Plant, built in 1958 and specializing in manufacture of knitting and automatic hosiery machines, and Wuchang Textile Machinery Plant in Heilungkiang, producing cotton spinning and synthetic fiber machinery. Numerous small and medium-sized textile machinery plants are located in Tientsin, Tsingtao, Chengchow, Hangshow and Shanghai.

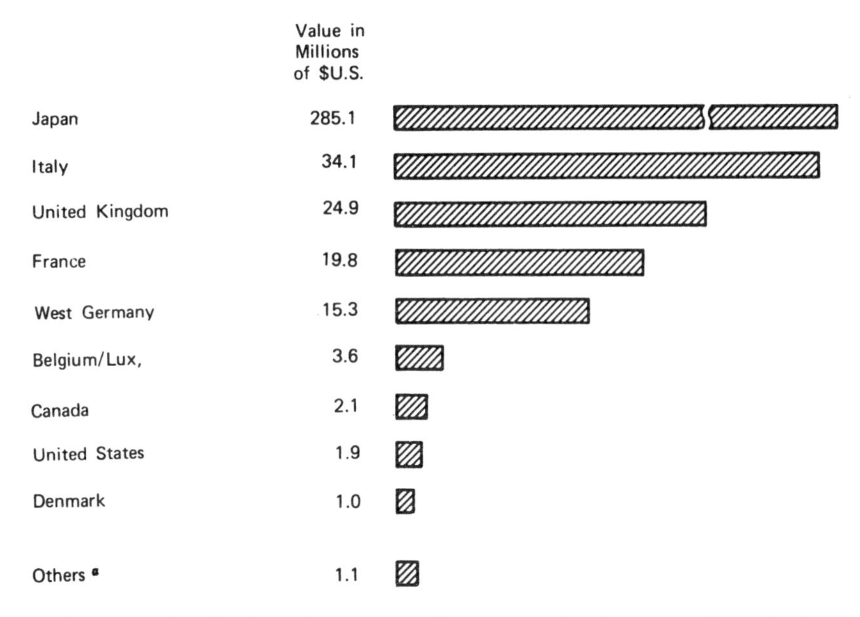

***Figure 16.1** Market shares of textile yarn and fabrics suppliers during 1970–1975.* (**Source: OECD Commodity Trade Statistics Series B** *for SITC code 65.* [[a] *Others include Switzerland (0.4), Yugoslavia (0.3), Austria (0.2), The Netherlands (0.1), Iceland (0.1).*]

During 1950–1965 textile machines represented the most developed area of China's machine-building industry. Some 10 million spindles were produced, and there was a diversification from cotton textile machinery. By 1965 about 1000 types of machines were being produced, including machines for wool, hemp, silk, knitting, printing and dyeing, and some synthetic fiber equipment.

Imports of textile machinery (SITC category 717.1) rose from $0.5 million in 1970 to $24.8 million in 1975. Japan was the main supplier, accounting for $16.1 million in 1975 and a total of $31.8 million during 1970–1975. The second largest supplier, Italy, exported only $3.48 million worth during that period. Imports from the United States in 1974 and 1975 came to only $0.1 million each year. The rise in imports of textile machinery coincides with the development of synthetic fiber production (see section on synthetic fibers production and import in Chapter 10). The major suppliers of textile machinery, Japan, Italy, the Netherlands, and the United Kingdom, are among the major suppliers of synthetic fibers to China (see Fig. 16.2). With the increasing production of synthetic fiber there may be more purchases of synthetic spinning plants, es-

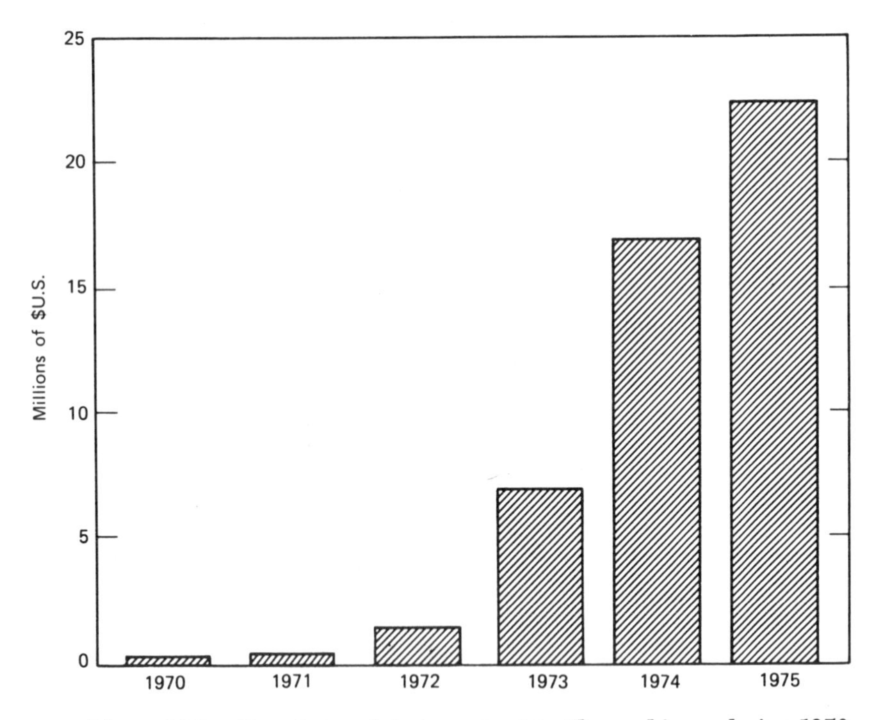

***Figure 16.2** Growth trends in imports of textile machinery during 1970–1975. (Source: Based on* **OECD Commodity Trade Statistics Series B** *and United Nations bulletins of statistics on world trade in engineering products for SITC commodity code 717.1.)*

pecially polyester staple and filament, and of automatic looms and double-knit machines.

PAPER AND PULP

In 1975 China was the world's fourth largest producer of paper and paperboard after the United States, Japan, and the Soviet Union, and the sixth largest producer of newsprint. It was also among the top ten producers of chemical and mechanical wood pulp. Its total paper production was estimated at 6.9 million metric tons, almost twice the output for 1965 of 3.6 million metric tons. It included almost 5 million tons of paper and 998,000 tons of newsprint (see Table 16.2).

Reportedly China became self-sufficient in paper production in 1954 and has since been exporting small amounts of high-quality news-

Table 16.2 Production of wood pulp and paper and paperboard in selected countries in 1975 (in thousands of metric tons)

			Wood Pulp (Dry Weight)	
Country	*Paper and Paperboard*	*Newsprint*	*Mechanical*	*Chemical*
China	4,983	998	628	965
United States	41,555	3,120	4,004	32,812
Japan	11,440	2,160	1,320	7,293
Soviet Union	6,862	1,334	1,842	6,340
West Germany	4,801	486	805	726
France	3,862	238	385	1,368
United Kingdom	3,296	319	179	140
Sweden	3,259	1,182	1,626	6,718
Italy	3,253	243	431	237
Canada	3,160	6,966	5,903	8,804
Finland	3,000	992	1,605	3,569
Brazil	1,518	128	65	1,236
Mexico	1,155	29	51	315
Australia	946	196	571	396
India	830	52	20	—
Norway	712	435	846	888
South Africa	624	210	181	601
South Korea	507	155	87	—
New Zealand	336	219	348	565

Source: United Nations Statistical Yearbook, 1976.

print to Australia, Cambodia, Japan, Sri Lanka, the Netherlands, and the United Kingdom. Its total exports of paper in 1975 were valued at $45 million. Nevertheless, China's per capita consumption of paper is among the lowest in the world, and much of the paper it produces is of low quality. This is mostly due to scarcity of suitable raw material. China is extremely short of timber; the forested area in China is about 0.13 hectares per person, or about one tenth of the 1.3 hectares per person in the United States. To overcome scarcity of timber China uses many substitute materials such as grass, reeds, straw, rags, bamboo, cotton stems, and waste products. Some timber is imported from the Soviet Union and Malaysia, but it is not clear whether it is used as raw material in paper making.

Most of the local paper demand is met by some 2200 small and medium paper plants located in all the administrative areas of China. Their daily capacity ranges from 500 kg to 5 tons, and they supplement the output of the larger plants, about forty of which have been identified. The ten paper plants listed in 1963 as the largest range in capacity from 33,000 to 105,000 tons. Two of them are located in the southern area of China, including the largest plant in Canton, one in the northern area, one in the southwestern area, and the remaining six in the northeastern area. The northeastern provinces, Kirin, Heilungkiang, and Liaoning, account for nearly one-third of China's paper production and contain nearly two-thirds of China's timber.

It is believed that China can produce all the required pulp and papermaking equipment it needs. It has also provided assistance in building paper mills in other Asian countries such as North Korea, Nepal, and Cambodia. There has been no record of Chinese purchase of paper plants since 1968. Known purchases include a pulp and paper plant in 1963 from Cellulose Developments of the United Kingdom for $1.4 million, a bank-note paper mill purchased in 1965 from ENSA of France for $424,000, and three plants from Finland—a straw cellulose plant with 62.5-ton daily capacity and a bleached sulfur cellulose plant with 80-ton daily capacity, both purchased in 1965, and a $4.2-million pulp plant purchased in 1968.

Domestic consumption of paper is growing rapidly with increasing industrialization, and China's demand for paper products may be reaching a point where insufficient raw materials make additional imports necessary. Reforestation programs and raw materials derived from petroleum are not as yet capable of providing significant alternative input for the industry. Exponential growth in future demands is expected as a result of rapid industrialization. This in turn may result in a continuing increase in pulp and paper imports until raw materials and additional plant capacity are resolved. Possibly the 1975 decline in imports of paper products and an increase in imports of wood pulp indicates the increased capability of the paper-making industry.

Imports of wood pulps and waste paper rose sharply in 1971 to $22.9 million from the $0.7 million level in 1970 and were maintained at a steady level of an average $25.6 million per year until 1975, when they increased to $50 million and again rose to $60 million in 1976. Sweden was the leading supplier during 1970–1975, with total sales of $67.3 million. However, pulp imports from Sweden dropped to $3.7 million in 1975, while imports from Canada rose to $17.9 million. Canada can provide needle leaved pulp which is in short supply in China and imports from Canada increased further to $33 million in 1976, although imports from Sweden continue and were valued at $8 million in that year (see Fig. 16.3).

China has also been importing paper and paperboard products as identified by SITC code 64. These imports reached a high of $88.6 million in 1974 but declined to $80 million in 1975 and even further to $45 million in 1976. During 1970–1975 Japan was the leading supplier, with

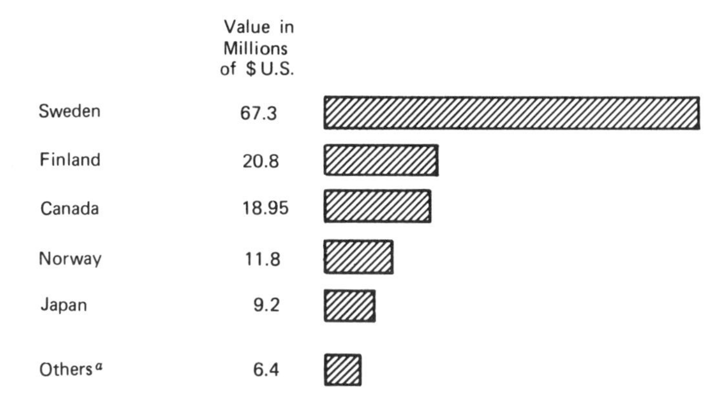

Figure 16.3 ***Estimated market shares of major suppliers of wood pulps and waste paper to China during 1970–1975. All data for SITC export commodity code 251 Source.*** **OECD Commodity Trade Statistics Series B.** **[*[a] Others include United States (6.1) and West Germany (0.3).*]**

$152.1 million. The second largest supplier was Finland, with $24.6 million. The United States exported a total of $10.8 million, of which $7.1 million was exported in 1974 when kraft paper and paperboard represented 0.9% of all the U.S. exports to China (see Fig. 16.4).

FOOD PROCESSING

There is relatively little information about China's food-processing industry, although food products represent a substantial share of Chinese exports. Processing of agricultural or sideline products is the principal industrial activity of the 50,000 or so communes, and much of tea curing, grain milling, sugarcane and edible oil pressing, and fruit canning is done at the commune or county level. Visitors to China are shown small rice or flour mills of 1.5 to 2.0-ton daily capacity operated by particular communes; compact flour mills with 25-ton/24 h capacity are produced by a large factory in Tsingtao.

Production of processed sugar has been increasing steadily. It rose from 1.8 million metric tons in 1970 to 2.3 million in 1975, and sugar refineries have been multiplying throughout the country, especially in the major sugar-producing provinces of Kwantung, Fukien, and Heilungkiang. Fukien is reported to have forty-six sugar refineries, Kwantung has 100 refineries, employing a total of 50,000 workers, and Heilungkiang re-

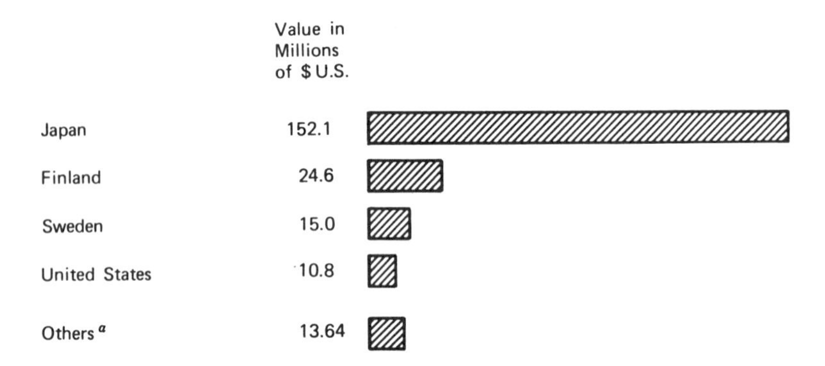

Figure 16.4** Estimated market shares of major suppliers of paper and paperboard products to China during 1970–1975; all data for SITC export commodity code 64.* (**Source: OECD Commodity Trade Statistics Series B.**) [[a] ***Others include France (3.0), West Germany (2.84), United Kingdom (1.9), Norway (0.8), Spain (0.8), Canada (0.7), Switzerland (0.2), Belgium (0.1), Denmark (0.1).]

cently built four large, three medium, and six small sugar refineries, all Chinese designed and equipped for multipurpose use of raw material. Besides granulated table sugar and soft white sugar they produce alcohol, citric acid, acetone, and butyl alcohol from beet residue. Heilungkiang also has a new Sugar Research Institute. Inner Mongolia, another major sugar-producing region, has ten sugar refineries, each with a capacity of several hundred tons. Reported capacities of individual refineries in various parts of China range from 500 tons per year to 2000 tons daily.

Food products constitute one of China's major export items. In the 1950s they accounted for nearly half the exports. During 1970–1975 they fluctuated between 30 percent and 32 percent. Exports of tea and spices are handled by China National Native Produce and Animal By-Products Import and Export Corporation, but most other foodstuff exports are the domain of China National Cereals, Oils and Foodstuffs Import and Export Corporation. Hong Kong is the largest importer of Chinese foods; about 50 percent of its imports from China are foodstuffs, especially live animals and frozen meat. The second largest importer is Japan, which in 1976 bought $236 million worth of foodstuffs from China.

China exported $2,050 million of foodstuffs in 1974 and $2,125 million in 1975. After elimination of grains exported mainly to the less developed countries and of live animals exported to Hong Kong, the export value of the remaining categories (meat and fish, eggs and dairy products, fruits and vegetables, tea and spices, and tobacco) was $915 million in 1974 and $960 million in 1975. Preliminary figures for 1976 indicate that exports of foodstuffs fell slightly to $1,945 million in that year.

Canned foods are an important item. China reportedly has 100 factories producing 400 varieties of canned goods. Canned Chinese products are popular with European buyers at the Canton Fair; however, exports to the United States present more of a problem because of stringent U.S. Government regulations and high-quality expectations on the part of the consumer. Recently, the Chinese filed registration and processing forms with the U.S. Food and Drug Administration for low-acid canned goods produced by Maling Canned Food Factory and Hangchow Canned Food Factory for pork and vegetable products, and Kwantung Cannery for canned melon and bamboo shoots. Frozen foods are another important export item. China uses blast freezing or freezing by refrigeration, which gives the food a frosted look, rather than the new IQF or nitrogen tunnel process used in the United States; however, this does not seem to discourage European or Japanese buyers.

In 1975 China had stepped up construction of meat-processing plants, cold-storage warehouses, and refrigerated trucks. A refrigeration plant capable of processing 10,000 chickens or rabbits daily for export has been put into operation in Tsinan, Shantung Province under the auspices of the local Foreign Bureau. However, it was estimated that inadequate refrigeration equipment may hamper further development of cold-storage plants. China has only about thirty plants manufacturing refrigeration equipment. *Business China* in February 1976 predicted a possible import market for refrigerated transport equipment or for the know-how and machinery to make it.

Another possible market related to food processing is for packaging equipment. China purchased $260,000 canning plant equipment from Toyo Canning of Japan in 1972, $119,000 biscuit and confectionery wrapping machinery in 1972, and $1,732,000 worth of tea-packing machinery in 1974, both from Rose Forgrove Ltd. of England. Chinese interest in packing and packaging equipment in general was demonstrated by the Japanese Printing and Packaging Machinery Exhibition held in Tientsin in November 1974, which featured automatic bag filling, vacuum packing and box-making machines, and the September 1976 Shanghai Exhibition of Italian Packing Machinery and Medical Appliances. In 1977 there were three foreign seminars of packaging held in China, and a delegation of the Chinese National Export Commodity Packaging Corporation (CHINA-

PACK) was scheduled to visit Sweden, Canada, and the United States to study packaging technology and equipment. At a July 1977 conference on foreign trade held in Peking, State Council Vice Premier Yu Chiu-li stated that Chinese should, among others, do a better job in packing to better meet the demands of foreign markets.

CONSUMER DURABLES

Ownership of consumer goods, especially the durable consumer goods such as bicycles, sewing machines, watches, clocks, and radios has been increasing in recent years, although some of these items continue to be not only rationed but expensive, and the demand far exceeds the supply. It is unlikely, however, that this demand will be met by any significant imports from abroad.

In the People's Republic of China the unsaturated market for the durable consumer goods provides a mechanism for additional capital formation. Retail prices of such durable consumer goods as watches or bicycles appear to be extremely high relative to average earnings, and their sales may be generating very high profits to the state that could be channeled for investment in other sectors with higher development priorities.

Bicycle and sewing-machine industries were already in existence in 1949 and in the early 1950s constituted the backbone of the consumer appliance industry. Although sewing-machine production originally grew more rapidly than production of bicycles, the actual number of units produced has been consistently lower than the number of bicycles since 1955. Output of watches was lower than either sewing machines or bicycles in 1965, but by 1971 it was above the output of either of these items. The radio industry, especially production of transistor radios, is a relative newcomer. For discussion of radio and television production, see Chapter 14 (see also Table 16.3).

Production and Import of Watches

Production of watches reached 7.8 million units in 1973. Prices vary from 85 yuan for the cheapest brand to over 200 yuan for the more expensive, Swiss-technology watches, made in Tientsin. In 1960 Shanghai, mainly Shanghai Watch Plant No. 1, accounted for 69 percent of total watch pro-

Table 16.3 Growth in production of selected durable consumer goods. In millions of units

Type of Product	*1965*	*1970*	*1971*	*1972*	*1973*	*1974*	*1975*
Bicycles	1.8	3.6	4.0	4.3	4.9	5.2	5.5
Sewing machines	1.6	2.4	3.0	3.3	3.9	NA	NA
Watches	1.2	NA	6.2	6.9	7.8	NA	NA
Radios	1.5	4.6	6.0	6.7	12.1	15.0	18.0
Televisions	0.005	0.015	0.02	0.04	0.075	0.115	0.205

Source: U.S. Government International Handbook of Economic Statistics, 1976; U.S. Government, "People's Republic of China Handbook of Economic Indicators," *Research Aid,* ER 76-10540, August 1976; U.S. Government, "China: Economic Indicators," *Reference Aid,* ER 77-10508, October 1977.

duction in China. In later years the proportion of Shanghai-produced watches diminished. Sian, another watch-producing center, has six watch factories, an industry research institute, and a watchmaking equipment plant that supplies over seventy enterprises in various parts of the country. In Peking, China's third largest wristwatch factory has 2000 workers and produces seventeen-jewel, three-hand watches and forty-jewel automatic watches.

Watches are one of the few consumer items imported by China in quantity and provide perhaps the most striking example of profits to the state generated by sale of consumer durables. While domestically produced watches are rationed, imported Japanese and Swiss models are available without rationing tickets but at prices of 250 to 800 yuan, compared to the 85 to 200-yuan price range for Chinese-made watches. Expensive Swiss watches are purchased by the Chinese often as personal investments at prices 500 to 1000 percent over their unit import value. Chinese citizens are not allowed to own gold or precious metals, and there seems to be a constant market for a certain amount of expensive foreign watches. Thus the import of 1 million Swiss watches valued at $10 million could bring in retail up to $100 million, providing the state with a very significant profit margin.

Chinese imports of Swiss watches began about 1955. During the late 1960s these imports reached almost 500,000 units annually, and in 1973 and 1974 over 1 million watches valued at about $10 million were imported each year. In 1974 total imports of watches reached $20 million, and in 1975 China imported $15 million worth of watches, with $1 million each from the United States and Japan and $13 million from

Switzerland. Imports in 1976 remained at the $15 million level, with Switzerland, Japan, and Italy being the major suppliers.

Bicycles

In 1976 there were some 50 million bicycles in use in China, and efforts were reportedly being made to further increase and improve production, which rose from 3.9 million units in 1970 to 5.5 million units in 1975. Currently produced lines include sports models, light- and heavy-duty roadsters, minibikes, collapsible and multigear bicycles, and mopeds.

Major bicycle-producing centers, namely, Shanghai, Tientsin and Tsingtao, together accounted for over 70 percent of total bicycle production in the 1960s. The Shanghai Bicycle Plant alone produced 33 percent of the total bicycles manufactured in China in 1961. It was making the "Permanent" brand bicycles, which were regarded as the best in China. Shanghai No. 3 Bicycle Plant, which was modernized in 1961 at the cost of 700,000 pound sterling, specializes in "Phoenix" bicycles, which are heavy-, standard-, light-, and miniature-wheel roadsters advertised as export items. It also makes sports and minibike models. Its output was over 1,000,000 units in 1973. The Tientsin Bicycle Plant was established in 1936 but has since been modernized and is the second largest bicycle plant in China. In 1974 over 70 percent of its production was automated, and it had at least ten mechanized production lines. Its output increased from 780,000 units in 1972 to 1 million in 1974. The plant produces "Flying Dove," bicycles which in 1972 were priced at 180 yuan. Other major bicycle plants include Tsingtao Bicycle Plant, the third largest in China, manufacturing "Dawn" bicycles, and Canton Bicycle Plant, with an output of 400,000 units per year.

The bicycle industry was already in existence in 1949, but all components had to be imported. Production of bicycles from Chinese-made parts rose steadily from 21,000 units in 1950 to 5,460,000 units in 1975. During 1955–1958 China imported 155,000 bicycles from Japan and the Soviet Union, about 5.2 percent of its total bicycle production during those four years.

Despite the steady increase in production, the demand for bicycles exceeds the supply, and bicycles are not only rationed but also expensive. The average price of a bicycle was 147.76 yuan in 1952, rose to 160.00 yuan by 1957, and was still at that level in 1972. Actual prices in 1972 ranged from 134.00 yuan for a "red Kapok" model to 176.00 yuan for the "Phoenix" or 180.00 yuan for the "Flying Dove" models. The maximum price was 200.00 yuan. Since an average worker earns about 50 yuan per month, purchase of a bicycle represents an average 3 months of work. In contrast, in the United States a typical model priced about $100 would represent only one week of work for an average worker. Nevertheless, there are no reports of recent bicycle imports, and in fact China exports bicycles to some fifty countries. This situation is expected to continue with increasing production of bicycles as well as mopeds, which may well become the highest personal luxury items in the not-too-distant future.

Sewing Machines

Production of sewing machines rose from 9360 units in 1949 to 3,894,000 in 1973, the last year for which production data are available. The sewing-machine industry grew rapidly during the 1950s. By 1960 there were eighteen sewing-machine plants in Shanghai alone and dozens of plants in Tientsin, Tsingtao, Harbin, and Shenyang, and the number of models produced rose from twenty in 1958 to thirty-five in 1962.

Shanghai No. 1 Sewing Machine Plant was the foremost producer, specializing in the JA 6–1 model machine for household use. Its production was 90,000 units in 1958 and double that in 1962. The second largest plant was Huanan Sewing Machine Plant in Canton, producing 300,000 machines per year. These two plants together accounted for 53 percent of the total output in 1959. Average prices of sewing machines in current yuan per unit were 190.00 yuan in 1952 and 108.50 yuan in 1957 exfactory, and 140.00 yuan in 1965 and 152.00 yuan in 1972 retail prices. In Peking average prices were 137 yuan in 1972 and 167.00 yuan in 1974. Some sewing-machine models are exported.

HANDICRAFTS

Handicrafts include a wide variety of utilitarian and purely ornamental objects hand made from wood, ivory, bronze, bamboo sticks, jade, porcelain, and other materials. They have been exported by China for over 2000 years.

In 1949 handicraftsmen represented the majority of nonagricultural workers and were largely self-employed. Cooperativization of individual handicraftsmen began with the communist administration and was virtually completed by 1956. Later the cooperatives were converted into factories. Today some handicrafts such as woven baskets, macrame, and simple embroidery are produced by commune workers as full-time or sideline occupation, but most traditional crafts such as art goods, pottery, cloisonné and enamel objects, ivory and jade carvings are produced in special factories.

Chinese handicrafts may be purchased from private companies in Hong Kong that have been designated by the Chinese government as the distributors of Chinese products. They are also available to foreign buyers during Canton Trade Fairs. In the past 2 years handicrafts have also been promoted at special minifairs held in different provinces of China and at the handicraft exhibitions in Hong Kong. Sales of handicrafts and antiques are handled by the China National Light Industrial Products Import and Export Corporation.

Demand for Chinese handicrafts is strong, especially among the industrialized countries. Recognizing the value of their handicraft exports, the Chinese have even adjusted the decorative themes to suit the foreign buyer. Since the early 1970s contemporary socialist themes have been replaced by "rejuvenated" traditional art, images of folk figures, landscapes, flowers, birds, insects, and fish.

In 1974 Chinese exports of handicrafts and light manufactures amounted to $210 million, of which $110 million was sold to the developed countries and $20 million to Hong Kong and Macao. In 1975 total exports amounted to $190 million, but the value of exports in 1976 increased sharply to a total of $320 million.

Because handicraft exports appear under several categories, it is difficult to assess the exact value of the U.S. imports of Chinese handicrafts. In 1972 the United States imported $1 billion worth of handicrafts, including $200 million worth from the Orient (China, Hong Kong, Japan, Korea, and Taiwan). Two of the products included in the handicraft and light-manufactures category are basketwork and pottery. In 1975 China's exports of basketwork (SITC code 8992) amounted to over $56 million, including $6 million to the United States, $14 million to Japan, and $11 million to Singapore. American imports of Chinese dinnerware and ornamental pottery have doubled from $767,149 in 1973 to $1,670,942 in 1975, most of the increases occurring in medium-priced dinnerware sets. However, Chinese products accounted only for 0.34 percent by value and 1 percent in terms of units of the household and hotel pottery products imported by the United States in 1975.

In the past 10 years there has been considerable expansion in the export of toys. Chinese toys are displayed regularly at the Canton Trade Fairs, but 90 percent are exported to Hong Kong for resale. Major toy factories located in Canton and Shanghai produce a variety of toys, sporting goods, and musical instruments designed specifically for export.

LIGHT-INDUSTRIAL EQUIPMENT PRODUCTION AND IMPORTS

The Production Control Bureau No. 1 of the First Ministry of Machine Building is believed to supervise the manufacture of most equipment required by light-industry manufacturing enterprises. Machinery for paper, rubber, printing industries, and some food- and sugar-processing equipment is also believed to be under the control of this ministry. The Ministry of Light Industry is believed to manufacture other equipment, notably textile machinery, which became its responsibility when the Ministry of Textile Industry was absorbed by the Ministry of Light Industry.

Except for textile machinery, the production of equipment for light industries did not enjoy priority in the earlier years, and its output value was not significant. Only 7000 tons of paper-making equipment and 9000 tons of sugar-refining equipment were produced in 1957, and the output value of light-industrial machinery was estimated at 2.8 percent of the total machinery production at that time. There is practically no information on production of specific equipment for the light industry, but China offers several products for export. These include over fifty types of textile machines, eight types of knitting machines, and almost thirty types of printing machines. There are also over twenty types of simple food-processing machines and over thirty types of rubber and plastic-making machines, all evidence of a well-developed light-industry manufacturing capability that is already looking for export markets.

Light-industry equipment represents a relatively small market for the foreign manufacturer. Early imports of food-processing machinery and light-industrial equipment from the Soviet Union reached a high of $571,000 in 1959, then declined, and ceased in the early 1960s. During the 1970s some textile and leather-making machinery and paper-making and printing equipment were imported from Japan, the United States, western Europe, and East Germany. The values of imports were higher during 1973–1975 than during the preceding 3 years. Imports of textile machinery (SITC code 717.1) totaled $51 million dollars during 1970–1975, averaging $8.5 million annually. However, during 1970–1972 the average annual import value of textile machinery was only $0.76 million. Imports of leather machinery (SITC code 717.2) and sewing machinery (717.3) averaged $1.2 million annually for 1970–1975 but only $0.03 and $0.24 million, respectively, for 1970–1972. Paper and pulp machinery (SITC code 718.1) and printing equipment (SITC code 718.2) totaled $10.6 million and $9.96 million, respectively, for 1970–1975, but these figures represent mainly the purchases made during 1973–1975. Imports of food-processing machinery were negligible, totaling only $1.78 million for 1970–1975 and $0.38 million for 1970–1972 (see Table 16.4).

Table 16.4 Imports of light-industrial equipment during 1970–1975 (Value in millions of $U.S.)

SITC Code	*Type of Equipment*	*1970*	*1971*	*1972*	*1973*	*1974*	*1975*
(717.1)	Textile machinery	0.5	0.5	1.3	6.9	17.0	24.8
(717.2)	Leather machinery	0.02	0.06	0.01	2.6	3.8	1.8
(717.3)	Sewing machinery	0.04	0.08	0.6	2.1	4.0	0.7
(718.1)	Paper & pulp machinery	0.05	0.17	0.18	0.2	4.6	5.4
(718.2)	Printing machinery	1.118	0.218	0.324	2.1	4.8	1.4
(718.3)	Food processing machinery	0.019	0.013	0.35	0.4	1.0	0.0

Source: U.S. Department of Commerce, *Market Share Reports,* "People's Republic of China 1969–73" and "People's Republic of China 1971–75"; *United Nations Bulletin of Statistics, World Trade in Engineering Products for 1973, 1974, and 1975.*

REFERENCES

Ashbrook, Arthur G., Jr. "China: An Economic Overview, 1975," in *China: A Reassessment of the Economy,* Joint Economic Committee, U.S. Congress, p. 20.

British Broadcasting Corporation. *Summary of World Broadcasts, part 3, Far East Weekly Economic Report:* FE/W877/A10, May 12, 1976; FE/W855/A13, December 3, 1975; FE/W899/A11, October 13, 1976; FE/W912/A4, January 19, 1977; FE/W919/A10, March 9, 1977; FE/W920/A3, March 16, 1977; FE/W911/A8, January 12, 1977; FE/W926/A11, April 27, 1977; FE/W930/A8, May 25, 1977; FE/W935/A18, June 28, 1977; FE/W949/A17, October 5, 1977; FE/W878/A10, May 19, 1976.

Chen, Nai-Ruenn. "China's Foreign Trade, 1950–1974," in *China: A Reassessment of the Economy,* Joint Economic Committee, U.S. Congress, July 1975, p. 617.

Cheng, Chu-yuan. *The Machine Building Industry in Communist China,* Edinburgh University Press, 1972.

China Machinery Corporation. *Export Catalogue 1973,* MACHIMPEX, Peking.

China Business Review, The. "China: Selected Economic Indicators as of 1977," September–October 1977, p. 12.

China Business Review, The. "China International Notes," September–October 1977, p. 50.

China Business Review, The. "Importer's Notes," July–August 1977, p. 24.

China Trade Report. "World Dateline: China," Far Eastern Economic Review Ltd., June 1977, Hong Kong.

China Trade Report. "World Dateline: Hong Kong," Far Eastern Economic Review Ltd., February 1977, Hong Kong.

China Trade Report. "Business Pipeline: Paper and Pulp," Far Eastern Economic Review Ltd., September 1975, Hong Kong.

China Trade Report. "Business Pipeline: Toys," Far Eastern Economic Review Ltd., March 1977, Hong Kong.

Current Scene. "Tachai and Economic Tasks," February 1977, Hong Kong.

Emerson, John Philip. "Employment in Mainland China: Problems and Prospects," in *An Economic Profile of Mainland China,* Vol. 2, Joint Economic Committee, U.S. Congress, February 1967, p. 403.

Field, Robert M. "Chinese Industrial Development 1949–1970," in *People's Republic of China: An Economic Assessment,* Joint Economic Committee, U.S. Congress, May 1972, p. 68.

Harmeling, Hope. "China's China," *The China Business Review,* May–June 1977, p. 27.

Jenkins, Brian Michael. "Arts and Crafts," in William W. Whitson, ed., *Doing Business with China,* Praeger, New York, 1974.

Kim, Young C. "Sino Japanese Commercial Relations," in *China: A Reassessment of the Economy,* Joint Economic Committee, U.S. Congress, July 1975, p. 600.

Marin, Minette. "China Industry Spotlight: Textiles," *China Trade Review,* Far Eastern Economic Review Ltd., September 1976, Hong Kong.

Oppenheimer, Arthur and Geoffrey McCarron. "Food Products," in William W. Whitson, ed., *Doing Business with China,* Praeger, New York, 1974.

Pomeranz, Y. "Food and Food Products in the People's Republic of China," *Food Technology,* March 1977.

Reynolds, Suzanne, R. "Foodstuffs from China," *U.S.–China Business Review,* January–February 1975.

Roberts, John. "The Textile Industry," in William W. Whitson, ed., *Doing Business with China,* Praeger, New York, 1974.

Roll, Charles Robert, Jr. and Kung-chia Yeh. "Balance in Coastal and Inland Industrial Development," in *China: A Reassessment of the Economy,* Joint Economic Committee, U.S. Congress, July 1975, p. 81.

Rossbach, Sara. "Handicrafts: Sales Rising," *China Trade Report,* Far Eastern Economic Review Ltd., February 1977, Hong Kong.

Sino British Trade Council. "China's Bicycle Production," *Sino-British Trade Review,* No. 139, April 1976.

Sino British Trade Council. "New Products at the Spring Fair," *Sino British Trade Review,* No. 141, June 1976.

Sino British Trade Council. "Chinese Industry," *Sino-British Trade Review,* No. 150, March 1977.

Strauss, Paul. "Fibres Future," *China Trade Report,* Far Eastern Economic Review Ltd., March 1976, Hong Kong.

South China Morning Post. "Watch Industry Thrives," October 24, 1974, Hong Kong.

United Nations. *Bulletin of Statistic on World Trade in Engineering Products,* issues 1973, 1974, and 1975, New York.

United Nations. *Statistical Yearbook, 1976,* New York, 1977.

U.S.–China Business Review. "China Economic Notes: Refrigeration," May–June 1976.
U.S. Government. "China: International Trade 1976–77," *Research Paper* ER77-10674, October 1977.
U.S. Government. *Handbook of Economic Statistics, 1976 and 1977.*
U.S. Government. "China: Role of Small Plants in Economic Development," *Research Aid,* A(ER)74-60, May 1974.
U.S. Government. "Production of Machinery and Equipment in the People's Republic of China," *Research Aid,* A(ER)75-63 May 1975.
U.S. Government. "Prices of Machinery and Equipment in the People's Republic of China," *Research Aid,* A(ER)75-64, May 1975.
U.S. Government. "People's Republic of China: Handbook of Economic Indicators," *Research Aid,* ER 76-10540, August 1976.
U.S. Government. "People's Republic of China: International Trade Handbook," *Research Aid,* ER 76-10610, October 1976; A(ER)75-73, October 1975; A(ER)74-63, September 1974.
U.S. Government. "People's Republic of China Timber Production and End Uses," *Research Aid,* 76-10493, October 1976.
Veilleux, Louis. "Paper Products," in William W. Whitson, ed., *Doing Business with China,* Praeger, New York, 1974.

17

Education and Research

The progress of education and research in the People's Republic of China has been characterized by alternating periods of expansion and consolidation, normal to any developing country that must achieve a balance between the demands of the growing economy and the supply of trained manpower. In China another factor complicates the issue; the ultimate goal of the communist ideology is to eliminate the difference between manual and mental labor and to abolish the elitism of the educated. The role of education is to create cultural laborers with social consciousness who are both ideologically aware (red) and technically trained (expert) and can move between the physical and intellectual aspects of a job.

The successive changes in the form and content of education since 1949 reflect the alternating shifts in the emphasis from "red" to "expert" and back (see Table 17.1). The most important shift to "red" came during the Cultural Revolution, which disrupted the existing education and research programs and produced some drastic reforms. In the 1970s the emphasis began to swing slowly away from the extreme "red" to a balance between the two concepts. The movement suffered a setback during the 1973 anti-Confucius campaign, but recent articles in the Chinese press and reports of visitors to China indicate that the new leadership directives to develop science and modernize science and technology augur well for the future development of research and higher education. For example, the graduate programs and university entrance examinations were revived in 1977, and the duration of college courses restored to 4 years for most subjects, and 5 years for medicine.

EDUCATIONAL SYSTEM

The educational goals and policies of the People's Republic of China are established by the central government, but their implementation is left

Table 17.1 Research and education in People's Republic of China, 1949–1977

Period	*Education*	*Research*	*Acquisition of Foreign Technology*
Liberation (1949)	elite of 36,300 higher intellectuals, including 1100 Ph.D.'s; most trained in western Europe, U.S., or Japan		
First Five Year Plan (1952–1957)	expansion of primary and secondary school system and establishment of specialized colleges and universities	Soviet model technocracy being formed (1957); Twelve Year Plan for science and research development	comprehensive technological transfer from U.S.S.R., Soviet advisers in China, Chinese students and scientists in U.S.S.R.
"Leap Forward" (1958–1960)	shift to "red"; decentralization, overexpansion; education to combine with production, workshops and small plants in educational establishment; decrease in state funds	stress on self-reliance, increase in state funds allocation	rejection of Soviet influence; Soviet aid withdrawn in 1960
Normalization (1960–1965)	shift to "expert"; substandard schools closed, advanced and graduate training in institutes and universities	return to centralized planning, research done mainly by the research institutes, basic and theoretical research	limited import of foreign equipment and technology, acquisition of scientific literature, prototype copying
Cultural Revolution (1966–1969)	revamping of the educational and research systems, colleges and universities shut down, scientists and educators sent to work in factories and field and to be reeducated ideologically		policy of self-reliance
1970–1977	colleges reopen (1970), educational process shortened, curricula and college admissions altered, open school policy; 1975 on—some normalization; In 1977 graduate studies and entrance examinations revived, college terms extended to 4–5 years	decentralization; basic research condemned; research directed at solving immediate industry problems; 1976 on—new emphasis on development and modernization of science & technology, more rational use of specialists	scientific exchanges begin; import of plants, systems, and equipment; foreign technicians in China; Chinese technicians abroad

largely to the discretion of the local authorities. Consequently, the level of primary, secondary, and higher education varies with the locality and the authority responsible for the particular school.

Primary and Secondary Education

Chinese youths enter the school system at the age of 7 years and leave it at 15 years, having completed some 9 years of primary–secondary education. Only a part of that time is spent in classroom activities. Students are expected to participate in production. Generally, the school year in a middle (secondary) school includes one month in a factory or a workshop and one month of farm work. The actual duration may vary. Since the Cultural Revolution primary and secondary schools are no longer operated by the Ministry of Education but are financed and organized by communes, enterprises, and urban neighborhoods, and their curricula are adjusted to meet local needs.

Middle-school graduates were expected to gain a minimum of 2 years practical experience in a commune, factory, or the army before continuing their education. Between 1968 and 1975 some 12 million urban youths were sent to work in the communes as part of the "down-to-the-countryside" youth program. This may now be changing. In December 1977 a mass university entrance examination was held throughout China. It was the first such examination since the Cultural Revolution, and for the first time since the revolution secondary school students were encouraged to apply to the university, directly after leaving school.

Higher Education

Higher-education institutions include colleges and universities under the Ministry of Education and a large number of specialized colleges and schools with a narrow range of courses run by individual industries, ministries, and local governments.

Most of China's scientific and technical contingent is trained in worker's universities, also called "July 21st colleges." They are patterned after the school set up by the Shanghai Machine Tool Plant and were commended by Mao Tse-tung on July 21st 1968. These colleges, run by a single enterprise or jointly by several enterprises, offer 1 to 2-year courses in vocational training and political education. Instructors include experienced workers and technicians as well as personnel from the universities or research institutes. Some 15,000 worker's universities were reported in existence in 1975, with a total enrollment of 780,000 students. During the 1977 Tachai Conference top officials called for further proliferation and development of these schools.

Similar education is provided by the state-run technical institutions. Provinces and municipalities with relatively well-developed industries are entrusted to train technicians for the less-developed areas. In most cases, however, graduate students return to their old work unit after completing the course. The "from commune or plant back to commune or plant" policy applies particularly to graduates from rural areas and those graduating from agricultural college equivalents of the worker's university. Medical, dental, and educational school graduates are sent to the areas where their skills are most needed.

Before the Cultural Revolution regular colleges and universities provided China with higher-level personnel and leading scientists and engineers. Their function was teaching, information, and research. The Cultural Revolution downgraded the educational status of these institutions to a level of trade schools, but they continued to function outside the worker's college system. Latest reports indicate that the new leadership recognizes the need for higher education: if China is to overtake the United States economically by the year 2006, it will need a core of highly trained people to absorb, develop, and disseminate new technologies.

Table 17.2 lists the universities known to have existed at the start of the Cultural Revolution. Some may no longer exist. Only about twenty of the universities offered a range of art and science courses comparable to that of a Western-type university. Today, the most publicized universities, that is, those mentioned by the New China News Agency reports and frequently exhibited to foreign visitors, include Chungshan in Canton, Futan in Shanghai, and Peking and Tsinghai Universities in Peking. Nanking, Lanchow, Kirin and Wuhan Universities are among the institutions visited by U.S. scientists.

Futan University has fourteen departments, forty-seven laboratories, three subsidiary research institutes, and five factories. It conducts research in electronics, laser beams, petroleum catalysis, electric light sources, crop cultivation and plant diseases, and application of seismological techniques in tapping oil. Basic sciences are also studied, and research is conducted in theoretical physics.

Peking University has seventeen arts and science departments. Its Electronic Plant produces medium and small computers. Scientific research projects include basic theoretical research in nuclear physics, geophysics, and mechanics as well as research into technological techniques and product manufacture. Although graduate programs in colleges were abolished, a graduation thesis was written by a Peking university graduate in 1973. Chou P'ei-yuan, vice chairman of the university's Revolu-

Table 17.2 **Universities of China existing prior to the Cultural Revolution**

Name of University	*Location*	*Province*	*Date Established*
Amoy	Amoy	Fukien	1921
Anhwei	Hofei	Anhwei	
Central			
Chekiang	Hangchow	Chekiang	1927
Chengchow	Chengchow	Honan	
China Scientific & Technical	Hofei	Anhwei	
Chinan	Canton	Kwantung	1958
Chinese Univ. Science & Tech.	Peking	Peking Municipality	1958
Chuanchow	Chuanchow	Fukien	
Chunking	Chunking	Szechwan	1957
Chungshan (Sun Yat-sen)	Canton	Kwantung	1924
Futan	Shanghai	Shanghai Municipality	1905
Foochow	Foochow	Fukien	1960
Hangchow	Hangchow	Chekiang	1958
Heilungkiang	Harbin	Heilungkiang	
Hopei	Tientsin	Hopei	
Hunan	Changsha	Hunan	
Hupeh			
Inner Mongolia	Huhehot	Inner Mongolia A.R.	1958
Kiangsi	Nanchang	Kiangsi	1958
Kirin (People's Univ. N.E. China)	Changchun	Kirin	1958
Kwangsi			
Kweichow	Kweiyang	Kweichow	
Lanchow	Lanchow	Kansu	1946
Liaoning	Shenyang	Liaoning	1958
Nankai	Tientsin	Hopei	1958
Nanking	Nanking	Kiangsu	1902
Ninghsia	Yingchwan	Ninghsia A.R.	
Northeastern			
Northwestern	Sian	Shensi	1937
Overseas Chinese	Peking	Peking Municipality	1960
Peking	Peking	Peking Municipality	1898
Peking Normal	Peking	Peking Municipality	1902
Shantung	Tsinan	Shantung	1926
Shanghai, Chiaotung	Shanghai	Shanghai Municipality	1816
Shansi			
Sian, Chiaotung	Sian	Shensi	
Sinkiang	Urumchi	Sinkiang	
South Anhwei			
Szechwan Comprehensive			
Szechwan	Chengtu	Szechwan	1931
Tientsin	Tientsin	Hopei	
Tsinghai	Hsining	Tsinghai	1960
Tsinghua (People's Univ.)	Peking	Peking Municipality	1908
Tungchi	Shanghai	Shanghai Municipality	1927
Science & Technology	Shanghai	Shanghai Municipality	1959
Wuhan	Wuhan	Hopei	1913
Yen Pien	unknown		
Yunnan	Kunming	Yunnan	1922

Source: Compiled by the 21st Century Research from sourecs listed at the end of this chapter.

tionary Committee and a noted physicist, was one of the strongest advocates of improving the academic quality of higher education during the education debate of the 1970s. He is a noted figure in international scientific activities and visited the United States in 1975.

RESEARCH

The central government of the People's Republic of China has always recognized the importance of scientific research. The science budget was the fastest growing budget between 1952 and 1965; with 117 percent increase compared to the overall increase of 16 percent. The largest single item was the nuclear development program. Goals for scientific research were initially laid down in the first Five Year Plan and later elaborated in the Twelve Year Plan (1956–1967). The plan identified 582 problems to be solved in the following areas: (1) peaceful use of atomic energy, (2) new electronic techniques, (3) jet propulsion, (4) automation in production and precision equipment, (5) surveying and prospecting for petroleum and other scarce materials, (6) exploration of mineral resources, (7) metallurgical studies, (8) development of fuels and heavy machines, (9) utilization of Yangtze and Yellow rivers, (10) agriculture, with emphasis on mechanization and use of chemicals, (11) prevalent diseases, and (12) basic natural sciences.

Overall research goals are established at the highest level, but their implementation is left to the local authorities. Since the Cultural Revolution industry has had a more direct control over the nature of research programs. Achievements in the civilian research during the past decade were mainly those in applied science affecting industry and agriculture. Other active research areas were meteorology, seismology, and oceanography, each under the direction of a special agency of the State Council. Basic research became restricted to a few vital areas, particularly nuclear physics.

Chinese Academy of Sciences

The Chinese Academy of Sciences (CAS) remains the single most important center for civilian research. Its director, Kuo Mo-jo, has held his post since 1949, the year in which the CAS was formed. The CAS is generally regarded as an independent agency of the State Council, although for several years it was also controlled by the State Scientific and Technology Commission, which was disbanded in 1971. It conducts advanced research, makes policy suggestions to higher levels, carries out government policy, and provides technical instruction to subordinate units. The CAS also organizes symposia and conferences and coordinates large-scale research projects such as the 1973–1977 comprehensive survey of Tibet, which involved some 400 scientists from fifty specialties.

In the wake of the Cultural Revolution the personnel of the CAS and the scope of its research and educational activities did become reduced. Of the original 114 CAS institutes only twenty-six, all in Peking, presently remain under direct CAS control, forty-three are under dual control of the CAS and local authorities, and the rest are no longer controlled by the CAS.

Research Institutes

Most of the research is carried out by research institutes. Some of these are run by CAS, and others are under direct control of industrial ministries or special agencies of the State Council. Universities and industrial enterprises may have their own subsidiary institutes. The institutes combine research with manufacture and either run their own pilot plants or work in cooperation with neighboring factories. They also train graduates from universities and middle schools, compile pertinent foreign literature, and publish magazines and reports. There is a close cooperation between the various institutes and between the institutes and industrial enterprises. Some research projects such as development of special equipment may require joint efforts of several institutes and plants (see Table 17.3).

A research institute consists of one or more laboratory section or department. There is one director and one or more deputy director, usually scientists with party affiliations, and a Revolutionary Committee consisting of scientists, technicians, and workers who make decisions concerning the research programs. A party structure parallels the administration.

Although institutes themselves make small claim on foreign technology, they play an important role in decisions regarding purchase of foreign technology. Representatives of the institutes participate in scientific exchange visits and in visits of industry study groups to foreign countries and are able to make recommendations based on these studies. They are often present at business negotiations with Chinese trade companies.

Acquisition and Dissemination of Technology

Educational and research institutions represent a small market for scientific instruments. Their chief interest in foreign technology, however, is

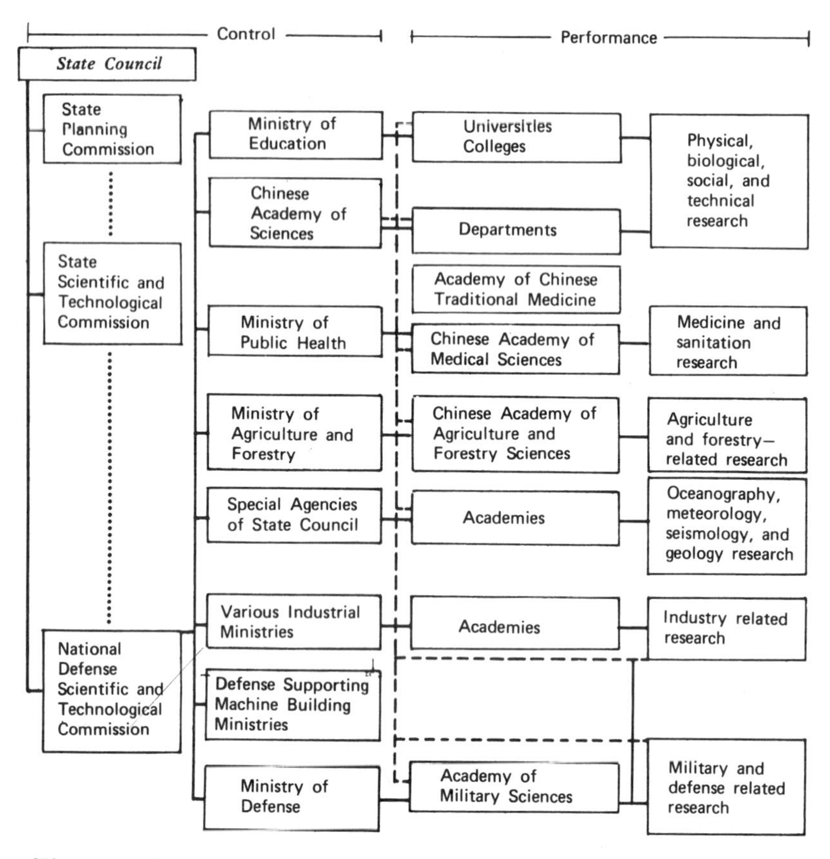

Figure 17.1 Organization of research in China. **Note:** ***State Planning Commission administrators are thought to be directly involved in scientific planning and technology transfer. The State Scientific and Technological Commission was disbanded in 1969–1971, but presence of provincial commissions suggests that a national-level organization may exist, perhaps under another name. The chart was compiled by 21st Century Research from various sources mentioned at the end of the chapter. It reflects presumed mechanics of operation rather than formal organizational structure.*** **Code:** ——— ***control and guidance; - - - - - - guidance and/or coordination; . . . . . . coordination.***

obtaining scientific information from literature and scientific exchange programs. Knowledge thus acquired is disseminated to other institutes and to the decision-making bodies in the government. The largest importers of foreign technology are the large-scale basic and military industries that acquire complete plants and systems, and medium to small manufacturing enterprises under provincial or municipal control that acquire mainly prototypes for adaptation and copying.

Technological know-how eventually filters down to the masses, mainly via numerous scientific exchange stations that display the latest techniques and organize mass scientific and technical activities involving specialists from different factories.

REFERENCES

Ashbrook, Arthur G., Jr. "China: Economic Overview, 1975," in *China: A Reassessment of the Economy,* Joint Economic Committee, U.S. Congress, July 10, 1975.

BBC Summary of World Broadcasts. FE/W798/A13, October 23, 1974; FE/W826/A7, May 14, 1975; FE/W842, September 3, 1975; FE/W893/A4, September 1, 1976; FE/W894/A12, September 8, 1976; FE/W947/A3, September 21, 1977.

Cheatham, T. E., et al. "Computing in China, a Travel Report," *Science,* **175,** 134, October 12, 1973.

Cheng, Chu-yuan. "Scientific and Engineering Manpower in Communist China," in *An Economic Profile of Mainland China,* Vol. 2, Part III, Joint Economic Committee, U.S. Congress, February 1967.

China Business Review, The. "China's Future," Vol. 4, pp. 3–5, January–February 1977.

China Business Review, The. "China's Grand Plan," **4,** 3–5, July–August 1977.

Current Scene. "Debate Over College Enrollment," **11,** 11–15, September 1973, Hong Kong.

Current Scene. "Defending Socialist New-Born Things," **12,** 22–24, July 1974, Hong Kong.

Current Scene. "New Formulas Sought in Youth Policies," **12,** 15–19, October 1974, Hong Kong.

Current Scene. "Shifts in Higher Education Policy," **13,** 25–26, September 1975, Hong Kong.

Current Scene. "Science and Technology Promoted," **13,** 17–19, December 1975, Hong Kong.

Current Scene. "Education Debate Renewed," **14,** 17–20, January 1976, Hong Kong.

Current Scene. "1975 Down-to-the-Countryside Program," **14,** 16–19, February 1976, Hong Kong.

Current Scene. "People in the News: New Education Minister," **15,** 23–24, March 1977, Hong Kong.

"Directory of Selected Scientific Institutions in Mainland China," *Hoover Institute Publication Series 96,* Hoover Institution Press, Stanford, California, 1970.

Esposito, Bruce J. "Science in Mainland China," *Bulletin of the Atomic Scientists,* January 1972, pp. 36–40.

Esposito, Bruce J. "The Cultural Revolution and China's Scientific Establishment," *Current Scene,* **12,** 1–12, April 1974, Hong Kong.

Far Eastern Economic Review. *Asia Yearbook, 1974.*

Ginsburg, Philip E. "Development and the Educational Process in China," *Current Scene,* **14,** 1–11, March 1976, Hong Kong.

Heymann, Hans, Jr. "Acquisition and Diffusion of Technology in China," in *China: a Reassessment of the Economy,* Joint Economic Committee, U.S. Congress, July 10, 1975.

New China News Agency September 15, 1974 report (transl. by JPRS #63442, November 1974).

Orleans, Leo A. "Communist China's Education: Policies, Problems, and Prospects," in *An Economic Profile of Mainland China,* Vol. 2, Part III, Joint Economic Committee, U.S. Congress, February 1967.

Orleans, Leo A. "Research and Development in Communist China; Mood Management and Measurement," in *An Economic Profile of Mainland China* Vol. 2, Part III, Joint Economic Committee, U.S. Congress, February 1967.

MacFarquhar, Emily. "Making China Work," *The Economist* December 31, 1977, pp. 28–30, London.

Phillips, W. H. "China's Tomorrow: How 3 Students Fare in a System of Education Undergoing Rapid Transformation," *Wall Street Journal,* November 3, 1972.

Prybyla, J. S. "Reports from China: Notes on Chinese Higher Education: 1974," *China Quarterly,* No. 62, pp. 271–296, June 1975.

Signer, Ethan and Arthur W. Galston. "Education and Science in China," *Science,* **175,** 15–23, January 7, 1972.

Sobin, Julian M. "New Doors into China: the Role of Technological and Commercial Exchanges," *Harvard Magazine,* **79,** 53, October 1976.

Suttmeier, R. P. "Science Policy Shifts Organizational Level Change and China's Development," *The China Quarterly,* No. 62, p. 207, June 1975.

Swannack, Nunn S. "Research Institutes in the People's Republic of China," *U.S.–China Business Review,* March–April 1976, pp. 39–50.

U.S. Government. "Directory of Officials of the People's Republic of China," *Reference Aid,* A (CR)75-16, April 1975.

U.S. Government. "Civilian Scientific Research Academies in the People's Republic of China," *Reference Aid,* CR 76-10211, January 1976.

Wu, Y. L. and Robert Sheeks. "The Organization and Support of Scientific Research and Development in Mainland China," Praeger, New York, 1970. Reviewed in Richard Baum, "Chinese Science" (book review), *Science,* **172,** 669, May 14, 1971.

18

Welfare and Medicine

Health policies outlined during China's First National Health Conference in 1950 emphasized the importance of preventive medicine, involvement of the masses in health programs, and integration of traditional and modern medical practice in bringing health to "workers, peasants and soldiers." By implementing these policies the Chinese were able to achieve outstanding results with minimal expenditure and to create a unique system of public health and welfare.

Despite modest budget allocation (only 2 percent of the national budget was spent on health and sanitation in the years 1953–1957), most parasitic and infectious diseases that once ravaged the country were brought under control by 1960. This was achieved through massive immunization programs and public involvement in "patriotic health movements" of basic hygiene and sanitation measures and destruction of animal vectors such as flies, rats, snails, and mosquitoes.

In the mid-1950s the integration of traditional medicine into the public health system enriched health services by about 500,000 workers and a number of remedies that provided an economical alternative to scarce and expensive Western-type drugs.

To solve an acute shortage of medical personnel the Ministry of Public Health originally set up a dual training program of "middle" medical schools for nurses, technicians, midwives, pharmacists, and "assistant physicians" and "higher medical schools" for physicians. The system based on a Soviet model, produced some 170,000 assistant physicians and 100,000 physicians by 1965, but it did not essentially improve the patient : doctor ratio and perpetuated the elitism of the medical profession. Medical care was still centered in urban areas, and the rural population (80 percent of the people) were without adequate medical help. The system came under severe criticism during the Cultural Revolution and has been since abolished.

The present system of health care and medical education is based on Mao Tse-tung's June 1965 directive, "In medical and health work put stress on rural areas." Since 1965 there has been a shifting of medical personnel and supplies to the countryside, a change in medical-schools curricula, and emergence of a new type of medical worker, the *chijiao yisheng,* or the barefoot doctor.

The barefoot doctor is primarily a peasant with some medical training. He is addressed as "comrade," works in the fields, and shares in the year-end distribution of profits. Barefoot doctors are selected by their community to undergo 3 to 6 months of training in a hospital or clinic, after which they return to serve the community. There are now some 1.5 million barefoot doctors, one third of them women. They treat minor ailments and injuries by drugs, herbal medicines, and acupuncture, immunize children, dispense birth-control information and supplies, supervise sanitary work and herb gathering, and launch patriotic health campaigns.

Together with their urban counterparts (worker doctor and the "Red Health Worker") they bridge the patient–doctor communications gap, educate people in basic health measures, and reduce the workload of hospitals and clinics.

PUBLIC-HEALTH NETWORK

The Ministry of Public Health administers health activities through governmental organizations that function at the province, district, and county levels. The health and welfare of an individual, however, are the primary responsibility of the group to which he or she belongs—a brigade, commune, or the industrial enterprise. The emphasis is on self-reliance rather than central funding and on active public involvement. Patients have the custody of their own records, thus eliminating much of the paperwork.

In rural areas over 85 percent of communes have a Cooperative Medical Service system under which each member of the commune (children included) contributes about 1 yuan annually and in return is entitled to free or partly free medical care. Minor ailments, immunization, and birth-control services are handled by health stations at the brigade level. These stations are sponsored collectively by the commune members. Patients requiring further medical attention go to the commune clinics or to county hospitals. The former are financed by communes and may be subsidized by the government. The latter are located in towns and funded by the financial department of the county. They have surgical, medical, and diagnostic facilities that satisfy the needs of most patients, but they may refer particular cases to larger city hospitals with the brigades paying part or most of the hospital costs.

A similar network exists in the urban areas. Urban districts are subdivided into "neighborhoods" (an urban equivalent of a commune), and each neighborhood consists of a number of lanes. A Lane Committee representing several lanes operates a health station staffed by Red Medical Workers, who are local housewives or retirees with few months of medical training. Health stations are supervised by the neighborhood medical facility. This is an outpatient clinic staffed by modern and traditional physicians, nurses, technicians, and trainees. Patients requiring hospitalization or special treatment are referred to the government-run district hospitals or special research centers. The city Bureau of Public Health coordinates all the health services in the city and is responsible for education of health workers other than physicians.

In urban areas there is no Cooperative Medical Service System, and the method of payment varies. Government workers, university students, certain minorities, and people in "jobs of national interest" (e.g., miners) receive free medical care, though their dependents may have to pay 50 percent of the costs. Job-related diseases and injuries are also treated free of charge. In other cases the patient may have to pay only the registration fee and all or part of the costs of medication. The payment system has great flexibility.

Industrial Health

In addition to rural and urban networks, individual enterprises have their own medical-care facilities. In smaller plants this may be only a health station staffed by worker doctors and health workers. A larger enterprise will have more elaborate medical services. For example, the Third Textile Mill in Peking has seventy full-time medical personnel, a central clinic, and a thirty-bed hospital for its 6000 workers and their families. Factories cover all or part of their employees' medical costs. Annual checkups done at the factory clinics and hospitals include x-ray, vaginal smear, and breast examinations. Cooks and other workers involved in food services are regularly screened for hepatitis, and their stools are examined for parasites. There is special protection for women workers during menstruation, pregnancy, labor, and lactation.

Measures have been introduced to reduce noise and air pollution in factories and mines. People working in high-temperature workshops or having contact with toxic materials receive protective clothing and special nutrition. Those working under specific hard conditions (miners, steel workers, etc.) may retire 5 to 10 years earlier than the normal retiring age.

According to the New China New Agency 1977 report, plans for improving working conditions are worked out at the same time as the annual production plans, and funds for these improvements are drawn from the economic income of the enterprise. Members of State Bureau of Labor inspect mines and factories to check labor safety, and the Peking Institute of Labor Protection sends its researchers to discuss latest developments in labor-protection techniques.

Availability of Medication

Medicines in China are considered relatively inexpensive compared with the United States or even the Soviet Union. Their price, which is standard throughout China has been reduced on several occasions and reportedly is now 20 percent of their cost in the 1950s.

Vaccines, drugs for children, and oral contraceptives are supplied free of charge. Western and herbal medications are available at pharmacies and in department stores, very few of them requiring doctor's prescription. They are also dispensed at clinics, hospitals, and health stations. A barefoot doctor has at his disposal some fifty Western-type drugs, including thirty that would require prescription in the United States.

Herbal medicines are usually dispensed in preweighed packages to be made into tea or brew, but some are available as sterile preparations for injections. In hospitals, decoctions of herbs are distributed to the inpatients in thermos flasks.

There is no agency similar to the U.S. Food and Drug Administration in China, and most pharmacists who prepare and dispense drugs would be regarded as pharmacist aides in the United States; they have 6 months to 3 years of training and no university degree.

Role of Hospitals

In-patient hospitals with advanced technology are found in towns and cities, but they also serve the rural areas. Each hospital acts as a referral center for two to four rural counties. It also sends medical teams composed of doctors, nurses, technicians, pharmacists, and third-year medical students to tour rural areas. Medical teams provide preventive and therapeutic services and supervision and on-the-job training of barefoot doctors and health workers.

In addition to teaching and health-care responsibilities, hospitals, especially those affiliated with research institutes, pursue specific research programs.

SOCIAL WELFARE

There are two types of welfare system in China: (1) the state social insurance system, which covers some 50 million nonagricultural workers, and (2) the cooperative system for rural areas and small enterprises such as street factories.

Type (1) is superior to type (2), as it provides assistance in case of sickness, disability, injury, maternity, and old age. It also has retirement benefits, funeral allowances, and pensions for surviving dependents. The mandatory retirement age is 60 years for men, 55 years for women staff, and 50 years for women workers. Under this system each enterprise sets aside 3 percent from its payroll for the welfare fund. Most of that money (70 percent) is paid out in benefits, and the remainder goes to central union organization for welfare programs. Early in 1977 posters in China called for further improvements of the welfare system for unemployed workers.

Under the cooperative system, each commune sets aside a small percentage of its total annual earnings prior to distribution of profits. This welfare fund is usually administered by production teams or by brigades. It covers all "households in difficulties" and offers them "five guarantees," namely, food, shelter, clothing, fuel, and medical care. In some cases the fund may also cover costs of cremation. The team's welfare fund may be supplemented from the reserve funds of production brigade of the commune, but the emphasis is on self-reliance. The aged and indigent are in the first place the responsibility of their own families, then of the team, and only as the last resort are funds from more central sources used to help them. The system is decentralized, and the quality of aid varies with the economic status of the commune.

CHINESE TRADITIONAL MEDICINE

Chinese traditional medicine is a well-organized body of knowledge that has served China for more than 2000 years. Prior to 1949 it was the only type of medical care available to the majority of people; modern medicine introduced in the early 1900s was practiced by fewer than 40,000 physicians concentrated mainly in the large urban areas.

Although it is not based on physiological methodology, as is Western medicine, traditional medicine does have a primitive science base of experiments and observations and stresses treating the whole patient, taking into account psychosomatic factors. According to traditional medicine, good health depends on maintaining a balance between the interacting forces within and outside the body.

Mao Tse-tung referred to the traditional Chinese medicine as the "great heritage for us to explore and promote," but the policy of integrating modern with traditional medicine was not popular among the modern-type physicians of the original Ministry of Public Health. At the insistence of political leaders, some move toward integration was made in the mid-1950s. The Chinese Medical Association started admitting traditional practitioners (over 1000 had joined by 1956), the Institute of Traditional Medicine in Peking was elevated to the status of Academy under the Ministry of Public Health, and by 1957 there were ten schools in Chinese traditional medicine and twenty-three refresher courses for its practitioners in the country and an integrated hospital in every county. Real progress in the unification of the two disciplines, however, did not come until after the Cultural Revolution.

Today both types of medicine are practiced in clinics, hospitals, and health stations, often on the same patient. For example, a patient undergoing complex surgery may receive Western-type anesthesia plus acupuncture analgesia. Cancer patients on chemotherapy regimens are given herbal remedies to improve general well-being and reduce severity of drug-induced side effects.

Western-style medical students receive instruction in traditional medicine, and Western-style practices are taught to the students of traditional medicine. There is scientific research into the most widely known traditional methods; namely, herbal medicine, acupuncture, and moxibustion (applying heated herbs at specific points of the body). Gymnastics, respiratory exercises, massage and tonics, the preventive measures of traditional practitioner, are in use throughout the country.

MEDICAL EDUCATION AND RESEARCH

The Cultural Revolution, which disrupted all research and education activities, had a particularly strong impact on medicine. The Ministry of Public Health was dismantled, its officials removed, and many top researchers and educators "sent into the countryside." Medical journals ceased publication or served as vehicles for ideological propaganda. No new classes were formed in medical colleges during 1966–1969, and the already enrolled students finished their education ad hoc.

When after a long debate schools reopened there was a drastic revision of the curricula and revision of medical textbooks. Some of the medical colleges failed to reopen, and a number of Western and traditional colleges merged to form schools of "new medicine." In research the direction was changed toward applied research with emphasis on practical application and short-term results.

Medical Education

The original medical education system was abolished during the Cultural Revolution. Today there is a deliberately decentralized system of "training of medical practitioners." Curricula vary in detail, but they all reflect the directives of the central government and emphasize merging of theory with practice, combining traditional and modern practices, and importance of rural health care.

The training period for physicians has been shortened to 3 years and stresses practical training. Students begin nursing duties a few weeks after entering college. In the first year they are assigned to health stations, and in the second year to hospitals and clinics at the commune level, and in the third year they train at tertiary hospitals and for several months tour the countryside with medical teams. They also serve in fields and factories both as ordinary workers and as health workers to gain experience in industrial and rural health problems and care. (In 1977 duration of medical college courses was extended to 5 years, according to the report by a British journalist who visited China that year. There are probably corresponding changes in the curriculum).

In addition to medical colleges there are alternate ways of becoming a doctor. They include vocational schools, apprenticeships under a traditional physician, and certification. Certification is applied to health workers with wide clinical experience and to nurses with many years of service (usually 20). The whole system is designed to "serve the people" and prevent reemergence of the medical professional elitism.

Medical Research

The main areas of medical research in China include acupuncture, herbal pharmacology, allergy, burns, and trauma. Infectious diseases having been brought under control, the priority diseases are now cancer, cardiovascular disease, and respiratory disorders. According to some American visitors, the Chinese may be ahead of the United States in treatment of severe burns and limb replantation, and their field studies and epidemiologic investigations are quite impressive. In the long run, however, lack of basic research and curtailing of medical schools curricula may have adverse effect on medical research.

The research is carried out by the institutes of the Chinese Academy of Medical Sciences, provincial research institutes under Ministry of Public Health, and by medical colleges and their affiliated hospitals. All these institutions work closely with the provincial and city organizations responsible for health services.

The Chinese Academy of Medical Sciences, under the Ministry of Public Health, coordinates medical research activities. It receives some guidance from the CAS and shares some of its top scientists with that organization.

The major vehicle for postgraduate training and information is the Chinese Medical Association. This government-funded organization is closely linked with the Ministry of Public Health and has branches in all the provinces and autonomous regions of China. It selects participants and programs for Chinese medical delegations traveling abroad and arranges scientific exchanges with other countries.

Since 1972, the United States has hosted several medical delegations from China, and several American medical groups have visited the P.R.C. The Chinese delegations included Medicine (1972), Biomedical Engineering and Pain (1973), Pharmacology (1974), Molecular Biology (1975), and Tumor Immunology (1976). Immunology in cancer therapy seemed to be of particular interest to members of the last three delegations.

Since there are no diplomatic relations and no student exchange program between the two countries, only a few Chinese–American scholars were allowed to work with the Chinese researchers. Most other visits are limited to short survey tours of 3 to 4 weeks.

Countries with diplomatic relations with China have progressed beyond the tour-group stage. The Canadian and Chinese Medical Associations are arranging for some extended trips and training of specialists. The Chinese seem interested in Canadian experience in kidney transplants, cardiac surgery, and immunology. The Scientific and Technical Cooperation Agreement between China and West Germany calls for intensified exchange in the fields of pharmacology, acupuncture, microsurgery, and cancer research.

Herbal Pharmacology

Clinical studies of herbal medicines are done in large hospitals and research institutes, including the Institute of Traditional Medicine in Peking. Clinical trials are not comparable to those in the United States Herbal medicines consists of several herbs, and the composition of a decoction may be changed from day to day. The Chinese are not bound by Western methodology, which demands isolation of variables. Moreover, the double-blind, placebo-controlled trial, which forms the basis of Western clinical pharmacology, conflicts with the Chinese principle of keeping the patient fully informed so that he can actively participate in the therapy.

The main centers of pharmacological research are the Materia Medica Institutes of Shanghai and Peking. They develop new drugs, investigate known ones, and raise utilization of herbal medicines. The institute in Shanghai specializes in anticancer and neurological drugs and has used leads from the National Cancer Institute, Drug Research and Development Program in the United States. The Shanghai institute succeeded in developing new process for manufacturing several anticancer drugs from plants.

According to American herbalists who visited China in 1974, development of a significant number of new synthetic drugs from isolated active principles from plants would require an intensive training program for biologists and chemists and the acquisition of extensive new instrumentation that is not likely to occur in the near future. The Chinese are not going to concentrate on isolation of active principles, except for the practical purpose of substituting those herbs that are in short supply. They are more interested in developing better dispensing forms (tablets, pills) and in improving the manufacturing processes.

ENVIRONMENTAL POLLUTION

In 1949 the immediate concern was improvement of human environment at its most basic level; sanitary disposal of human and animal waste. This was achieved through Patriotic Health Campaigns and construction of sewage system in the cities. Today all large cities have sewage-disposal systems, if only in the newer districts, and although human excrement is still widely used as fertilizer, handling of this material is under supervision of health workers.

With the development of industry, inorganic waste has replaced organic as the chief source of pollution. In the earlier years, pollution control was a health project and was handled by the Ministry of Public Health, but since the mid-1960s it also became the responsibility of the economic departments and production units. Workers were exhorted to utilize waste gases, liquids, and slag as valuable sources of raw material.

Environmental research has now (in the 1970s) become an area of rising activity and priority. Industrial enterprises have been pressured to stop water pollution. The Office of Environmental Protection was set up in 1974 and the CAS Institute of Environmental Chemistry was created in 1975. Chinese groups interested in basic research in environmental sciences and application of chemical analysis to monitoring of environmental pollution visited the United Kingdom, West Germany, Sweden, and the United States. The Japanese had an Environmental Protection and Hydraulic Pneumatic Techniques exhibition in Peking in October 1976 and hope to increase their sales of pollution-measuring equipment.

MEDICAL EQUIPMENT PRODUCTION AND IMPORTS

According to reports by foreign visitors, virtually all diagnostic, therapeutic, and research equipment in medical institutions is Chinese-made. Main centers of medical instrument industry are the Peking and Shanghai Municipalities, other centers exist in Hunan Province and Ninhsia Hui Autonomous Region, and a Medical Instrument Factory was identified in Canton. Other electronic equipment or instrument plants may also produce medical equipment.

Shanghai's precision medical-instrument industry produces cancer scanners and orthopedic surgical instruments. It also trial-produced some 300 kinds of advanced apparatus, including the rotary cobalt-60 (^{60}Co) machine, microsurgical instruments, laser therapeutic units and telemetric electrocardiographs. Shanghai has a special Operating Theater and Hospital Equipment Research Center to aid its industry.

Peking produces equipment for export as well as for domestic use. Its output quadrupled between 1965 and 1975. Products include automatic cell counters, high-resolution scintillation cancer scanners, surgical lasers, and laser ophthalmological units.

In 1975, Ninghsia Hui A.R. had two main and four subsidiary plants, producing thirty kinds of products, including incubators, thermostats, operating tables, operating-room lights, osteopathic tractions, and acupuncture needles. In Hunan Province the output of medical instruments his risen twelve times since the Cultural Revolution. The province produces 30-mA x-ray machines, high-pressure sterilizers, cardiographs, surgical instruments, and ultrasonic dental equipment.

Imports and sales of medical apparatus are handled by the China National Chemicals Import and Export Company; however, although there have been some purchases, the total volume involved is fairly insignificant. During 1970–1975 combined imports from fourteen countries surveyed by the Department of Commerce Market Share reports came to about $18 million. Japan seemed to be the largest single supplier, followed by West Germany and the United Kingdom, Italy, France, and Sweden. Imports from the United States totaled only $111,000 during 1972–1975 (see Fig. 18.1).

Total Chinese imports of medical equipment doubled from $4.2

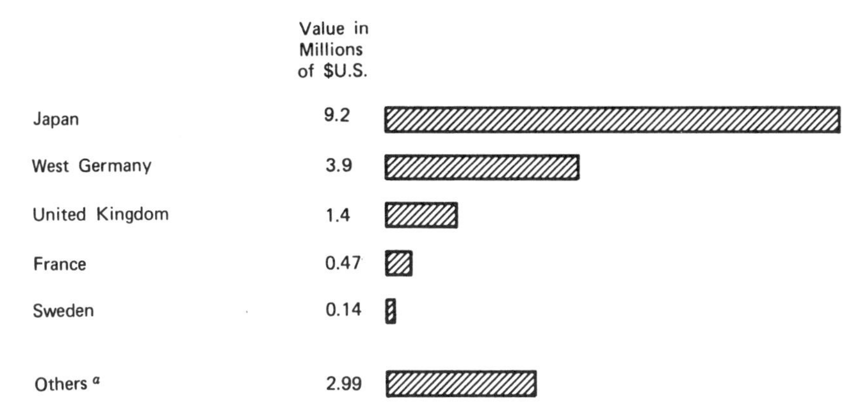

***Figure 18.1** Estimated market shares of major suppliers of medical equipment products to China during 1970–1975; all data for SITC export commodity code 726.1 and 726.2.* **(Source:** *U.S. Department of Commerce, "People's Republic of China 1969–73"* **Market Share Reports** *country Series; "People's Republic of China 1971–75,"* **Market Share Reports** *country series.*) ([a] *Others include Belgium/Luxembourg, the Netherlands, and the United States.*)

million in 1973 to $9.1 million in 1974, but dropped to $5.6 million in 1975, according to the United Nations statistics. Recently, China also signed a trade agreement with Hungary and plans to purchase some of its medical instruments.

REFERENCES

American Medical Writers Association, Metropolitan N.Y. Chapter, "Chinese Medicine Advanced at AMWA," *Scrivener,* March 1974.

American University, Foreign Area Studies. *Area Handbook for People's Republic of China,* 1972, pp. 182–189.

BBC Monitoring Service, Reading England. *Summary of World Broadcasts (SWB). Part 3. The Far East, Weekly Economic Report,* several issues: October 16, 1974; May 21, 1975, April 7, 1976, March 9, 1977; March 16, 1977; May 11, 1977; July 7, 1976.

Blakeslee, A. "Chinese Medicine, a Truly Great Leap Forward," *Saturday Review/World,* October 23, 1973, pp. 70–72.

Champeu, Harold C. "Five Communes in China," *Current Scene,* **14,** 1–16, January 1976, Hong Kong.

Chemical Engineering News. "Environmental Research," p. 32, March 15, 1976.

Committee on Scholarly Communication with the PRC. *China Exchange Newsletter,* 1974–1977.

Foreign Language Press. *New Women in China, 1972,* Peking.

Lampton, David M. "Trends in Health Policy," *Current Scene,* June 1974, Hong Kong.

Lasagna, L. "Herbal Pharmacology and Medical Therapy in the People's Republic of China," *Annals of Internal Medicine,* **83,** 887–893, December 1975.

Li, C. P. "Chinese Herbal Medicine," Department of Health, Education, and Welfare Publication, (NIH) 75-732, 1974.

Li, Victor H. "Health Services and New Relationship between China Studies and Visits to China," *China Quarterly* No. 59, p. 566, July–September 1974.

Orleans, L. A. "China's Environomics: Backing Into Ecological Leadership," in *China a Reassessment of Economy,* Joint Economic Committee, U.S. Congress, July 10, 1975.

Petitt, George R. "View of Cancer Treatment in the People's Republic of China," *China Quarterly* No. 68, pp. 789–796, December 1976.

Przybyla, Jan S. "Growing Old in China," *Current History,* pp. 57–59, 80, September 1976.

Sidel, Victor and Ruth Sidel. *Serve the People,* Josiah Macy, Jr. Foundation, New York, 1973.

Sidel, Victor and Ruth Sidel. "The Delivery of Medical Care in China," *Scientific American,* **230,** 19–27, April 1974.

Tobin, J. "The Economy of China; a Tourists View," *Current Scene,* **12,** 1–11, May 1974, Hong Kong.

Wen, Chi-pang and Charles W. Hays. "Medical Education in China in the Postcultural Revolution Era," *New England Journal of Medicine,* **292,** 998–1005, May 8, 1975.

Worth, R. M. "Health and Medicine," in Yuan-li Wu, ed., *China, a Handbook,* Praeger, New York, 1973, Chapter. 25.

Index